AF290129

Physical Optics

Physical Optics

Alan Rolf Mickelson

Electrical and Computer Engineering
University of Colorado at Boulder

SPRINGER SCIENCE+BUSINESS MEDIA, LLC

Library of Congress Catalog Card Number 91-42420
ISBN 978-1-4613-6566-2

I⦿P   Van Nostrand Reinhold is an International Thomson Publishing company.
ITP logo is a trademark under license.

16 15 14 13 12 11 10 9 8 7 6 5 4 3 2

Library of Congress Cataloging-in-Publication Data
Mickelson, Alan Rolf, 1950-
   Physical optics / Alan Rolf Mickelson.
      p.   cm.
   Includes bibliographical references and index.
   ISBN 978-1-4613-6566-2   ISBN 978-1-4615-3530-0 (eBook)
   DOI 10.1007/978-1-4615-3530-0
   1. Optics, Physical. I. Title.
QC395.2.M53   1992
535.2—dc20   91-42420
                                        CIP

Contents

Preface

This present text has emerged from the lecture notes for a one semester, first year, graduate level course which has been offered yearly since fall 1985 here in the Electrical and Computer Engineering Department at the University of Colorado at Boulder. Enrollment in the course, however, has not been limited to first year graduate electrical engineering students, but has included seniors, as well as more advanced students, from a variety of disciplines including other areas of engineering and physics. Although other Physical Optics texts exist, the most up-to-date ones are written primarily for undergraduate courses. As is discussed in slightly more depth in the introduction in the beginning of Chapter 1, up-to-dateness is important in a Physical Optics text, as even classical optics has been greatly rejuvenated by the events of the last 30 years, since the demonstration of the laser.

The perception of this author is that the needs of a graduate level text are quite different from that of an undergraduate text. At the undergraduate level, one is generally pleased if the student can qualitatively grasp a portion of the concepts presented and have some recollection of where to look them up if need be later in his/her career. A deeper insight is necessary at the graduate level and is generally developed through qualitative analysis of the problems within the subject area. This text represents an attempt to present a unified treatment of quantitative approaches to the analysis of a relevant set of problems in the general area of physical optics. The development is therefore somewhat formal and theoretical, yet examples and problems are included in an attempt to improve the pedagogy.

It is not conceivable that a text could cover the complete body of material that makes up physical optics. The field is all encompassing and would require multiple encyclopediae for complete coverage. The topics treated here, which are discussed slightly less perfunctorally in the last section of Chapter 1, are the ones that the author chose to present, those which the author felt would be of the most use to the optics graduate students here in our department in Boulder. The majority of students here are involved in experimental laboratory bench setups involving guided wave and/or optical processing systems. The topics covered here are meant to give the student the background necessary to understand the polarization, dispersion, ray optical system, interference and diffraction effects he/she will encounter in practice and quite possibly will need to understand in some depth in order to make his/her experiment work. This course may also serve as a first graduate course in optics for those interested in pursuing theoretical aspects of optics in more advanced courses.

Although, as explained above, there has been no attempt made to be comprehensive in treatment, the material is more than sufficient to cover a one semester course. Quite generally, some picking and choosing of topics is necessary in order to present a course with a reasonable pace.

Acknowledgments

I wish to acknowledge all the students who have taken this course over the last six years for their contributions and comments, which have been a guide in numerous revisions of the manuscript. I further acknowledge many useful discussions with my colleagues, especially Ed Kuester, John Dunn, and Zoya Popović. I particularly thank a number of my present and former graduate students who have materially contributed to the manuscript, including Walter Charczenko, Chow Quon Chon, Dag Roar Hjelme, Indra Januar, Holger Klotz, Paul Matthews, Marc Surette, Sandeep Vohra, Mike Yadlowsky, Shao Yang, and Peter Weitzman, as well as everyone else who has helped with the manuscript in any way. I am grateful to all of you.

List of Symbols Used

a, a_x, a_y	electric field amplitudes,
a	guide radius
$\mathbf{a}$	acceleration
A	area,
$\mathbf{dA}$	unit area vector,
A_{ij}	Einstein coefficient
$\mathbf{b}$	microscopic $\mathbf{B}$,
$\hat{\mathbf{b}}$	unit binormal
$\mathbf{B}$	magnetic induction vector,
B	magnitude of $\mathbf{B}$,
B_{ij}	Einstein coefficients
c	velocity of light
$c_i(t)$	mode amplitude
C	degree of circular polarization,
d	distance between apertures
$\mathbf{D}$	electric displacement vector,
$\hat{\mathbf{e}}_i$	basis vectors,
e	negative charge of an electron,
$\mathbf{e}$	microscopic electric field vector
$\mathbf{E}$	electrical field amplitude,
E	magnitude of $\mathbf{E}$,
$f(x)$	a function,
f	focal length,
$f(x, p, t)$	density in phase space
$\mathbf{F}$	force vector,
F	force magnitude,
F	F number,
$\mathbf{F}$	finesse
$g^{(i)}(\tau)$	i^{th} degree of temporal coherence
h	Planck's constant, $\hbar = h/2\pi$
$\mathbf{H}$	magnetic field intensity,
H	Hamiltonian
$i_d(t)$	detector current
I	optical intensity,
I	identity matrix,
$I(\mathbf{r}, \Omega)$	specific intensity,
$\mathbf{j}$	microscopic $\mathbf{J}$,
$\mathbf{J}$	current density vector,
J	Jones matrix
J_{ij}	components of Jones matrix
k_B	Boltzmann's constant,
k_1	arbitrary constant,
$\mathbf{k}$	propagation vector,

k	magnitude of propagation vector	t	time,
K	an arbitrary constant	t	transmission coefficient
l	degree of linear polarization	T	transmissivity,
L	Lagrangian,	T_{opt}	period of an optical wave,
$\mathbf{L}$	length	$\boldsymbol{T}$	transmission matrix,
m	mass,	T	temperature,
m	degree of polarization,	$T(x, y)$	transmission function
m_p	proton mass	$u_i(\mathbf{r})$	spatial wave function
$\boldsymbol{M}$	Mueller matrix,	U	a scalar wave function
$\boldsymbol{M}$	characteristic matrix,	$\mathbf{v}$	velocity vector,
$\boldsymbol{M}$	matrix in general	v_p	phase velocity
$\boldsymbol{M}$	magnification,	$\mathbf{v}_g$	group velocity
n	index of refraction,	$\mathbf{v}_E$	energy velocity
$\boldsymbol{n}$	index tensor,	V	volume,
$\hat{\mathbf{n}}$	unit normal	$\mathbf{V}$	Jones vectors,
N_i	level occupancy,	V_i	Jones vector elements,
N	number density,	$V(\mathbf{r}, t)$	potential function,
NA	numerical aperture	V_{ij}	matrix elements of potential
$\hat{O}$	an operator	V_{th}	thermal velocity,
$\mathbf{p}$	microscopic $\mathbf{P}$	$V(\mathbf{r}, t)$	analytical signal
p	momentum	w_0	spot size,
$\mathbf{P}$	polarization vector,	w_{ij}	transition rates
$\boldsymbol{P}$	intrinsic dipole vector,	W_e, W_m	energy densities,
$\mathbf{P}$	loss function	$W(w)$	spectral density
q	electric charge,	x	coordinate,
q_a	electric monopole moment, electric quadripole moment	$\dot{x}$	time derivative of coordinate
		x'	longitudinal derivative of x
		X	a spatial period
$\mathbf{r}$	radius vector,	y	coordinate,
r	reflection coefficient	$\dot{y}$	time derivative of y,
$\boldsymbol{R}$	rotation matrix,	y'	longitudinal derivative of y
R_{ij}	matrix elements,		
$R(\tau)$	autocorrelation function,	z	coordinate,
R	reflectivity	$\dot{z}$	time derivative of z,
s	distance along a ray path,	z'	longitudinal derivative of y
$\hat{s}$	unit vector along a ray path		
$\mathbf{S}$	Poynting vector,		
$\mathbf{S}$	Stoke's vector,	$\boldsymbol{\alpha}$	polarizability tensor,
S	eikonal	α	microscopic polarizability,

GREEK

α exponent of complex degree of coherence

β second order polarizability tensor,

β propagation constant

γ complex polarization scalar,

$\hat{\gamma}_{12}$ complex degree of coherence,

$\hat{\gamma}$ complex coherence function,

γ damping constant,

γ third order polarizability tensor

Γ antisymmetric index tensor,

Γ retardance,

Γ_{ij} coherence functions

δ a phase difference

Δ index contrast

ϵ_p photon energy,

ϵ dielectric constant,

ε dielectric tensor

η impedance,

η inverse of ε

θ angle from x to y in cartesian coordinate system

Θ areal dilation

κ composite wavenumber or ray period in a guiding medium,

κ radius of curvature

λ wavelength

μ permeability,

μ_{ij} complex degree of coherence

ν frequency

ξ coordinates of ray surface

ρ charge density,

$\rho(\omega)$ blackbody radiation density,

ρ ray density

σ conductivity,

σ conductivity tensor

τ_f, τ_b phase functions,

τ_d time constant,

τ_c coherence time,

τ torsion

φ phase angle

χ ellipticity angle on Poincaré sphere or ellipticity of polarization state,

χ susceptibility

χ a coupling constant

ψ angle on Poincaré sphere or angle to major axis

ψ wave function

ω angular frequency

Ω solid angle

1

Introduction

1.1. ABOUT PHYSICAL OPTICS

The twentieth century has seen two revolutions in the field of physical optics. The first was comprised of the genesis of quantum theory and, subsequently, quantum field theory (see, for example, Heitler 1954). The second was the demonstration and subsequent development of the high power coherent source referred to as the laser (review material is given Sargent, Scully, and Lamb 1974 and Haken 1984). The laser suddenly made many of the predictions of the pre-1960 quantum theory amenable to experimental investigation, and very mature fields, such as spectroscopy, changed radically overnight. For the present exposition, however, the effect of the laser on classical optics is perhaps more relevant. As the historical review in Born and Wolf (1975) evidences, pre-1900 optics was already a mature field in that everyday effects were already explained, and the many more exotic effects in the fields of polarization, interference, diffraction, and so on, had at least been touched on. However, pre-laser sources were incoherent and could be made to exhibit coherence only through stringent filtering, which made the sources weak in the sense that the number of photons per time period per steradian per unit area was small. In the post-laser era, arbitrary coherence functions (see, for example, Loudon 1983) could be generated with high powers. This allowed for not only the demonstration of exotic optical effects, but the application of these effects to diverse areas of endeavor. Classical physical optics in this era is no longer a mature field that catalogs some optical realities and curiosities, but it is a dynamic field that serves as a hotbed of ideas for wholly new application areas of a large number of engineering and scientific fields.

Physical optics is an all-encompassing subject that includes all aspects of the interaction of light and matter. For purposes of writing a book on the topic, some limitation must be placed on the subject. The present book will concentrate on the concepts involved in classical physical optics. The only quantum-mechanical argument used in the work is a semiclassical one involved with explaining the frequency dependence of the refracture index. All other material in the book should follow directly with assumed constitutive relations. It will be seen that this approach is sufficient for the explanation of practically all everyday phenomena as well as the field of highly accurate measurement by interferometry and spectroscopy, and the description of strange and intricate diffraction patterns whose manipulation has application in areas as diverse as mechanical strength testing and parallel processing.

1.2 THE ELECTROMAGNETIC SPECTRUM

Some discussion is needed here about how optical phenomena differ from other electromagnetic (EM) phenomena. Figure 1.1 depicts a frequency line on which various regions of the spectrum are denoted by their center wavelength, generic title, frequency, photon energy, and something like their effective temperature, where h is Planck's constant and k Boltzmann's constant.

More consideration will be given to the properties of the matter/radiation interaction for the different portions of the spectrum in Chapter 3, where material dispersion will be discussed. It will suffice here to give only some observations based on the wavelength, period, photon energy, and effective temperature of the optical radiation. First, one can see that the wavelength of the optical radiation, λ_{opt}, satisfies the inequality

$$3 \ \mu\mathrm{m} < \lambda_{\mathrm{opt}} < .7 \ \mu\mathrm{m} \tag{1-1}$$

The most salient observation to be made from this is that the wavelength is much smaller than "everyday" distances. It is much easier to picture a 1 cm

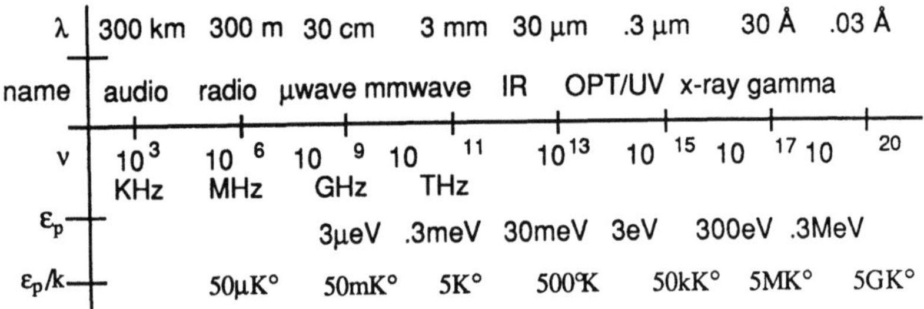

FIGURE 1.1.   A frequency line that gives the wavelengths λ, the frequencies ν, the photon energies $\epsilon_p = h\nu$ and effective temperatures $h\nu/k$ for the various regions of the frequency spectrum named on the line.

wavelength than a 1 μm wavelength, as one cannot optically resolve an optical wavelength. Whereas audio waves correspond to intercity distances, radio waves to spacings of buildings, microwaves to everyday objects, and X-rays and gamma rays to interatomic spacings, optical waves correspond to micrometers. Among other things, this opens up the possibility of the use of optics for very accurate length measurement.

A second fact to be noted from the frequency line would be that the optical time period, T_{opt}, satisfies the inequality

$$2 \times 10^{-15} \text{ sec} \geq T_{opt} \geq 10^{-15} \text{ sec} \qquad (1\text{-}2)$$

Perhaps the most obvious observation to be made from these values is that this period is very short—in fact, so short that despite recent progress in ultrafast phenomena there is no way that one can resolve a single optical period. Therefore, when one speaks of optical measurement, especially with ''usual'' detectors, one speaks of taking a long time average over the optical intensity. The detector, therefore, functions as a collector of energy. Information is included as a modulation of the radiant energy incident on the detector's front facet. This is in contradiction to the way that lower frequency detectors actually respond to the changes in phase and amplitude of the incident waveform itself.

The third observation to be made from Figure 1.1 concerns the optical photon energy. It can be surmised from the figure that the optical photon energy, ϵ_p, roughly satisfies

$$4 \text{ eV} \geq \epsilon_p \geq 2 \text{ eV} \qquad (1\text{-}3)$$

If one recalls that the ionization energy of the most tightly bound outer valence electron, that of hydrogen, corresponds to 1 Ryberg, which is 13.6 eV, one then will rapidly realize that optical photon energies correspond closely to electronic transition energies. One thus can conclude that optical constitutive relations are going to be strongly dependent on the details of energies, line shapes, and oscillator strengths of the various electronic transitions of the material in question. Although there are other transitions (vibrational, rotational, etc.) to worry about at infrared and millimeter wavelengths, from the microwave spectrum down one can generally take the dielectric constant to be frequency-independent. Because the photon energy is high enough at optical frequencies to actually free an electron from the valence to the conduction band the possibility of actually detecting a single photon (photon counting) exists. Lower frequency radiation can be detected only in masses of photons that combine to form a classical wave. Although this simple photon detection sounds nice, as if it were very efficient, it leads to a noise problem. The quantum noise that plagues photon statistics becomes directly observable in the output of photon counting detectors and leads to a form of noise known as shot noise, which limits minimally detectable signals to some number of photons per information bit.

Because a photon carries with it a discrete quantity of energy, $h\nu$, where h is Planck's constant and ν the photon frequency, one can also associate a quantity of heat with the photon. This quantity of heat would be the thermal energy that would be released were the photon energy to be converted to heat. Now, one could think of light propagating through a material at thermal equilibrium at a temperature T. One could speak of the light being in thermal equilibrium[1] with the matter if an effective temperature of the light corresponded to the equilibrium temperature of the matter. Such a temperature can be defined (apart from a proportionality constant) by

$$k_B T \approx h\nu \qquad (1\text{-}4)$$

where k_B is Boltzmann's constant. Perusal of Figure 1.1 indicates more than all else that light is hot. Room temperature falls within the infrared part of the spectrum, and indeed, infrared detectors "see" hot spots and other such temperature changes in the normal environment. Optical frequencies correspond to thousands of Kelvins, and indeed, stars, which have photo- and chromospheres that are in local thermal equilibrium at thousands of Kelvins, are great emitters of optical radiation. Light bulbs are incandescent sources where current within a lossy conductor can generate effective temperatures like those on stars. It would be hard to imagine that there could be thermal sources of either X rays or γ-rays, as materials at those temperatures could not be in thermal equilibrium but must be in some transient (exploding) state. Although there are extraterrestrial generators of microwaves and even radio waves, these emissions are generally atomic and not thermal in nature with the exception of the cosmic blackbody spectrum at $3°K$, which is, perhaps, a remnant of the big bang. Although there are regions of space cold enough to correspond to microwave frequencies, these regions are generally quite devoid of matter, and, further, a thermal radiator radiates a total energy that is proportional to T^4. Cold, vacuum-like regions, therefore, are poor radiators. For these reasons, we can see that most thermal radiation is either infrared or optical. Conversely, nonthermal radiators at infrared and optical frequencies are forced to compete with thermal radiators, as well as thermal effects.

1.3 OVERVIEW OF THE FOLLOWING CHAPTERS

The organization of this book is as follows: After this introduction, the properties of plane wave solutions of Maxwell's equations are presented, first for

---

[1]This is not really quite true. To be in thermal equilibrium the light would have to take on the blackbody distribution. This concept will be discussed along with the coherence properties of light in the next chapter.

monochromatic waves in free space, then for finite linewidth sources, and finally for monochromatic waves propagating in layered dielectric media. Discussion then naturally turns to the nature of the dielectric constant, including both its atomic origin and its frequency dependence. Propagation in a dispersive but isotropic medium is the topic that completes the third chapter before the case of anisotropic media is taken up in the fourth. More complicated inhomogeneous dielectric structures are taken up in Chapter 5, on geometrical optics, where both waveguides and imaging systems are introduced. Questions of coherence are addressed in Chapter 6, on interference, which also includes discussion of interferometer measurement techniques. The closing chapter, on diffraction, begins with discussion of scalar diffraction theory and proceeds to propagation of partial coherence, followed by discussion of spectrometry using various types of grating spectrometers.

References

Born, M. and E. Wolf, *Principles of Optics*, Fifth edition, Pergamon Press, Elmsford, NY (1975).

Haken, H., *Laser Theory*, Springer-Verlag, New York (1984).

Heitler, W., *The Quantum Theory of Radiation*, Third edition, Clarendon Press, Oxford (1954).

Loudon, R., *The Quantum Theory of Light*, Second edition, Clarendon Press, Oxford (1983).

Maimon, T. H., "Stimulated optical radiation in ruby masers," *Nature 187*, 498 (1960).

Sargent, M., M. O. Scully, and W. E. Lamb, *Laser Physics*, Addison-Wesley, Reading, MA (1974).

Problems

1. What is the frequency, wavelength, and period of an EM source that produces photons with energy of
   (a) 50 meV
   (b) 1 eV
   (c) 40 eV
   (d) 1 keV
2. In order to measure the atomic layer spacing of an unknown crystal, a photon with an energy greater than 50 keV must be used. What wavelength photons will satisfy such a requirement? What is the interatomic spacing?
3. A source of wavelength λ is radiating onto a detector held at a temperature of 0 K. If 0.01% of the incident energy is converted to heat, calculate the rise in temperature if the wavelength of the radiation is
   (a) 1 m
   (b) 1 mm
   (c) 0.6328 μm
   (d) 3.39 μm
   (e) 5 nm

4. Consider a system in which a source is radiating onto a detector, and 0.01% of the incident energy is converted to heat, 100% of which is passed through a metal tube whose outer and inner walls are initially at a temperature T_1 (Figure 1.2). Calculate the total heat flow (current) at steady state if the incident wavelength is 0.83 μm, the tube inner diameter is 2 mm with an outer diameter of 5.44 mm, and $T_1 = 300\%$. Assume that the tube is made of copper which has a thermal conductivity of 385 Joules $\cdot$ sec^{-1} $\cdot$ m^{-1} $\cdot$ K^{-1}. The tube length is 1 m.

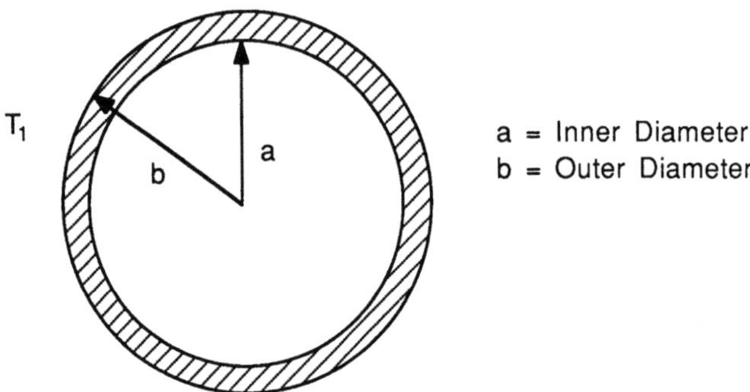

FIGURE 1.2.   A sketch of the structure discussed in problem 4.

2

Maxwell's Equations and Plane Wave Propagation

2.1 INTRODUCTION

In this chapter, many of the equations and concepts fundamental to later developments are presented. The discussion begins with a statement of Maxwell's equations and a short discussion of what additional information is necessary to find solutions to them. This discussion leads to the concept of constitutive relations. The presentation turns to the discussion of wave propagation in the medium with the simplest constitutive relations, that of free space. Monochromatic plane waves are taken up as the simplest of solutions to Maxwell's free space equations. Energy propagation is introduced through the construct of Poynting's vector. This vector then is used in a discussion of the physical reality of plane wave solutions, which by nature must be truncated. The discussion then turns to that of polarization and its representation in terms of complex numbers. The following section takes up the more realistic case of the propagation of polychromatic plane waves. The section's presentation is motivated by a discussion of the nature of thermal electromagnetic sources and the concepts of stimulated and spontaneous emission. The Poynting vector is then generalized to the polychromatic case. The introduction of the Wiener-Khintchine theorem allows discussion to turn to spectra of nonthermal light sources. A brief discussion of quasi-monochromatic Jones vectors is then used to introduce the concept of polychromatic polarization and its description in terms of Stokes parameters. The following section first discusses the propagation of Jones and Stokes vectors by means of Jones and Mueller matrices, respectively. The discussion is followed by a presentation of various examples of polarizers, wave plates, and more complex compound polarization structures, as well as the case of a microwave optical sampling head. The final section of this chapter is dedicated to the discussion of striated media. This presentation begins with a discussion of transverse electric and transverse magnetic modes, followed by considerations of energy conservation. A matrix method for solution of a boundary value problem for a dielectric stack then is introduced. This method is applied to various examples, including the case of the simple beam splitter, polarizing stacks, and frequency filters. The chapter concludes with a discussion of how the matrix method can be applied to guided wave problems.

2.2 SOME PRELIMINARIES

Maxwell's equations can be written in MKSA units, in the form

$$\nabla \times \mathbf{E} = -\frac{\partial \mathbf{B}}{\partial t} \qquad \text{(a)}$$

$$\nabla \times \mathbf{H} = \mathbf{J} + \frac{\partial \mathbf{D}}{\partial t} \qquad \text{(b)} \qquad \text{(2-1)}$$

$$\nabla \cdot \mathbf{B} = 0 \qquad \text{(c)}$$

$$\nabla \cdot \mathbf{D} = \rho \qquad \text{(d)}$$

where $\mathbf{E}$ is the electric field intensity in volts per meter, $\mathbf{H}$ is the magnetic field intensity in amperes per meter, $\mathbf{D}$ is the electric displacement vector in coulombs per meter squared, $\mathbf{B}$ is the magnetic induction vector in webers per meter squared, $\mathbf{J}$ is the current density vector in amperes per meter squared, and ρ is the volume density of charge in coulombs per cubic meter. Unfortunately, in units other than MKSA, such as the CGS units which are employed in various other physical optics texts (Born and Wolf 1975; Klein and Furtak 1986; various others), constants can appear, and many of the dimensions of the electromagnetic quantities in Maxwell's equations can change. This state of affairs really arises from the fact that, in the Coulomb force law (for the force between two charges labeled q_1 and q_2 located at a distance r from each other),

$$F_e = k_1 \frac{q_1 q_2}{r^2} \qquad \text{(2-2)}$$

the measurables of the electric force F_e and the distance r serve to define the particle's charge, but only through an arbitrary constant k_1. By changing the dimensions of k_1, one can therefore change the dimensions of the charge. The same comments must apply to the magnetic force law. For the novice in optics, these differences can prove to be quite confusing. Fortunately, however, there is a very clear and complete discussion of these matters in Appendix B of Jackson (1975).

Some interesting observations can be made immediately about Maxwell's equations simply from the form of (2-1). One can immediately see that $\mathbf{E}$, $\mathbf{H}$, $\mathbf{D}$, $\mathbf{B}$, $\mathbf{J}$, and ρ comprise 16 unknowns to be solved by the eight equations comprised by (2-1) (a–d). A bit of extra observation shows that the situation is still worse than this, in that the divergence of (2-1) (a) gives one (2-1) (c) because the divergence of a curl is zero, and the divergence of (2-1) (b) together with (2-1) (d) gives one

$$\nabla \cdot \mathbf{J} + \frac{d\rho}{dt} = 0 \qquad (2\text{-}3)$$

which is just a statement of conservation of charge, which one would hope was built into the system of equations. This implies that (2-1) (c) and (d) are not independent of the first six equations (2-1) (a) and (b), and therefore Maxwell's equations actually comprise six equations in 15 unknowns. It is clear, therefore, that the "exact" system defined by Maxwell's equations cannot be solved without something extra.

Maxwell's equations are a mathematical set of equations relating electromagnetic quantities to elements of charge. The solution of these equations requires that one augment the number of relationships with a set of nine relations that describe the physics of the situation. These relations are generally written in frequency space, that is, they relate the temporal Fourier transforms of the quantities in equation (2-1). A possible form of these so-called constitutive relations is

$$\mathbf{J}(\mathbf{r}, \omega) = \mathbf{f}_J(\mathbf{E}(\mathbf{r}', \omega').\ \mathbf{B}(\mathbf{r}', \omega')) \qquad \text{(a)}$$

$$\mathbf{D}(\mathbf{r}, \omega) = \mathbf{f}_D(\mathbf{E}(\mathbf{r}', \omega'), \mathbf{B}(\mathbf{r}', \omega')) \qquad \text{(b)} \quad (2\text{-}4)$$

$$\mathbf{H}(\mathbf{r}, \omega) = \mathbf{f}_H(\mathbf{E}(\mathbf{r}', \omega'), \mathbf{B}(\mathbf{r}', \omega')) \qquad \text{(c)}$$

where the $\mathbf{f}_J$, $\mathbf{f}_D$, and $\mathbf{f}_H$ are, in general, nonlinear operators in space and frequency. The first of these relations describes the motion of the charges under the influence of the electromagnetic fields. In general, this relation requires one to solve the quantum mechanical equations of motion and takes a complicated integro-differential form. The last two relations express the properties of the electric and magnetic polarizabilities of the medium. With these relations, one can reduce Maxwell's equations to a system of six equations in six unknowns, $\mathbf{E}$ and $\mathbf{H}$, but cannot in general solve them. In a linear lossless, homogeneous, isotropic medium, (2-4) reduces to

$$\mathbf{J}(\mathbf{r}, \omega) = 0 \qquad \text{(a)}$$

$$\mathbf{D}(\mathbf{r}, \omega) = \epsilon(\omega)\mathbf{E}(\mathbf{r}, \omega) \left(\begin{array}{l}\text{linear, lossless,}\\ \text{homogeneous,}\\ \text{isotropic medium}\end{array}\right) \qquad \text{(b)} \quad (2\text{-}5)$$

$$\mathbf{H}(\mathbf{r}, \omega) = \mathbf{B}(\mathbf{r}, \omega)/\mu(\omega) \qquad \text{(c)}$$

where ϵ is the medium permittivity, and μ is the medium permeability.

2.3 MONOCHROMATIC PLANE WAVES

Now, with some knowledge of the structure of Maxwell's equations, we can attempt to construct the simplest solutions to them, the plane wave solutions, in the simplest medium, that of free space. The plan here is to introduce the free space constitutive relations, apply them to a monochromatic form of Maxwell's equations, and then investigate the solutions of these equations that have no transverse variation, that is, are totally plane along the direction of propagation. The meaning of the direction of propagation then is discussed in terms of energy flow concepts. These energy flow concepts are used to investigate what happens to the field structure when one truncates a plane wave. The resulting discussion vindicates the plane wave concept by indicating that in many practical cases it is a very good approximation. Although it is only touched on in the discussion, it should be pointed out here that there is a still more important reason for the great utility of the plane wave concept. The reason stems from the basic tenet of Fourier analysis that it is possible to expand arbitrary functions in terms of complex exponentials, and, therefore, that it is possible to describe arbitrary, spatial wave patterns in terms of three-dimensional plane wave expansions (see, for example, Clemmow 1966). Therefore, it can be said that any solution of Maxwell's equations can be expanded in terms of a plane wave expansion. As will be mentioned, however, in the original discussion of the Poynting vector in this section, and again in Section 2.6 when the Fresnel relations are discussed, not all of the plane waves in the expansion need to be the "nice" homogeneous kind. There are also inhomogeneous plane waves that exhibit an evanescent behavior and complex polarization. Such plane waves describe nonpropagating (nonradiative) fields and are necessary to discuss many phenomena, especially phenomena that occur in the vicinity of sources, and guided wave structures. A majority of the applications in this text, and in "free space" optics in general, are radiative in nature and can be described without the use of the inhomogeneous plane waves. Following the truncated plane wave discussion, polarization will be discussed in some depth. Although the Poynting vector gives the direction of energy flow, it is the Lorentz force law that gives the effect of the field on a particle. It is, indeed, the energy associated with this force that leads to a sloshing back and forth of energy between the field and the host material, which leads to the $\mathbf{E} \cdot \mathbf{P}$ term in the medium-dependent form of Poynting's theorem, and which is one of the most important themes of Chapter 3. This section will conclude with a discussion of the representation of monochromatic polarization states by Jones vectors.

Now a specific homogeneous, source-free region is free space, where (2-5) reduces to

$$\mathbf{J}(\mathbf{r}, \omega) = 0 \tag{a}$$

$$\mathbf{D}(\mathbf{r}, \omega) = \epsilon_0 \mathbf{E}(\mathbf{r}, \omega) \tag{b}$$

$$\mathbf{B}(\mathbf{r}, \omega) = \mu_0 \mathbf{H}(\mathbf{r}, \omega) \tag{c}$$

(2-6)

where ϵ_0 is the permittivity of free space, and μ_0 is the free space permeability. As one is dealing with monochromatic phenomena here, one can take

$$\mathbf{E}(\mathbf{r}, t) = \mathrm{Re}\,[\mathbf{E}(\mathbf{r})e^{-i\omega t}] \qquad \text{(a)}$$
$$\mathbf{H}(\mathbf{r}, t) = \mathrm{Re}\,[\mathbf{H}(\mathbf{r})e^{-i\omega t}] \qquad \text{(b)}$$

(2-7)

where $\mathbf{r}$ is the position vector. Using (2-6) and (2-7) in (2-1), one finds the following equations for $\mathbf{E}(\mathbf{r})$ and $\mathbf{H}(\mathbf{r})$:

$$\nabla \times \mathbf{E} = i\omega\mu_0\mathbf{H} \qquad \text{(a)}$$
$$\nabla \times \mathbf{H} = -i\omega\epsilon_0\mathbf{E} \qquad \text{(b)} \quad \text{(2-8)}$$
$$\nabla \cdot \mathbf{H} = \nabla \cdot \mathbf{E} = 0 \qquad \text{(c)and (d)}$$

where the prescription of (2-7) must be used to find $\mathbf{E}(\mathbf{r}, t)$ and $\mathbf{H}(\mathbf{r}, t)$. Applying the curl operator to (2-8) (a) and (b) and using a vector identity and (c) and (d), one finds the wave equations

$$\nabla^2\mathbf{E} + k^2\mathbf{E} = 0 \qquad \text{(a)}$$
$$\nabla^2\mathbf{H} + k^2\mathbf{H} = 0 \qquad \text{(b)}$$

(2-9)

where $k^2 = \omega^2\mu_0\epsilon_0 = \omega^2/c^2$, as $\mu_0\epsilon_0 = 1/c^2$. The solutions to (2-9) can be written in the form

$$\mathbf{E} = \mathbf{E}_f e^{i\mathbf{k}\cdot\mathbf{r}} + \mathbf{E}_b e^{-i\mathbf{k}\cdot\mathbf{r}} \qquad \text{(a)}$$
$$\mathbf{H} = \mathbf{H}_f e^{i\mathbf{k}\cdot\mathbf{r}} + \mathbf{H}_b e^{-i\mathbf{k}\cdot\mathbf{r}} \qquad \text{(b)}$$

(2-10)

where $\mathbf{k}$ is a vector of magnitude k which points in the direction of propagation. $\mathbf{E}_f$, $\mathbf{E}_b$, $\mathbf{H}_f$, and $\mathbf{H}_b$ are vector constants that are determined from (2-8) and the boundary conditions. To gain more insight into the nature of the solution, it is instructive to look at the time-varying $\mathbf{E}$-fields, which can be derived from (2-10) through the use of (2-7), to obtain

$$\mathbf{E}_f(\mathbf{r}, t) = \mathbf{E}_f \cos\,(\omega t - \mathbf{k}\cdot\mathbf{r} + \varphi_f) \qquad \text{(a)}$$
$$\mathbf{E}_b(\mathbf{r}, t) = \mathbf{E}_b \cos\,(\omega t + \mathbf{k}\cdot\mathbf{r} + \varphi_b) \qquad \text{(b)}$$

(2-11)

Now, in turn these could be written in the form

$$\mathbf{E}_f(\mathbf{r}, t) = \mathbf{E}_f \cos\,(\tau_f(\mathbf{r}, t)) \qquad \text{(a)}$$
$$\mathbf{E}_b(\mathbf{r}, t) = \mathbf{E}_b \cos\,(\tau_b(\mathbf{r}, t)) \qquad \text{(b)}$$

(2-12)

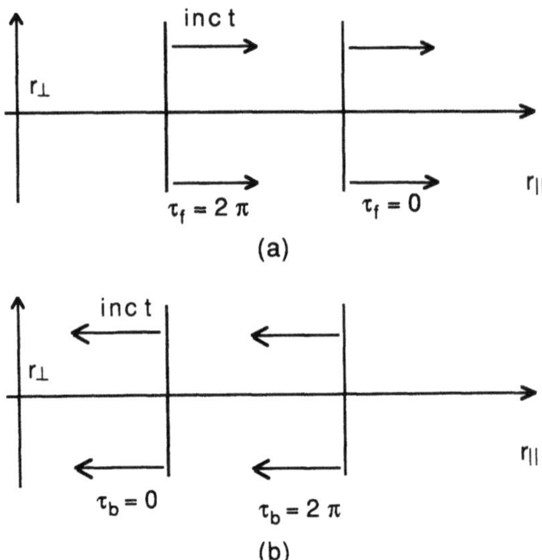

FIGURE 2.1.   Schematic depiction of planes of constant $\tau_f(a)$, and $\tau_b(b)$, and how they propagate with increasing time where $\mathbf{r}_\perp$ ($\mathbf{r}_\parallel$) denotes the component of the radius vector $x\hat{e}_x + y\hat{e}_y + z\hat{e}_z$ perpendicular (parallel) to the wave propagation vector $\mathbf{k}$.

where $\tau_f = \omega t - \mathbf{k} \cdot \mathbf{r} + \varphi_f$ and $\tau_b = \omega t + \mathbf{k} \cdot \mathbf{r} + \varphi_b$ are phase functions. If one were to draw a line of constant phase and trace its behavior as a function of increasing time, one would obtain a figure like that of Figure 2.1. In the figure, the $r_\perp$ represents the direction transverse to the propagation direction which is denoted $r_\parallel$ that is parallel to $\mathbf{k}$. It is clear from the figure that the first lines of equations (2-11) and (2-12) represent forward-traveling waves, and the second terms represent backward-traveling waves.

To learn more about the relations between the amplitudes and phases of the constants $\mathbf{E}_f$, $\mathbf{H}_f$, $\mathbf{E}_b$, and $\mathbf{H}_b$, one can restrict one's attention to forward-going waves, as it is clear that forward and backward-going waves will not couple in free space and, further, one can always choose boundary conditions such that either the forward or the backward wave is excited. So, choosing the forward-going wave,

$$\mathbf{E} = \mathbf{E}_f e^{i\mathbf{k} \cdot \mathbf{r}} \qquad \text{(a)}$$

$$\mathbf{H} = \mathbf{H}_f e^{i\mathbf{k} \cdot \mathbf{r}} \qquad \text{(b)}$$

(2-13)

and plugging back into (2-8), one finds that

$$\hat{e}_k \times \mathbf{E}_f = \eta_0 \mathbf{H}_f \qquad \text{(a)}$$

$$\hat{\mathbf{e}}_k \times \mathbf{H}_f = -1/\eta_0 \mathbf{E}_f \qquad\qquad \text{(b)} \quad (2\text{-}14)$$

$$\hat{\mathbf{e}}_k \cdot \mathbf{H}_f = \hat{\mathbf{e}}_k \cdot \mathbf{E}_f = 0 \qquad\qquad \text{(c) and (d)}$$

where $\hat{\mathbf{e}}_k$ is the unit vector in the $\mathbf{k}$ direction, defined by $\hat{\mathbf{e}}_k = \mathbf{k}/|k|$, and η_0 is the impedance of free space, defined by $\eta_0 = \sqrt{\mu_0/\epsilon_0}$. The situation is as illustrated in Figure 2.2. The field directions are mutually orthogonal and, further, are both orthogonal to the propagation direction. Further, the two field magnitudes must satisfy the relation

$$|\mathbf{E}_f| = \eta_0 |\mathbf{H}_f| \qquad\qquad (2\text{-}15)$$

which, from the dimensions of $\mathbf{E}$ and $\mathbf{H}$, should make it clear why η_0 is identified as an impedance. Equation (2-15) is of the form of Ohm's law.

Equations (2-14) and (2-15), however, still do not completely nail down the solutions for $\mathbf{E}_f$ and $\mathbf{H}_f$, as the requirements of mutual perpendicularity and perpendicularity to the propagation direction only require that they lie in a plane. To simplify the exposition, assume that $\hat{\mathbf{e}}_k$ coincides with $\hat{\mathbf{e}}_z$. In practically all optical problems, the z-axis is assumed to be the so-called optic axis or, in other words, the direction of propagation. With this choice of axes, one can pick two independent solutions for (2-14) to be

$$E_x = \eta_0 H_y \qquad (x\text{-polarized}) \qquad\qquad \text{(a)}$$
$$E_y = -\eta_0 H_x \qquad (y\text{-polarized}) \qquad\qquad \text{(b)} \quad (2\text{-}16)$$

where this situation is schematically illustrated in Figure 2.3. Clearly many other solutions were possible, but as the solutions of (2-16) satisfy orthogonality, that is, $\mathbf{E}_x \cdot \mathbf{E}_y = 0$, where $\mathbf{E}_x$ is the x-polarized electric field vector, and $\mathbf{E}_y$ is the y-polarized electric field vector, then all solutions of (2-14) can be

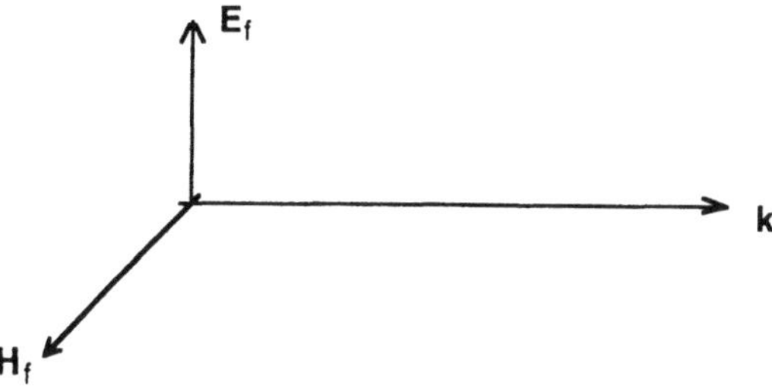

FIGURE 2.2.   A depiction of the relative directions of the propagation vector, and the $\mathbf{E}$ and $\mathbf{H}$ fields for a plane wave.

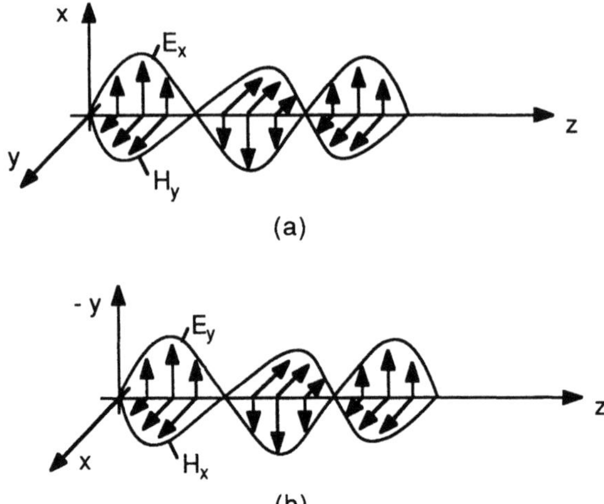

FIGURE 2.3.   Illustration of the spatial distribution (at a given time) of the **E** and **H** fields of a
z-propagating plane wave.

built up from linear combinations of the solutions of (2-16). This situation soon
will be taken up in more detail.

Perhaps the best way to continue this discussion is by giving some attention
to the details of the energy flow and detection processes as the concept of time
averaging of square law detected signals will be used extensively here. Consider
the Poynting vector $\mathbf{S}(t)$, defined by (see, for example, the discussion in Johnk
1973, Chapter 7):

$$\mathbf{S}(t) = \mathbf{E}(t) \times \mathbf{H}(t) \tag{2-17}$$

Taking a divergence of (2-17) and using Maxwell's curl relations (2-1) (a) and
(b) for a free space medium leads to the expression

$$\nabla \cdot \mathbf{S}(t) = -\frac{1}{2} \mu_0 \frac{\partial}{\partial t} (\mathbf{H}(t) \cdot \mathbf{H}(t))$$

$$-\frac{1}{2} \epsilon_0 \frac{\partial}{\partial t} (\mathbf{E}(t) \cdot \mathbf{E}(t)) - (\mathbf{E}(t) \cdot \mathbf{J}(t)) \tag{2-18}$$

If one now integrates the divergence of $\mathbf{S}(t)$ over a finite volume V, one can
use the divergence theorem

$$\int_V \nabla \cdot \mathbf{S}(t) \, dV = \int_A \mathbf{S}(t) \cdot \mathbf{dA} \tag{2-19}$$

where A is the (closed) surface of the (finite) volume, and $\mathbf{dA}$ is the unit vector normally pointing outward from the surface, to obtain the following relation:

$$\int_A \mathbf{S}(t) \cdot \mathbf{dA} = -\frac{\partial W_m}{\partial t} - \frac{\partial W_e}{\partial t} - \int_V \mathbf{E}(t) \cdot \mathbf{J}(t)dV \qquad (2\text{-}20)$$

where

$$W_m = \frac{\mu_0}{2} \int_V \mathbf{H}(t) \cdot \mathbf{H}(t)dV \qquad (a)$$

$$(2\text{-}21)$$

$$W_e = \frac{\epsilon_0}{2} \int_V \mathbf{E}(t) \cdot \mathbf{E}(t)dV \qquad (b)$$

The physical interpretation of (2-20) becomes clear from the diagram of Figure 2.4. The left-hand side of (2-20) represents the amount of the Poynting vector flowing out across the surface of A. The first two terms on the right represent the amount of loss of internally stored magnetic and electric energy due to this flow [at least as far as the representations of (2-21) really are energies, which they certainly are dimensionally]. The last term on the right-hand side represents the amount of energy that is transferred from electromagnetic energy to mechanical (particle flow) energy. It seems reasonable from this argument to associate the Poynting vector $\mathbf{S}(t)$ with some sort of energy flow. Some care must be used in doing so, as the derivation of (2-20) used a divergence of $\mathbf{S}(t)$. As a divergence of a curl is zero, the quantity $\mathbf{S}(t)$ appearing in (2-17) is not unique, as it can be modified by a circulant term (that is, a term which is a curl), which indicates, from Stokes theorem, an equal flow in and out of the surface. A second problem that can arise in such interpretation is involved with standing waves. Equation (2-20) involves only $\mathbf{S}(t) \cdot \mathbf{dA}$, which

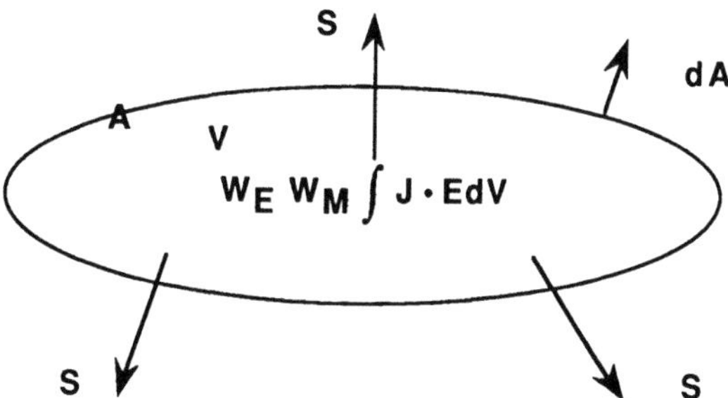

FIGURE 2.4.   Schematic depiction of the volume V within which there is storage of electric, magnetic, and kinetic energy. The stored energy is tapped by an outward flow of the Poynting vector $\mathbf{S}$ over the surface A.

is an operation on $\mathbf{S}(t)$ which picks out the forward propagating part of the field at the surface A. Therefore, to use the Poynting vector $\mathbf{S}(t)$ as the direction of energy flow, we must be able to pick out the forward-going from the backward-going part of the wave, and this is not always possible. Fortunately, for plane unidirectionally propagating waves, it is. For inhomogeneous plane waves, it is not, in general, possible.

Let us consider the Poynting vector for a forward propagating plane wave of arbitrary polarization. We will assume that the wave is propagating in the z-direction toward a detector surface of area A oriented perpendicular to the z-axis, a situation that is depicted in Figure 2.5. In this case, the transverse fields can be expressed by

$$E_x = a_1 \cos (\omega t - kz)$$

$$H_y = a_1/\eta_0 \cos (\omega t - kz) \quad \text{(a)}$$

$$\hspace{5.5cm}\text{(2-22)}$$

$$E_y = a_2 \cos (\omega t - kz + \varphi)$$

$$H_x = -a_2/\eta_0 \cos (\omega t - kz + \varphi) \quad \text{(b)}$$

Using (2-17) to calculate the Poynting vector gives

$$\mathbf{S}(z,\, t) = \left\{ \frac{a_1^2}{2\eta_0} (1 + \cos (2\omega t - 2kz)) \right.$$

$$\left. + \frac{a_2^2}{2\eta_0} (1 + \cos (2\omega t - 2kz + 2\varphi)) \right\} \hat{\mathbf{e}}_z \qquad \text{(2-23)}$$

One sees that the energy flows in the two polarization states simply are additive, and that there are no cross terms. The time dependence of the energy in one polarization state for a fixed z-value is plotted in Figure 2.6. As one can see

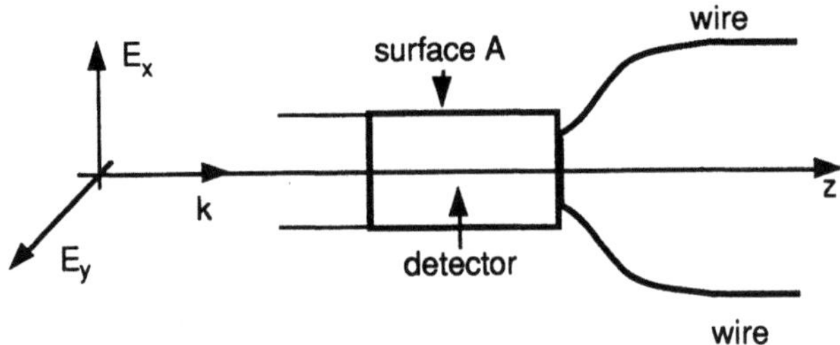

FIGURE 2.5.   Diagram illustrating the positioning of a (optical to electrical) detector relative to an incoming plane wave that propagates along the optical axis.

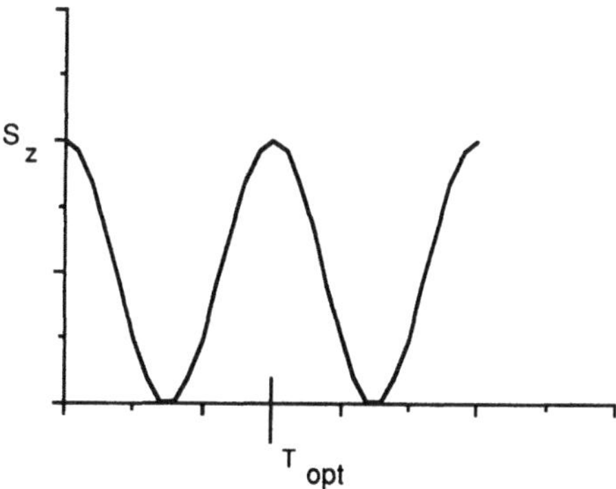

FIGURE 2.6.   Plot of the time variation of the Poynting vector evaluated at a fixed z-value, assumed to be the detector surface.

from the figure, the flow across the detector surface will oscillate between a maximum and zero during an optical period. At optical frequencies, there are no detectors that can react as rapidly as the period, and therefore what will actually be seen in the detector current (assuming 100% quantum efficiency) will be an effect driven by the average energy crossing A:

$$\langle \mathbf{S}(z, t) \rangle = \frac{1}{2\tau_d} \int_{t-\tau_d}^{t+\tau_d} \mathbf{S}(z, t)\, dt \tag{2-24}$$

where τ_d is the detector response time. As the detector response time corresponds to many optical periods, one could replace the detector time average with an infinite time average to obtain the averaged Poynting vector:

$$\langle \mathbf{S}(z, t) \cdot \hat{\mathbf{e}}_z \rangle = \frac{a_1^2}{2\eta_0} + \frac{a_2^2}{2\eta_0} \tag{2-25}$$

and indeed, it is seen that the energies in the two polarization states simply add, independently of phase, within the detector. Further, it is clear that the energy is directly proportional to the time average of $\mathbf{E}(t) \cdot \mathbf{E}(t)$. One can easily show that for a propagating wave

$$\mathbf{S}_{av} = \langle \mathbf{S}(t) \rangle = \tfrac{1}{2}\mathrm{Re}\,[\mathbf{E} \times \mathbf{H}^*] \tag{2-26}$$

where

$$\mathbf{E}(t) = \mathrm{Re}\,(\mathbf{E}e^{-i\omega t}) \tag{a}$$

$$\mathbf{H}(t) = \mathrm{Re}\,(\mathbf{H}e^{-i\omega t}) \tag{b}$$

$$\tag{2-27}$$

In line with the above definitions, the optical intensity is often defined by

$$I = \frac{1}{2\eta_0}\, \mathbf{E} \cdot \mathbf{E}* \tag{2-28}$$

For a plane wave, this optical intensity is directly proportional to the detectable energy. Unfortunately, for more general waves this is not the case, as could occur in the near-field of an aperture. More will be mentioned on this point later in this chapter, when the Fresnel relations are discussed.

After all this discussion of plane waves, one might wonder about the observability of such abstract solutions, which have perfectly flat, infinite wave fronts. So this appears to be a good time to digress a little and discuss the degree to which an obtainable wave can approximate a plane wave. Consider the laser collimation system of Figure 2.7. Let us assume that the electric field directly outside the laser has the form

$$\mathbf{E} \cong \hat{\mathbf{e}}_x E_x e^{-r^2/2w_0^2}\, e^{ikz} \tag{2-29}$$

where the w_0 is called the spot size or beam waist and indicates that the field intensity must gradually fall off away from the optic axis (z-axis), which is defined by $r = 0$ where $r^2 = x^2 + y^2$. Now if the aperture of the first lens subtends an area for which $r \ll w_0$, one would think that the field transmitted by the collimator would be roughly a truncated x-polarized plane wave. For such a plane, one should have $E_y = H_x = E_z = H_z = 0$ and $E_x = \eta_0 H_y$. However, one should also have

$$\nabla \cdot \mathbf{E} = \frac{dE_x}{dx} + \frac{dE_y}{dy} + \frac{dE_z}{dz} = 0 \tag{2-30}$$

But (2-30) cannot be satisfied by $E_y = E_z = 0$, as the x derivative of E_x cannot be zero if the wave is to truncate. From the symmetry of the collimator system,

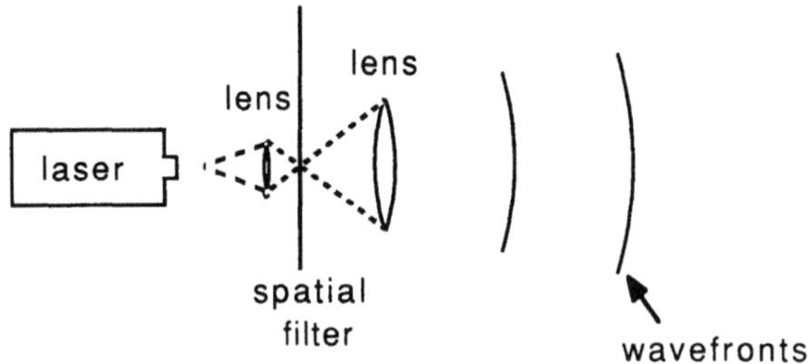

FIGURE 2.7.   Schematic illustration of a laser collimation system.

it would be impossible to explain how an E_y would arise from an E_x, so the conclusion must be that a truncated plane wave must have both E_x and E_z. If one assumes that the Poynting vector $\mathbf{S}$, defined by equation (2-17) represents the direction of energy flow, and writes the polarization vector for the truncated plane wave in the form

$$\mathbf{E}_0 = E_x\hat{\mathbf{e}}_x + E_z\hat{\mathbf{e}}_z \qquad (a)$$
$$\mathbf{H}_0 = H_y\hat{\mathbf{e}}_y + H_z\hat{\mathbf{e}}_z \qquad (b)$$

$$(2\text{-}31)$$

one comes to the unmistakable conclusion that there must be radial components to the Poynting vector. This effect is exactly what is known as diffraction spreading, and is illustrated in Figure 2.8 and will be discussed at greater length in Chapter 7.

As is well known, the diffraction spreading angle θ is roughly proportional to the wavelength, λ, divided by the aperture size, d. As optical wavelengths are small (<1 μm) and apertures can be large (meters), diffraction effects can be made small. For example, in the case of laser ranging from the McDonald Observatory at the University of Texas at Austin, a ruby laser is collimated through a 2.7 m telescope. For a ruby wavelength of roughly 0.7 μm and a distance to the moon of roughly 200,000 miles (about 300,000 km), one finds that the spot size (full width half maximum) on the moon is roughly 80 m, which indeed is that measured (Silverberg 1974). The diffraction spreading angle is this case of 2.5×10^{-7} rad certainly is not significant for terrestrial applications.

An interesting case where diffraction spreading is suppressed occurs in the case of guided waves. Consider the metallic parallel plate waveguide of Figure 2.9. (We could have considered a square or circular waveguide, but the physics would be the same as in this simple one-dimensional case). Here the possible solutions will break up into two polarizations, the transverse magnetic (TM) polarization which has components E_x, H_y, and E_z, and the transverse electric (TE) polarization with components H_x, E_y, and H_z. The argument for free space

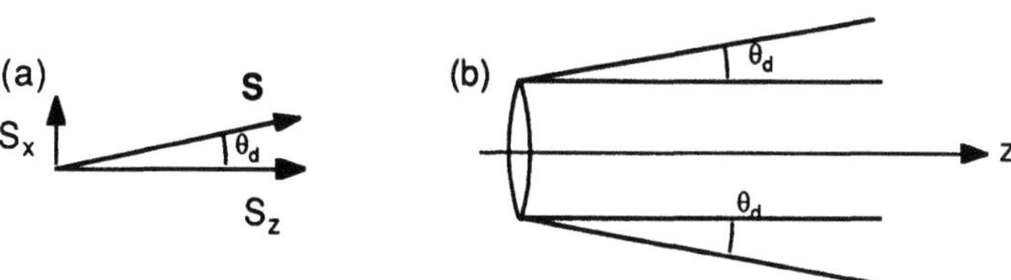

FIGURE 2.8. Illustration of diffraction spreading indicating (a) direction of the Poynting vector and (b) the spreading of a beam as it emerges from a collimator.

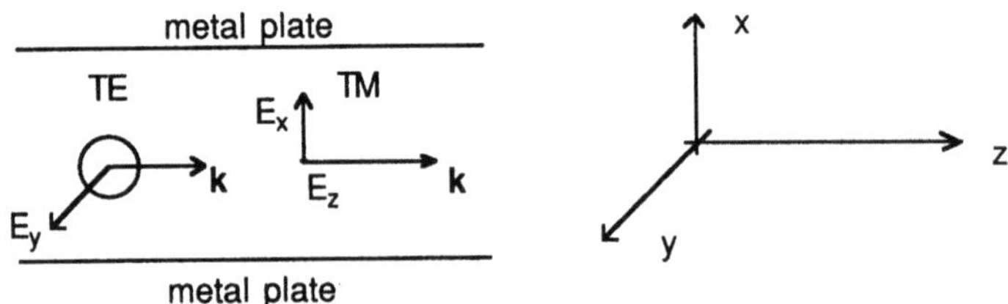

FIGURE 2.9.   Illustration of a z-directed, y symmetric parallel plate waveguide, showing the **E** vectors for the TE and TM polarizations.

diffraction spreading that was presented above no longer applies here, as the divergence **E** equation is now driven by a charge density ρ, and the curl **H** equation now contains a **J**. These extra terms cause the waves to bounce back and forth between the plates and remain bounded.

We now return to the question of the general state of polarization, having assured ourselves that we can generate plane waves and therefore can think of measuring polarization properties. Consider a total electric field of a z-propagating plane wave to be of the form

$$\mathbf{E}_t(\mathbf{r},\ t) = E_x(\mathbf{r},\ t)\hat{\mathbf{e}}_x + E_y(\mathbf{r},\ t)\hat{\mathbf{e}}_y \tag{2-32}$$

where

$$E_x(\mathbf{r},\ t) = a_x \cos\ (\omega t - kz + \varphi_x) \tag{a}$$
$$E_y(\mathbf{r},\ t) = a_y \cos\ (\omega t - kz + \varphi_y) \tag{b}$$
$$\text{(2-33)}$$

Note that in the complex notation of equation (2-7), a minus sign is associated with ωt and here, in the cos, a plus sign shows up. This does not have a physical consequence. However, the relative sign between ω_t and φ does. the convention used here is that of references (Papas 1965; Johnk 1975; Kraus 1984). Unfortunately, the widely used texts of references (Shurcliff 1962; Shurcliff and Ballard 1964; Jackson 1975) use just the opposite. We will soon see that this sign convention problem is compounded by the existence of contradictory electromagnetic and optical definitions of polarization handedness.

One could think of representing the total state of polarization of (2-32) as a Lissajous pattern, which would be traced out by letting the time vary in (2-33). One can actually find an analytical form for that pattern by using the following argument (Papas 1965). First, let $\tau = \omega t - kz$, and then use the sum angle formula to obtain

$$E_x/a_x = \cos\ \tau \cos\ \varphi_x - \sin\ \tau \sin\ \varphi_x \tag{a}$$
$$E_y/a_y = \cos\ \tau \cos\ \varphi_y - \sin\ \tau \sin\ \varphi_y \tag{b}$$
$$\text{(2-34)}$$

One can eliminate the τ from (2-34) to obtain

$$\left(\frac{E_x}{a_x}\right)^2 + \left(\frac{E_y}{a_y}\right) - 2\,\frac{E_x}{a_x}\frac{E_y}{a_y}\cos\varphi = \sin^2\varphi \tag{2-35}$$

where $\varphi = \varphi_y - \varphi_x$. Equation (2-35) is, in general, the equation of an ellipse and represents the locus that the tip of the electric polarization traces out during an optical period. We will presently consider some special cases of this ellipse. If one recalls the Lorentz force law

$$\mathbf{F}_q = q(\mathbf{E} + \mathbf{v} \times \mathbf{B}) \tag{2-36}$$

where $\mathbf{F}_q$ is the force felt by a particle of charge q traveling with a velocity $\mathbf{v}(t)$ under the influence of $\mathbf{E}$ and $\mathbf{B}$ fields, one can find the meaning of the path which the tip of the E-field traces out from the following argument. First, let us approximate the ratio of the magnitudes of the two terms of the right-hand size of (2-36) for z-propagating incident plane wave. Here, one can write that

$$\frac{|\mathbf{v} \times \mathbf{B}|}{|\mathbf{E}|} \leq \frac{|\mathbf{v}|\mu_0|\mathbf{E}|/\eta_0}{|\mathbf{E}|} \approx \frac{|\mathbf{v}|}{c} \tag{2-37}$$

If the particle is exhibiting nonrelativistic motion, therefore, the force from the B-field is much smaller than the force due to the E-field. Therefore, the force on an electron of charge $-e$ could be written as

$$\mathbf{F}_e = -e\mathbf{E} \tag{2-38}$$

Now using Newton's law $\mathbf{F} = m\mathbf{a}$ and writing a general form for $\mathbf{E}$ for a z-directed plane wave, one finds that

$$\frac{\partial^2 x}{\partial t^2} = -\frac{e}{m}\,E_x \cos\omega t \tag{a}$$

$$\tag{2-39}$$

$$\frac{\partial^2 y}{\partial t^2} = -\frac{e}{m}\,E_y \cos(\omega t + \varphi) \tag{b}$$

which for particles initially placed at the origin of the x–y coordinate system with zero velocity, has the solution

$$x(t) = \frac{eE_x}{m\omega^2}\,(\cos\omega t - 1) \tag{a}$$

$$\tag{2-40}$$

$$y(t) = \frac{eE_y}{m\omega^2}\,[\cos(\omega t + \varphi) - \cos\varphi] + \left[\frac{eE_y}{m\omega}\sin\varphi\right]t \tag{b}$$

which indicates that the motion of an electron under the influence of an incident plane wave approximately mimics the motion of the tip of the electric field of the plane wave incident, apart from a shift in both position and velocity induced by the initial wave interaction.

Let us now consider some specific motions that the tip of the polarization vector trace out, as defined by (2-35). For example, consider the case where $\varphi = 0$, which is to say $\varphi_x = \varphi_y$. Here, one finds that

$$\frac{E_x}{a_x} = \frac{E_y}{a_y} \tag{2-41}$$

and therefore the pattern traced out is a straight line, as is sketched in Figure 2.10(a). Now, let us consider the case where $\varphi = \pi/2$. In this case, equation (2-35) reduces to

$$\left(\frac{E_x}{a_x}\right)^2 + \left(\frac{E_y}{a_y}\right)^2 = 1 \tag{2-42}$$

and the resultant figure is an ellipse whose major axes coincide with the x and y coordinates, as is sketched in Figure 2.10(b). The figure shows the direction of rotation of the polarization vector. This circulation can be referred to either as left-handed or right-handed, depending on whether one is facing a source or sitting at the source and looking after the wave. Both of these conventions are used in practice, the first viewpoint by persons working in optics and the second by radio engineers and persons working with standards (see the discussion and footnote in Kraus 1984, 497). As was previously stated, the convention used here will be that of classical optics, which corresponds to the motion of the tip of the polarization vector as seen when one is standing in front of the wave and looking back toward the source, as one might do in a polarization experiment. Recall that in the figure, the wave causing the motion is traveling out of the paper as given by the right-hand rule, and therefore, by the optical convention, the state in Figure 2.10(b) is left-hand circularly polarized.

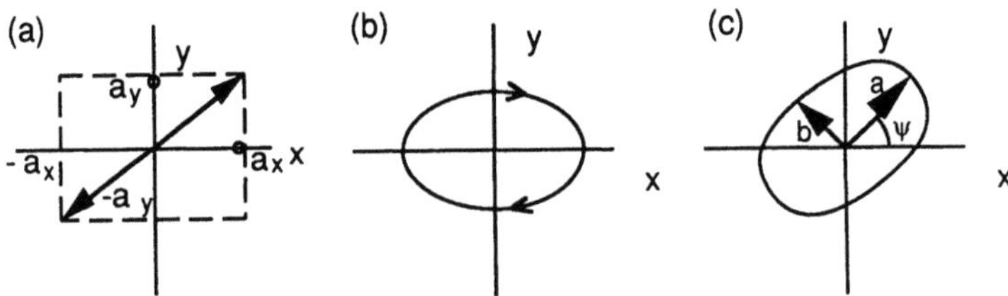

FIGURE 2.10.   Sketches of the pattern traced by the tip of the electric field vector **E** during an optical period.

Now we are ready to treat the general case, as sketched in Figure 2.10(c).
The parameters of the ellipse are determined from the set of equations

$$a^2 + b^2 = a_x^2 + a_y^2 \qquad \text{(a)}$$

$$b/a = \pm\tan \chi \qquad \text{(b)}$$

$$\sin 2\chi = [2a_x a_y/(a_x^2 + a_y^2)] \sin \varphi \qquad \text{(c)}$$

$$\tan 2\psi = [2a_x a_y/(a_x^2 - a_y^2)] \cos \varphi \qquad \text{(d)}$$

$$(2\text{-}43)$$

where a and b are half of the major and minor axes of the ellipse, respectively,
ψ is the angle that the major axis makes with plus x, and χ is an auxiliary angle
that defines the ellipticity and is often referred to as the ellipticity.

Figure 2.11 illustrates a set of Lissajous plots for $a_x = a_y = a$ for various
values of the phase difference φ varying from 0 to 2π. Such a set of plots is

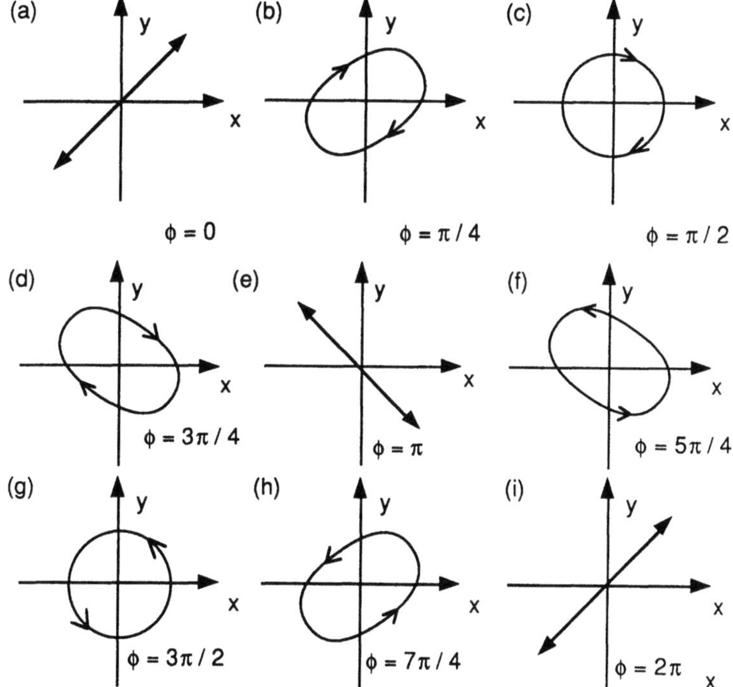

FIGURE 2.11.   Illustration of a set of Lissajous patterns, one for each $\pi/4$ increment of φ going
from 0 to 2π, for equal-amplitude polarization states. This set resembles the evolution of the polarization state for 45° excitation of an anisotropic crystal.

especially relevant to propagation in anistropic materials where the phase difference φ is a slowly varying (compared to a wavelength) function of the propagation distance. Such propagation will be taken up in some detail in Chapter 4, but it was deemed sufficiently important that one understand the effect of the phase angle φ upon the polarization state that such a figure was included here.

Now, clearly there must be some energy involved in setting about changes in particles' states of motion. This energy would have to be removed from the incident Poynting vector. Also, if the charges interacting with the incident plane wave were really free charges, they would be endowed with electromagnetic fields, which would be static electric fields before the wave interaction, but which would become both dynamic and radiative during the field interaction. These created fields would act back on the incident field by returning energy to it, although not necessarily in the same direction. Although the situation is really even more complicated inside a material, the following phenomenological picture is a little simpler to write down and much more illuminating than an accurate theoretical picture. Let us say that the material has no free charges, but only electrons bound to nuclei. There are no macroscopic fields in this case until the wave arrives. As will be discussed in much more detail in Chapter 3, the primary effect of the field in this case is to cause a small displacement of the electrons from their respective nuclei in a direction determined by the Lorentz force law. The energy required to do this will cause a reduction in the incident field value. However, simultaneously it leads to an increase in the displacement field value with respect to the free space constitutive relation value, and the new constitutive relation is expressible as

$$\mathbf{D} = \epsilon_0 \mathbf{E} + \mathbf{P} \tag{2-44}$$

where, in general, $\mathbf{P}$ does not have to be collinear with $\mathbf{E}$ (a case to be discussed in Chapter 4.) Using (2-44) in Poynting's theorem (2-20), one finds a new energy term W_{mat}:

$$W_{\text{mat}} = \int \frac{\partial \mathbf{P}(t)}{\partial t} \cdot \mathbf{E}(t) dV \tag{2-45}$$

which corresponds to the electromagnetic energy stored in the material. This energy will be exactly offset by the energy removed from the electric field because of the reduction in the electric field strength. As will be seen, this effect is the origin of the refractive index, which causes a slowing of energy flow in a material, and is the cause of dispersion.

Now, an alternative representation of the polarization vector of equation (2-32) would be a complex column vector representation such as

$$\mathbf{E}(\mathbf{r}) = e^{ikz} \begin{bmatrix} a_x e^{i\varphi_x} \\ a_y e^{i\varphi_y} \end{bmatrix} \tag{2-46}$$

When one ignores the common propagation factor, one obtains a representation originally due to Jones (1941a, b, 1942, 1947, 1948a, b; Hurwitz and Jones 1941), that of

$$\mathbf{E} = \mathbf{E}(0) = \begin{bmatrix} a_x e^{i\varphi_x} \\ a_y e^{i\varphi_y} \end{bmatrix} \tag{2-47}$$

which is referred to as the Jones vector representation, or, simply, the Jones vector. In what follows, we will try to give a review of some of the useful properties of this representation. [There are many books reviewing the properties of Jones vectors (see, for example, Azzam and Bashara 1989).]

Now the polarization states that we discussed in conjunction with Figure 2.10 are easily expressible as Jones vectors. The state of Figure 2.10(a) can be expressed as

$$\hat{\mathbf{e}}_{\psi 1} = \begin{bmatrix} \cos \psi \\ \sin \psi \end{bmatrix} = \cos \psi \hat{\mathbf{e}}_x + \sin \psi \hat{\mathbf{e}}_y \tag{2-48}$$

and the state orthogonal to this by

$$\hat{\mathbf{e}}_{\psi 2} = \begin{bmatrix} \sin \psi \\ -\cos \psi \end{bmatrix} = \sin \psi \hat{\mathbf{e}}_x - \cos \psi \hat{\mathbf{e}}_y \tag{2-49}$$

where $\hat{\mathbf{e}}_x$ and $\hat{\mathbf{e}}_y$ are, respectively, the unit amplitude (normalized) linear polarization states along the x and the y axes. They are normalized in the sense that

$$\hat{\mathbf{e}}_x^\dagger \hat{\mathbf{e}}_x = \hat{\mathbf{e}}_y^\dagger \hat{\mathbf{e}}_y = 1 \qquad \text{(a)}$$
$$\hat{\mathbf{e}}_x^\dagger \hat{\mathbf{e}}_y = \hat{\mathbf{e}}_y^\dagger \hat{\mathbf{e}}_x = 0 \qquad \text{(b)} \tag{2-50}$$

where the dagger denotes Hermitian transpose of the vector. That such a normalization is reasonable follows from the fact that the optical intensity I is given by equation (2-38), which, for the present column vectors, would be expressible as

$$2\eta_0 I = \mathbf{E}^\dagger \mathbf{E} \tag{2-51}$$

Now, one might look at (2-51) and conclude that the vectors of (2-48) and (2-49) should include $\sqrt{2\eta_0}$'s if they are to be considered normalized to unity intensity of power flow, and then equation (2-50) should contain a $2\eta_0$ on its right-hand side. One could counter that one is generally not concerned with absolute intensity. This is reasonable for free space propagation, as the normalization does not change with position, and in what follows we will stick to the normalization of (2-50). It should be pointed out, however, that for propagation in materials, or multimode waveguide propagation, relative intensities will be adversely affected by such a normalization, and one must use more care in such cases.

Now, in line with the discussion following Figure 2.10, the unit vectors of the left (as in Figure 2.10b) and right-hand circularly polarized states will be given as

$$\hat{e}_r = \frac{1}{\sqrt{2}} \begin{bmatrix} 1 \\ -i \end{bmatrix} \qquad \text{(a)}$$

$$\hat{e}_\ell = \frac{1}{\sqrt{2}} \begin{bmatrix} 1 \\ i \end{bmatrix} \qquad \text{(b)}$$

$$(2\text{-}52)$$

As before, these states are normalized such that

$$\hat{e}_r^\dagger \hat{e}_r = \hat{e}_\ell^\dagger \hat{e}_\ell = 1 \qquad \text{(a)}$$

$$\hat{e}_r^\dagger \hat{e}_\ell = \hat{e}_\ell^\dagger \hat{e}_r = 0 \qquad \text{(b)}$$

$$(2\text{-}53)$$

This is to say that, just as the x and y axes can be considered as orthogonal, so can the right and left axes as defined by (2-52). The situation is analogous, if a little bit more complicated, for general polarization states. For the sake of argument it is perhaps best to begin with principal axis system elliptical states in which the angle ψ of Figure 2.10 is zero. Now, just as one could consider the state of equation (2-49) as a rotated generalization of the state of equation (2-48) one could consider the state

$$\hat{e}_{xr} = \begin{pmatrix} \cos \chi \\ -i \sin \chi \end{pmatrix} \qquad (2\text{-}54)$$

as a right-hand rotating, elliptically generalized version of (2-52) (a), as can be verified by finding the real parts and plotting a Lissajous pattern. The parameter χ is referred to as the ellipticity of the state. Now, if one state $\hat{e}_1$ of an orthonormal pair is known, the two components of the other state $\hat{e}_2$ can always be determined from the two equations

$$\hat{e}_1^\dagger \hat{e}_2 = \hat{e}_2^\dagger \hat{e}_1 = 0 \qquad (2\text{-}55)$$

and applying these conditions to (2-54), one finds that one can complete this pair with its left-handed mate

$$\hat{\mathbf{e}}_{\psi\ell} = \begin{bmatrix} \sin \chi \\ i \cos \chi \end{bmatrix} \tag{2-56}$$

To complete the discussion of a general elliptical state, it can be of use to give some consideration to the case of coordinate rotations. Considering Figure 2.12, one an easily convince oneself that the components of a vector $\mathbf{V}$ in a primed coordinate frame are given by

$$V'_x = V_x \cos \psi + V_y \sin \psi \qquad \text{(a)}$$
$$V'_y = -V_x \sin \psi + V_y \cos \psi \qquad \text{(b)} \tag{2-57}$$

which can be expressed in a matrix form:

$$\mathbf{V}' = \mathbf{R}(\psi)\mathbf{V} \tag{2-58}$$

or as:

$$\mathbf{V} = \mathbf{R}(-\psi)\mathbf{V}' \tag{2-59}$$

where

$$\mathbf{R}(\psi) = \begin{pmatrix} \cos \psi & \sin \psi \\ -\sin \psi & \cos \psi \end{pmatrix} \tag{2-60}$$

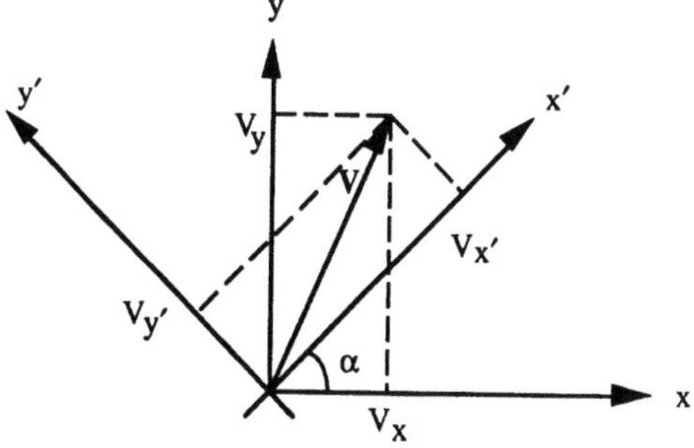

FIGURE 2.12.   Schematic depiction of the effect of rotating coordinates from an x-y to an x'-y' coordinate frame.

Perusing equation (2-57), one immediately sees how equations (2-48) and (2-49) were obtained from the basis states of (2-50)—by a straight inverse coordinate rotation, that is, by matrix multiplying $R(-\psi)$ by each of the basis vectors $\hat{e}_x$, $\hat{e}_y$. There is more than abstract meaning to such a rotation. If a component is designed to, for example, change the phase of one linear polarization state with respect to the other, then the axis of this component is important. If the component is rotated, that rotation will have the same effect as rotating the input states, as in equation (2-59). There will be more on this later. Here, the important point is to find the general elliptical states. Equations (2-54) and (2-56) give these states in a major axis coordinate system. To put these states into a system at an angle ψ to the x-y system, one need only apply $R(-\psi)$ to these states to obtain

$$\hat{e}_{xr\psi} = R(-\psi)\hat{e}_{xr} = \begin{bmatrix} \cos\psi\cos\chi + i\sin\psi\sin\chi \\ \sin\psi\cos\chi - i\cos\psi\sin\chi \end{bmatrix} \quad \text{(a)}$$

$$\hat{e}_{x\ell\psi} = R(-\psi)\hat{e}_{x\ell} = \begin{bmatrix} \cos\psi\sin\chi + i\sin\psi\cos\chi \\ \sin\psi\sin\chi - i\cos\psi\cos\chi \end{bmatrix} \quad \text{(b)}$$

$$(2\text{-}61)$$

where the normalization of the vectors is unaffected because

$$\det R(-\psi) = 1 \quad \text{(a)}$$

$$R^{\dagger}(\psi) = R(-\psi) \quad \text{(b)} \quad (2\text{-}62)$$

$$R^{\dagger}(\psi)R(\psi) = I \quad \text{(c)}$$

where det denotes the matrix determinant and I the identity matrix, and, again, the dagger denotes Hermitian transpose. The states of (2-61) now should represent the most general representations of the two orthonormal polarization states whose major axes lie at angle ψ to the x-y coordinate system, and which exhibit ellipticity χ.

If one is not interested in the full knowledge necessary to reconstruct the propagating wave, then the Jones vector is really more information than is needed. The vector really consists of four pieces of information, namely the amplitude and phase of each of two complex components. If we could factor out a given phase and amplitude, as in essence we have already done with the basis states through normalization, there should be only two degrees of freedom left, which therefore should be representable as a single complex quantity. Indeed, if we take the ratio of the two components of a Jones vector, we obtain just such a complex number. Defining the complex number γ by

$$\gamma = \frac{E_y}{E_x} \qquad (2\text{-}63)$$

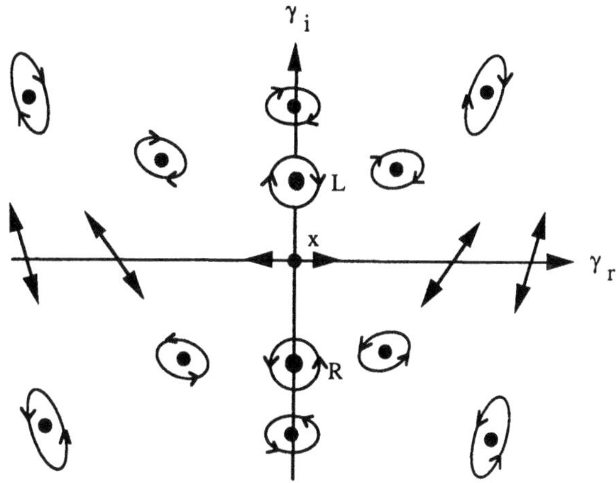

FIGURE 2.13.   Schematic depiction of the various monochromatic polarization states and where they are located in the complex plane defined by the γ of equations (2-63) and (2-64) (a).

for an arbitrary Jones vector, and then particularizing to our "general" states of equation (2-61) (a) and (b), we find that

$$\gamma_1 = \frac{\tan \psi - i \tan \chi}{1 + i \tan \psi \tan \chi} \qquad \text{(a)}$$

$$\gamma_2 = \frac{\tan \psi - i \cot \chi}{1 + i \tan \psi \cot \chi} \qquad \text{(b)}$$

(2-64)

where the subscripts 1 and 2 denote the (arbitrary) number of that elliptical basis state.

Concentrating attention on the γ_1 of equation (2-64) (a), we can plot points in the complex plane to indicate given polarization states, which is done in Figure 2.13. As can easily be verified, the linear states lie on the $\gamma_i = 0$ axis, with the x state at the origin and the y state at infinity. Left and right states both lie on the $\gamma_r = 0$ axis reflected across the γ_i axis. For a given polarization state, its orthogonal state is given by (2-64) (b), but is not necessarily easy to find as in the x–y and right–left cases. More discussion of orthogonal states is given in Section 2.5, where some practical device applications are considered.

2.4 POLYCHROMATIC PLANE WAVES

In this section, discussion will be given to generalizing the salient characteristics of monochromatic waves to the more realistic polychromatic case. The section begins with a discussion of thermal sources. This discussion shows that thermal sources are the most prevalent ones at optical frequencies because of the attendant efficiency possible. The discussion of rate equations shows that optical sources will always be plagued with spontaneous emission noise, when compared with comparable size, comparable temperature microwave sources. The monochromatic approximation is then seen to be more valid for the lower frequency regime. This discussion then turns to generalizing the Poynting vector to the polychromatic case, as well as finding better, more physical, representations of the polychromatic electromagnetic field. With these polychromatic results in hand, it is possible to give a plausibility argument for the Wiener-Khintchine theorem, which then is used to discuss the linewidths (extents of polychromaticity) of nonthermal sources, including gas and semiconductor lasers. The next topic of the section is again polarization and, indeed, it is immediately seen that the Jones vector representation of Section 2.3 is no longer valid, at least, in a time-averaged sense, that is, as seen by a ''normal'' detector. Discussion then naturally turns to Stokes parameters and the Poincaré sphere representation of partial polarization.

Before the demonstration of the laser in 1960 (Maimon 1960), essentially all optical sources were thermal (with the exception of particle accelerators, if one is to consider them as pre-1960 light sources). Even laser sources exhibit varying degrees of polychromaticity which induce partial polarization and partial coherence. Thermal sources in their unfiltered state, however, exhibit almost no coherence at all, and coherence and partial polarization of such sources are induced only through filtering operations. As the following discussion will show, the ubiquity of thermal optical sources is tied to their efficiency, which in turn is closely tied to their coherence (polychromatic) properties. By contrast, thermal microwave sources are neither efficient nor especially incoherent, to a large extent because of the ambient conditions in which we live. Hopefully, the following discussion will help to clear up these issues, as well as to convince one of the necessity of studying the properties of partially coherent light.

To begin the discussion of thermal sources, we will consider the concept of a blackbody. A blackbody is an ideal thermal electromagnetic body that, by absorbing all the radiation incident on it and reemitting enough to remain in thermal equilibrium with its surroundings, comes to equilibrium (or remains at equilibrium) at a temperature. The idealization comes into the problem in that the object can only interact radiatively with its surroundings and not through conduction or convection. In practice, however, one could (closely) approximate such a material by placing a highly conductive element inside an evacuated chamber and running a current through it. The vacuum in the chamber precludes convection, the distance to the walls of the vessel precludes conduction, and the current, which is really comprised of an electromagnetic field dragging electrons into material collision, acts as an almost totally absorbed incident electromagnetic radiation stream. Of course, such a configuration would lead to poor radiation efficiency, and light bulb as well as arc lamp makers have learned to backfill the tubes with a gas that radiates a strong line spectrum upon heating.

However, let us go back to the blackbody idealization. To calculate what radiation must be emitted in order to keep the material at temperature T, one needs to place the radiator in a large resonator whose walls will be allowed to recede to infinity. To calculate the energy spectrum one needs to find an expression for the average energy per mode at a given frequency, and then to scan over the mode spectrum. Now, if each mode is a plane wave labeled by its wave vector $\mathbf{k}$ and polarization state, then the number of modes per volume in an isotropic space must be

$$\frac{\text{Modes}}{\text{Volume}} = \frac{2 \text{ polarizations} \times d^3k}{(2\pi)^3} = \frac{k^2}{\pi^2}\, dk \qquad (2\text{-}65)$$

where the d^3k is the number of modes between k and $k + dk$, and the $(2\pi)^3$ is a unit volume. Therefore, the energy per volume per frequency interval $\rho(\omega)$, must be given by

$$\rho(\omega) = \frac{\omega^2}{\pi^2 c^3} \langle E(\omega) \rangle_m \qquad (2\text{-}66)$$

where it has tacitly been assumed that the average energy per mode $\langle E(\omega) \rangle_m$ can be written as a function of the mode angular frequency $\omega = kc$ alone.

The solution of the blackbody energy problem is one that relates directly to the roots of quantum mechanics (see, for example, Hermann 1971, Chapters 1 and 2). The classical solution to the problem was to use equipartition of energy and assume that the energy per mode was the Boltzmann constant k_B times the temperature T, and to obtain an expression that, when integrated over frequency, gave an infinite energy/volume. Planck solved the problem by assuming that the electromagnetic radiation was quantized into quanta of energies $h\nu$, where h is Planck's constant and $\nu = \omega/2\pi$ is the frequency, and then calculated the average energy assuming that the probability of occupancy of an energy level satisfied a Boltzmann distribution. (For details, see any elementary text on statistical thermodynamics, such as Kittel and Kroemer 1980, Chapter 4.) The result for the energy density $\rho(\nu)$, written in terms of the frequency ν is

$$\rho(\nu) = \frac{8\pi h \nu^3}{c^3(e^{h\nu/k_B T} - 1)} \qquad (2\text{-}67)$$

Now, expression (2-67) well explains the emissions of heated filaments, light bulbs (excepting the strong line emissions), and stars, but we want to learn a little more about such things as efficiency and coherence of the source. Equation (2-67) itself suffices for an efficiency discussion. The shape of the curve of

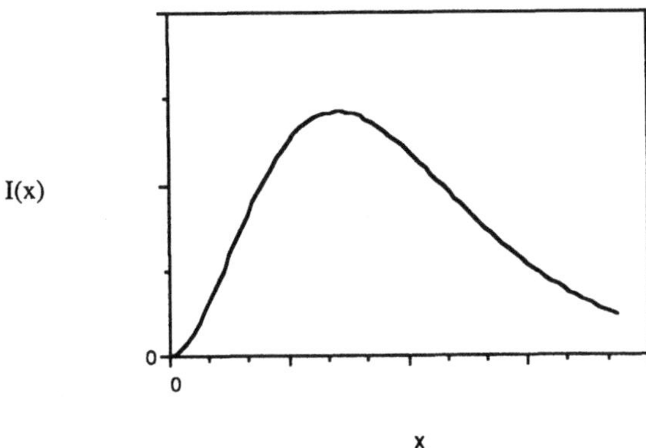

FIGURE 2.14.   A sketch of the Planck distribution of equation (2-67).

(2-67) is sketched in Figure 2.14. As is seen from Figure 2.14 (or could be derived by solving for the maximum of (2-67)) the peak of the curve lies in the vicinity of the frequency:

$$\nu_{MAX} = 3\,\frac{k_B}{h}\,T \approx 6 \times 10^{10}\,T\ \ \text{Hz} \tag{2-68}$$

With reference to the frequency line of Chapter 1, Figure 1.1, it is readily seen that for temperatures greater than a few kelvins, the microwave spectrum lies on the extreme left-hand part of the curve, and is only weakly radiated. For room temperature, where $T \sim 300°K$, the peak lies in the infrared at roughly a 15 μm wavelength. For temperatures such as those of molten metals or stars, where T is in tens of thousands, the peak lies in the visible or even the ultraviolet portion of the spectrum. Indeed, the reason why we can make efficient optical thermal sources is that the temperature range of interest is easily available to us.

Investigation of the coherency properties of optical sources requires a further argument beyond the result of (2-67). Consider a hollowed-out blackbody, where the hollow enclosure within forms a cavity that we will assume is large compared to the wavelength in which we are interested. We will further assume that something outside of this cavity is keeping the blackbody walls at a temperature T. Now, for a large-enough resonator, one would expect a mode density like that of equation (2-65). Our blackbody enclosure is not a resonator, in that the walls are totally absorbing and not totally conducting as they should be in a resonator. But the idea that something is totally absorbing is in itself an abstraction, and it thus cannot be true independent of frequency. But let us say that this condition can nearly be. Still, to retain equilibrium the walls must also

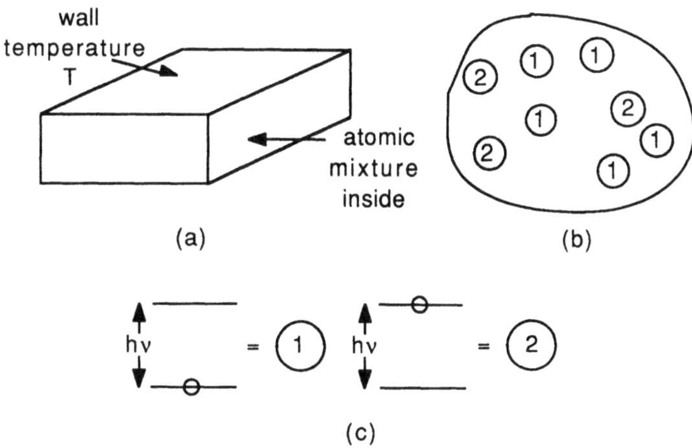

FIGURE 2.15.   Descriptive depiction of the blackbody enclosure and its contents. (a) The enclosure, filled with two-level atoms at equilibrium with the walls and outside reservoir at a temperature T, (b) a blow-up of a sphere filled with two-level atoms, and (c) definitions of the upper and lower states in terms of an energy-level diagram.

reradiate as much as they absorb and, on the average, in the same direction that they absorbed it in. This enclosure looks like a resonator, except that the reflected radiation is not coherent with the incident. So let us assume that the radiation density inside agrees pretty well with (2-67). Now let us fill the enclosure with two-level atoms, as is depicted in Figure 2.15. (By a two-level atom, we mean one in which there are only two states, and these states are separated by an energy $h\nu$). These atoms within the cavity must interact with the radiation in the cavity in such a manner that they also come to thermal equilibrium with the walls and thereby the outside surroundings.[1] It should be pointed out here that this attainment of equilibrium can take quite some time. The arguments used in what follows, however, are valid during this approach period; and because they are valid during approach, the equations derived when equipped with time derivatives can describe even situations that may be far from equilibrium. After the following argument is completed, we will investigate several nonequilibrium situations.

There are three processes (depicted schematically in Figure 2.16) by which the atoms and radiation can interact. The first is the well-known process of absorption, whereby an atom in its lower state absorbs a photon and therefore is raised to its upper state. A second process is spontaneous emission, whereby

---

[1]The original treatment of induced and spontaneous transitions was given in Einstein (1917). Summaries of the treatment can be found in various textbooks on lasers, as, for example, Yariv 1989, Section 8.5.

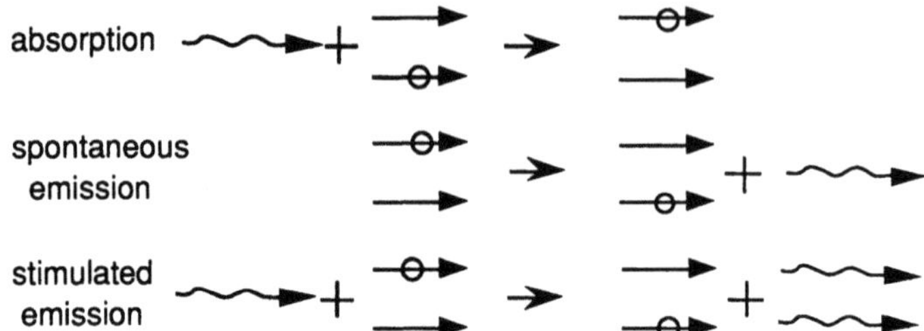

FIGURE 2.16.   Schematic depiction of the three processes involved in bringing and keeping an atomic system in equilibrium with a radiation field. The wavy lines indicate photons, the straight lines mark the direction of the interaction, and the energy-level diagram gives the atomic state.

an atom in its upper state emits a photon of frequency ν and thereby relaxes to its lower state. A less well-known interaction, although the one that is basic to laser action, is stimulated emission, whereby a photon of energy $h\nu$ interacts with an atom in its upper state and causes the atom to relax to its lower state while emitting a photon that is exactly "locked" to the frequency, phase, and direction of propagation of the incident photon. These radiation/matter interactions must settle down to constant rates of energy exchange for the composite system to obtain equilibrium. The situation is schematically depicted in Figure 2.17, where the upward transition rate is denoted by w_{12} and the downward rate by w_{21}. Among other things, thermal equilibrium must imply that the rate of upward transitions must exactly equal the rate of downward transitions, and therefore

$$w_{12}N_1 = w_{21}N_2 \text{ (equilibrium)} \tag{2-69}$$

where N_1 and N_2 are the number of atoms in states 1 and 2, respectively. If one denotes the absorption constant by B_{12}, the spontaneous emission rate by A_{21}, and the stimulated emission constant by B_{21}, one can write

$$w_{21} = B_{12}\rho(\nu) \tag{a}$$
$$w_{21} = B_{21}\rho(\nu) + A_{21} \tag{b}$$

$$(2\text{-}70)$$

Now, what is to be done here is to determine at what frequencies spontaneous emission dominates stimulated emission, and vice versa. It is clear that which

$$w_{12} \Big(\boxed{} \Big) w_{21}$$

FIGURE 2.17.   Schematic depiction of the upward and downward transition rates.

process dominates will have a profound effect on the classical properties of the source, as spontaneous emission is essentially a noise process and stimulated emission is essentially a noise-free quantum mechanical amplification process. To continue, recall that in thermal equilibrium one can write that (Kittel and Kroemer 1980, Chapter 3):

$$N_2/N_1 = e^{-h\nu/k_B T}\text{(equilibrium)} \tag{2-71}$$

Using (2-70) (a) and (2-71) in (2-69), gives

$$w_{21} = B_{12}\rho(\nu)e^{h\nu/k_B T} \tag{2-72}$$

Using (2-72) in (2-70) gives

$$A_{21} = \rho(\nu)(B_{12}e^{h\nu/k_B T} - B_{21}) \tag{2-73}$$

Substituting (2-67) in (2-73) yields

$$A_{21} = \frac{8\pi h\nu^3}{c^3}\frac{(B_{12}e^{h\nu/k_B T} - B_{21})}{e^{h\nu/k_B T} - 1} \tag{2-74}$$

Now, clearly we are not going to be able to solve for the three quantities in (2-74), that is, A_{21}, B_{21}, and B_{12}, all uniquely without including complete knowledge of the dynamics of the specific atomic system in which we are interested. We should (and will see that we can) go further than equation (2-74), however, and interrelate all of the atomic coefficients A_{21}, B_{21}, and B_{12} so that they can all be expressed in terms of a single one. That single one (whichever one chooses to make it), therefore, must be expressible in terms of the atomic constants. But atomic constants must be independent of temperature, as temperature is a macroscopic quantity and requires averages over multitudes of atoms before it can become meaningful. Any relationship between A_{21}, B_{21}, and B_{12} must, then, be independent of temperature. The only way that (2-74) can be independent of temperature is for

$$B_{21} = B_{12} \tag{2-75}$$

yielding the result that

$$\frac{A_{21}}{B_{21}} = \frac{8\pi h\nu^3}{c^3} \tag{2-76}$$

which was derived for equilibrium, but is in general true. For spontaneous emission to dominate stimulated emission it is necessary that A_{21} exceed $B_{21}\rho(\nu)$, which in turn requires that $e^{h\nu/k_B T} - 1 > 1$, which gives

$$\nu > \frac{.7 k_B T}{h} \tag{2-77}$$

Using the previously quoted values for k_B and h, that is, values $k_B T \sim 4.2 \times 10^{-14}$ erg (room temperature or $T = 300°$K) and $h = 6.6 \times 10^{-27}$ erg sec, one finds that the spontaneous emission regime is defined by

$$\nu > 4.5 \times 10^{12}/\text{sec} \qquad \text{(a)}$$
$$\lambda = c/\nu < 67 \ \mu\text{m} \qquad \text{(b)} \tag{2-78}$$

Before we see what consequences (2-78) has when discussing specific radiation systems, it can be of interest to try to understand why such a relation exists. One could rewrite equation (2-76) in the form

$$A_{21} = (C)(h\nu)(\nu^2)B_{21} \tag{2-79}$$

where C is a constant with respect to the frequency ν, and $h\nu$ is the photon energy. Equation (2-79) is a telling form, in that the factor ν^2 that shows up there is just the mode density factor of equation (2-65). This essentially tells us that as the frequency increases, there are more modes that a photon can spontaneously emit into, and therefore the spontaneous emission rate increases with respect to the stimulated rate. Thus, for a given cavity size, the lower the frequency, the lower the spontaneous emission. Although the derivation was made by assuming that the cavity was very multimoded, the relation of stimulated emission rate to frequency holds all the way down to single-mode cavities. The fact is that the original masers were built in single-mode cavities to avoid the spontaneous emission problem, and it was four years after the demonstration of the maser that Schawlow and Townes (1958) realized that laser action might be possible despite spontaneous emission. It was another two years before Maimon (1960) was able to demonstrate the principle that a high-enough gain could overcome the difficulties associated with a high spontaneous emission rate. In a later section, the laser linewidth, which is primarily due to spontaneous emission, will be derived.

The above thermal emission discussion can now be applied to a discussion of the nature of electromagnetic sources. At radio frequencies and below, most electromagnetic generation is done with lumped elements, and the radiation is guided on wires. In such cases the wavelength is large, and although the cavity is large (infinite perhaps) there is essentially no spontaneous rate, as there are no atomic transitions involved, and if there were any, the thermal rate would be very low because of the position of the frequency regime on the blackbody curve. Masers are the lowest-frequency (microwave) generators to use atomic transitions. As mentioned above, masers are often housed in single-mode cavities, as are other microwave sources such as klystons, if only to assume single-mode generation with the attendant spectral purity. Masers are often cooled, not to shift the blackbody curve to a lower peak, but rather to limit the total spontaneous emission, as the energy emission of a blackbody is proportional to T^4. It would be very hard to make single-mode lasers, as the total size of the cavity would have to be roughly one cubic micron. Therefore, in the laser, one tries to achieve a high-enough stimulated gain to overcome the spontaneous loss. There are ongoing efforts to lower spontaneous rates through the use of carrier confinement (quantum wells, quantum dots, etc.) and even few-mode cavities. The majority of lasers, however, are grossly multimoded and have a spectral purity defined by the spontaneous emission rate. This is a major reason why partial coherence theory is even more important in the laser age than it was in the thermal source age. X-ray lasers pose a complicated technological problem. They have been demonstrated (see, for example, Trebes et al. 1987) but are still a long way from existing in a hand-held continuous wave (CW) form. Spontaneous emission in the X-ray regime is a serious problem in the room-temperature world. In the interior of stars ($T \approx 50,000$ K°) it might be easier to make one, but there are no volunteers ready to set up such an experiment.

It has already been mentioned that monochromatic calculations can be simplified by removing the time dependence of the field through the prescription

$$\mathbf{E}(\mathbf{r},\ t) = \text{Re}\ [\mathbf{E}(\mathbf{r})e^{-i\omega t}] = \text{Re}\ [\mathbf{V}(\mathbf{r},\ t)] \tag{2-80}$$

with similar definitions for other quantities of interest. Clearly, if one knew the response of the system for each ω of a given spectral range, one should be able to reconstruct a polychromatic response by summing over these elemental responses. Indeed, this is generally done in the following manner. One can define the Fourier spectrum of $\mathbf{E}(\mathbf{r},\ t)$ by

$$\mathbf{E}(\mathbf{r},\ \omega) = \frac{1}{2\pi} \int_{-\infty}^{\infty} \mathbf{E}(\mathbf{r},\ t)e^{i\omega t}dt = \frac{1}{2\pi}\ \text{Re}\ \left[\int_{-\infty}^{\infty} \mathbf{V}(\mathbf{r},\ t)e^{i\omega t}dt\right] \tag{2-81}$$

and then one finds that $\mathbf{E}(\mathbf{r},\ t)$ can be expressed as

$$\mathbf{E}(\mathbf{r},\ t) = 2\text{Re}\ \left[\int_{0}^{\infty} \mathbf{E}(\mathbf{r},\ \omega)e^{-i\omega t}d\omega\right] \tag{2-82}$$

where the quantity between the brackets on the right-hand side of (2-82) is often referred to as $\mathbf{V}(\mathbf{r},\ t)$, the analytical signal representation of the real signal

$E(\mathbf{r}, t)$. This representation will be used later when more general partially coherent signals are considered.

Although (2-82) often is a very good formal representation for a polychromatic wave, one often can get by with a simpler, more physically motivated form, at least for the case of laser light, or strongly bandlimited light, for that matter. Let us consider one component of the electric field of a wave whose spectrum is strongly peaked about an average frequency ω. It would stand to reason that such a disturbance would have a temporal variation something like that depicted in Figure 2.18. Here the wave, on the average, has the form of a cosinusoidal amplitude variation. Both the phase and the amplitude of the wave, however, vary randomly around the φ_0 point and the a_0 value, respectively. A possible representation of this single polarization state would be

$$E_x(t) = \mathrm{Re}[a_x(t)e^{-i\overline{\omega}t}e^{-i\varphi_x(t)}] = a_x(t) \cos{(\overline{\omega}t + \varphi_x(t))} \qquad (2\text{-}83)$$

where $a_x(t)$ and $\varphi_x(t)$ are random processes of mean a_0 and 0 respectively. Plugging (2-83) into a Fourier transform relation of the form of (2-81), one finds

$$E_x(\omega) = \frac{1}{2\pi} \mathrm{Re}\left[\int_{-\infty}^{\infty} a_x(t')e^{-i\varphi(t')}e^{i(\omega - \overline{\omega})t'}dt'\right] = \mathrm{Re}\,[V_x(\omega)] \qquad (2\text{-}84)$$

Unfortunately, (2-84) has little computational meaning, as one must perform (2-84) for an infinite number of realizations for it to represent the process described by (2-83). One can remedy this problem somewhat by finding a power spectrum, defined by $V_x^*(\omega)V_x(\omega)$. Performing the $V_x^*(\omega)V_x(\omega)$ operation with

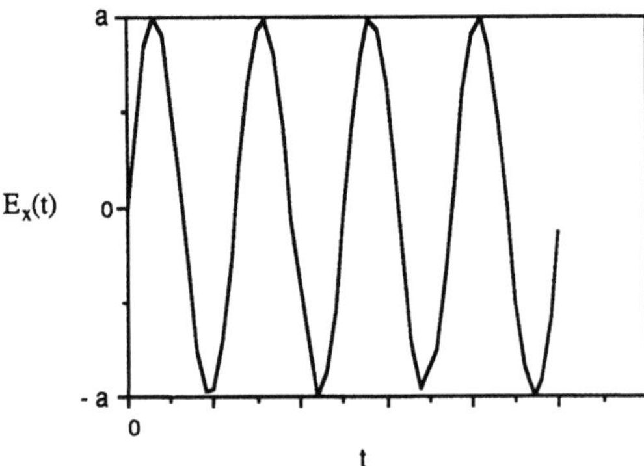

FIGURE 2.18.   Plot of a possible realization of the equation $E_x(t) = a_x(t) \cos{(\omega t + \varphi(t))}$ where $a_x(t)$ is Gaussian, and $\varphi(t)$ is folded Gaussian.

some self-evident simplifications, one finds

$$V_x^*(\omega)V_x(\omega) = \frac{1}{4\pi^2} \int_{-\infty}^{\infty} dt'' e^{i(\omega - \bar{\omega})t''}$$

$$\cdot \int_{-\infty}^{\infty} dt' a_x(t')a_x(t' + t'')e^{i(\varphi_x(t');\varphi_x(t' + t''))} \qquad (2\text{-}85)$$

which is a Fourier transform, centered about $\bar{\omega}$, of the autocorrelation of the analytical signal representation of the field. This is almost a form of the Wiener-Khintchine theorem. It should be noted here that $V_x(t)$ is assumed to be mean-squared-integrable, and that suitable time averaging would be necessary were the function not to be time-limited. It should also be noted here that the a_x's and φ_x's must be defined such that the autocorrelation is a symmetric function of t'', as this is the condition for the transform to be real, as it must be.

Perhaps the best way to understand (2-85) is by considering a few specific cases. For example, say that the temporal variation of φ is much slower than that of a. This will be the case in essentially any laser. The point is that phase fluctuations due to spontaneous processes are not damped out but tend to grow slowly because of stimulated amplification. The main effect of spontaneous emission is to generate an almost white component to the amplitude spectrum. The situation is as depicted in Figure 2.19, where, as indicated, the phase noise determines the low frequency part of the spectrum, which in turn determines the linewidth, whereas the amplitude noise contributes a pedestal upon which the phase noise sits. In a gas laser, the lasing medium is rarefied, and therefore the spontaneous emission rate will be low, especially in comparison with the photon lifetime, which can be very large because of high-reflectivity mirrors and the low absorptivity of the rarefied gaseous medium. As the photons are long-lived, the phase within the cavity can remain constant for a relatively long

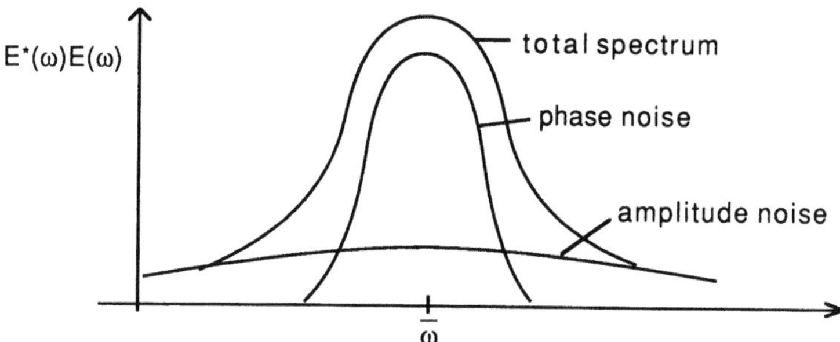

FIGURE 2.19.   A sketch of a possible spectrum of a laser source, where the phase noise and intensity noise spectra are plotted both as separate entities and as a sum.

period (inverse linewidth). The amplitude, however, will not. Again, as the medium is rarefied, the gain will be low, and each time that a photon is stimulatedly amplified, a significant change of the cavity photon number will take place, causing a granular change in the output amplitude. A normal gas laser, however, will exhibit some degree of multiple-cavity-mode operation (as will be discussed in Chapter 5). As all the modes will be at slightly different frequencies, their phases will be uncorrelated, and thus, depending on the total number of modes, will contribute an amount of phase noise. As the modes can also have amplitude coupling, not only the total linewidth of the laser but also the width of each spectral line can be greatly expanded when going from single to multimode operation.

One can calculate the form of a single-mode laser's line shape from the above considerations, and, indeed, it is an instructive exercise to carry out. (This case is treated in, for example, Yariv 1989, Chapter 21.) Oftentimes, the Wiener-Khintchine theorem is written in the form

$$W(\omega) = \frac{1}{2\pi} \int_{-\infty}^{\infty} R(\tau)e^{i\omega\tau}d\tau \qquad (2\text{-}86)$$

where $W(\omega)$ is the spectral density function (which is proportional to an average of $E*E$), R is the autocorrelation function of V_x, defined by

$$R(\tau) = \mathcal{E}[V_x^*(t + \tau)V_x(t)] \qquad (2\text{-}87)$$

where the $\mathcal{E}$ operator is an expectation operator. Now in order for $R(\tau)$ to be independent of time, one must assume the V_x to have time-stationary statistics. In such a case, however, $V_x(t)$ is not mean-squared-integrable, and so the form of (2-85) is no longer valid as the inner integral must diverge. This problem can be taken care of by expressing the expectation operator of (2-87) in the form

$$\mathcal{E} = \mathcal{E}_{ax}\mathcal{E}_{\varphi x} \lim_{T \to \infty} \frac{1}{2T} \int_{-T}^{T} dt' \qquad (2\text{-}88)$$

where $\mathcal{E}_{ax}$ and $\mathcal{E}_{\varphi x}$ are the expectation operators for the amplitude and phase, defined by

$$\mathcal{E}_{ax}(f(a_x)) = \int da_x p(a_x) f(a_x) \qquad \text{(a)}$$
$$\qquad (2\text{-}89)$$
$$\mathcal{E}_{\varphi x}(f(\varphi_x)) = \int d\varphi_x p(\varphi_x) f(\varphi_x) \qquad \text{(b)}$$

where $p(a_x)$ $(p(\varphi_x))$ is the probability density of $a_x(\varphi_x)$, f is an arbitrary function, and the last operator on the right-hand side of (2-88) is the time-average operator, often denoted by closed angular brackets. Now, in the preceding paragraph it was argued that the amplitude fluctuations were unimportant for linewidth determination. Ignoring the amplitude fluctuations, denoted $\Delta\varphi(t, \tau) = \varphi(t) - \varphi(t + \tau)$, one can rewrite (2-86) with (2-87) and (2-88) as

$$W(\omega) = \int d\tau \, \frac{a_x^2}{\pi} \, \langle \mathcal{E}_{\Delta\varphi(t,\tau)}(e^{i\Delta\varphi(t,\tau)}) \rangle \, e^{i(\omega - \bar{\omega})\tau} \tag{2-90}$$

Now, assuming that the fluctuation term $\Delta\varphi$ is due to multiple independent spontaneous emission events, one can invoke the central limit theorem to express the distribution of $\Delta\varphi(t, \tau)$ in the Gaussian form:

$$p(\Delta\varphi) = \frac{1}{\sqrt{2\pi\langle(\Delta\varphi)^2\rangle}} \, e^{-(\Delta\varphi)^2/2\langle(\Delta\varphi)^2\rangle} \tag{2-91}$$

to simplify (2-90) to the form

$$W(\omega) = \frac{a_x^2}{2\pi} \int d\tau e^{-i(\omega - \bar{\omega})\tau} e^{-\langle(\Delta\varphi)^2\rangle/2} \tag{2-92}$$

As we have already assumed $\langle(\Delta\varphi)^2\rangle$ to be time-independent, without loss of generality one can take

$$\tfrac{1}{2}\langle(\Delta\varphi)^2\rangle = C|\tau| \tag{2-93}$$

where C is some constant, to obtain the Lorentzian lineshape

$$W(\omega) = \frac{K}{(\omega - \bar{\omega})^2 + C^2} \tag{2-94}$$

where K is another constant.

A second interesting example of a coherent optical source is that of the semiconductor laser. Here the host medium is dense, and therefore the gain and spontaneous emission rate are high. As a semiconductor laser is short and the facet mirrors are cleaved and thus of low reflectivity, the photon lifetime is short, making spontaneous emission an even more important intercavity effect. Two salient features of spontaneous emission are illustrated in Figure 2.20. Figure 2.20(a) illustrates the field due to spontaneous emission at high absorption, that is, below threshold. Although each event has a fixed amplitude, cor-

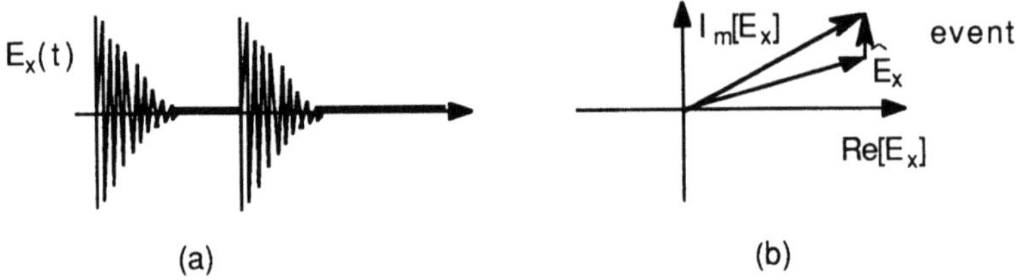

FIGURE 2.20.   Illustration of effects of spontaneous emission. (a) the electric field of a low-rate spontaneous source; (b) the effect of a spontaneous event on the electric field phasor of an operating source.

responding to one photon, the phases are completely random. In a thermal source, such photons are the only light carriers, and so, although a thermal source can be reasonably well amplitude-stabilized, the phases will always be completely uncorrelated between adjacent time periods. Because of the high gain in the semiconductor laser, however, a single-mode semiconductor laser can achieve a nominally fixed phase output. However, as is illustrated in Figure 2.20(b), each spontaneous event, and the rate of such events, is as large as the photon lifetime and will cause the phase and amplitude to stray. The situation is even more serious than has been explained, as the small change in amplitude causes a change of refractive index in the cavity and therefore causes the phase to shift still more, and so on, leading to strong amplitude phase coupling in the semiconductor laser. As in the gas laser, the slowly varying phase fluctuations will determine the laser linewidth although here there will be a linewidth enhancement factor due to the phase amplitude coupling. Operationally, this means that the C in equation (2-93) is enhanced in the semiconductor laser case over the gas laser case, but the form is unaltered. In this sense, the semiconductor laser is an intermediate case between the single-mode gas laser and the arc lamp. In general, however, the arc lamp is so broadband that without narrow band filtering the representation of (2-83) is not very accurate although it can be used in calculations as an indication of the way in which quantities vary.

After the above discussion, one should be convinced that one can approximately represent the electric field vector of a z propagating polychromatic electromagnetic plane wave in free space in the following form:

$$\mathbf{E}(r,\ t) = E_x \hat{\mathbf{e}}_x + E_y \hat{\mathbf{e}}_y \tag{2-95}$$

where

$$E_x = a_x(t) \cos(\tau + \varphi_x(t)) \tag{a}$$

$$E_y = a_y(t) \cos(\tau + \varphi_y(t)) \tag{b}$$

$$(2\text{-}96)$$

where τ is defined (as previously) by

$$\tau = \overline{\omega}t - \mathbf{k} \cdot \mathbf{r} \tag{2-97}$$

As was discussed above, generally a_x, a_y, φ_x, and φ_y will be random variables. Perhaps now is a good time to give some discussion of the possible correlations between these variables. As was discussed above, all optical sources will have some degree of spectral width and, therefore, some degree of randomness. This can vary over many orders of magnitude. For a thermal source like a light bulb, it is probably safe to assume that all the amplitudes and phases of the two components of the field vector are completely uncorrelated. In addition, a_x and a_y should have identical statistics and φ_x and φ_y should be uniformly distributed. Such light is known as unpolarized light. Any effect that will correlate the amplitudes and phases or make the distributions of the amplitudes unequal will polarize the light. Semiconductor lasers will not be unpolarized, in that there is amplitude and phase coupling. Further, a semiconductor laser cavity will not be symmetric in the transverse plane due to technology considerations, and thus the two polarization states will not have equal gain (gain being defined as stimulated gain minus cavity losses). In gas lasers quite generally, one tries to make the stimulated gain different for the two polarization states in an attempt to cause a linearly polarized output. Optical sources are not the only cause of polarization or nonpolarization of light. Scattering events generally involve both absorption and a shift of phase and also, in general, will not affect the two polarization states in the same manner. With all this in mind, one sees that the question of statistical correlation is one that requires insight into both the generation and the propagation processes.

Now, one could define a Jones vector for the polarization state defined by equations (2-95) through (2-97). Indeed, the Jones vector would be of the form of equation (2-47), where the a_x, a_y, φ_x, and φ_y would all be functions of time. The value that γ would take would be

$$\gamma = \frac{a_y(t)}{a_x(t)} \, e^{i\varphi(t)} \tag{2-98}$$

Whereas in the monochromatic case a polarization state could be represented by a single complex number, we see that here, in the polychromatic case, γ becomes a time function and could wander all over the complex plane. If γ were to wander all over the complex plane, one would say that the light was unpolarized. Were it to wander through a fixed region, one would say that the light was partially polarized. Only in the case where γ became constant would one say that the light was fully polarized. One could use the statistical properties of γ to determine the state of polarization, but there are better ways to investigate partial polarization.

To get a grasp of the degree of partial polarization, one needs a tool. Such a tool was developed by Sir George Gabriel Stokes during the nineteenth century (see, for example, Shurcliff 1962 or Shurcliff and Ballard 1964). The tool involves the use of the so-called Stokes parameters, defined as follows:

$$S_0 = \langle a_x^2(t) \rangle + \langle a_y^2(t) \rangle \qquad \text{(a)}$$

$$S_1 = \langle a_x^2(t) \rangle - \langle a_y^2(t) \rangle \qquad \text{(b)}$$

$$S_2 = 2 \langle a_x(t) a_y(t) \cos \varphi(t) \rangle \qquad \text{(c)}$$

$$S_3 = 2 \langle a_x(t) a_y(t) \sin \varphi(t) \rangle \qquad \text{(d)}$$

$$(2\text{-}99)$$

where $\varphi(t)$ is defined as before as $\varphi(t) = \varphi_y(t) - \varphi_x(t)$. It is not surprising that it takes four parameters to describe the state of partially polarized light. Polarized light required the three parameters a_x, a_y, and φ. In the partially polarized case, one additionally needs to carry along information about the degree of polarization, which will require at least an additional piece of information. But, indeed, these considerations should be made clearer by the following examples.

Example 1. Monochromatic source

Here we have

$$a_x(t) = a_x(0) = a_x \qquad \text{(a)}$$

$$a_y(t) = a_x(0) = a_y \qquad \text{(b)} \quad (2\text{-}100)$$

$$\varphi(t) = \varphi(0) = \varphi \qquad \text{(c)}$$

and therefore one can ignore the time-averaging brackets. Forming the sum $S_1^2 + S_2^2 + S_3^2$, one finds that

$$S_1^2 + S_2^2 + S_3^2 = S_0^2 \qquad (2\text{-}101)$$

which illustrates that in the monochromatic case the four Stokes parameters are not independent, as was discussed above.

One could note that (2-101) is the form of the equation for the surface of a three-dimensional sphere defined by axes S_1, S_2, and S_3. Indeed, this was noted by Poincaré and the resulting sphere is referred to as the Poincaré sphere. As we will see in what follows, each possible (partial) polarization state corresponds to a point on (in) the sphere.

Example 2. Thermal source

As was discussed abore, here one would have identical statistics for $a_x(t)$ and $a_y(t)$ and a uniform distribution for φ, with all parameters uncorrelated. Writing

$$\langle a_x^2 \rangle = \langle a_y^2 \rangle = a^2 \tag{2-102}$$

one finds that

$$S_0 = 2_a^2 \tag{a}$$
$$S_1 = S_2 = S_3 = 0 \tag{b}$$

(2-103)

and it is readily seen that S_0 represents the energy in the wave, as it did in the monochromatic case. This suggests that S_0 always represents the total energy in the wave. One could (correctly) abstract that $S_1^2 + S_2^2 + S_3^2$ represents the quantity of polarized energy in the above equation, and thereby one comes to the definition of the degree of polarization of the wave, m:

$$m = \frac{\sqrt{S_1^2 + S_2^2 + S_3^2}}{S_0} \tag{2-104}$$

It should be noted here that monochromatic light will always have $m = 1$, whereas polychromatic light can have any degree of polarization between 0 and 1. This implies that polychromatic polarization states correspond to states inside the Poincaré sphere defined by (2-101). As will be discussed in more depth as we proceed, this implies that polychromatic light can be polarized without changing its spectrum. Monochromatic light, however, cannot be depolarized without spectral broadening. This can be an important consideration in many areas of optical measurement.

Example 3. x-Polarized monochromatic light

What is meant by x-polarized light is essentially that

$$a_x^2 = a^2 \tag{a}$$
$$a_y^2 = 0 \tag{b}$$
$$\varphi = 0 \tag{c}$$

(2-105)

and therefore one finds the Stokes parameters

$$S_0 = a^2 \qquad \text{(a)}$$

$$S_1 = a^2 \qquad \text{(b)} \quad \text{(2-106)}$$

$$S_2 = S_3 = 0 \qquad \text{(c)}$$

with a degree of polarization of 1, as it should be. This is the polarization state of a point on the Poincaré sphere located on the equator (S_1–S_2 plane) along the Greenwich Meridian (West Africa?). As x-polarized light is fully polarized, one can also use a Jones representation, which here gives a $\gamma = 0$, or the origin point of the complex plane.

Example 4. Linearly polarized monochromatic light

What is meant here can be defined by

$$a_x^2 + a_y^2 = a^2 \qquad \text{(a)}$$
$$\varphi = 0 \qquad \text{(b)} \qquad \text{(2-107)}$$

leading to the set of Stokes parameters

$$S_0 = a^2 \qquad \text{(a)}$$

$$S_1 = a_x^2 - a_y^2 \qquad \text{(b)}$$

$$S_2 = 2a_x a_y \qquad \text{(c)} \qquad \text{(2-108)}$$

$$S_3 = 0 \qquad \text{(d)}$$

One notes that, here again, the degree of polarization is again 1. One might further abstract from the following two examples that the linearly polarized energy is represented by $S_1^2 + S_2^2$ and that therefore one could define the degree of linear polarization, ℓ, by

$$\ell = \frac{\sqrt{S_1^2 + S_2^2}}{S_0} \qquad \text{(2-109)}$$

We note here that linear states are all those states located on the equator of the Poincaré sphere. In the complex plane, these linearly polarized states correspond to the real axis of the complex plane.

Example 5. Circularly polarized monochromatic light

As we have seen, what is meant by circularly polarized light is

$$a_x^2 = a_y^2 = a^2 \tag{2-110}$$

$$\varphi = \pm \frac{\pi}{2}$$

which gives for the Stokes parameters

$$S_0 = 2a^2 \tag{a}$$

$$S_1 = S_2 = 0 \tag{b} \quad (2\text{-}111)$$

$$S_3 = \pm 2a^2 \tag{c}$$

Analogous to the definition of the degree of linear polarization, one can define a degree of circular polarization C by

$$C = \frac{S_3}{S_0} \tag{2-112}$$

Circular states, therefore, correspond to the poles of the Poincaré sphere. In the complex plane, these states correspond to the points $\pm i$.

2.5 PROPAGATION IN POLARIZING OPTICAL SYSTEMS

Having seen the forms that the Stokes parameters take for the most common polarization states, we now investigate the propagation of partially polarized light through an optical system. A common method of analyzing the propagation of monochromatic light through an optical system involves using the complex valued 2×2 Jones matrices. Where possible we will do this. Unfortunately, though, these matrices cannot be used for partially polarized light, as they do not carry enough information. Here one must use the real-valued 4×4 Mueller matrices to propagate the Stokes vectors, as we will see. These Mueller matrices can be thought of as corresponding to rotations and other motions on the Poincaré sphere. It should be noted here that the use of matrices to represent the components of a polarizing element requires that the components have a response that is linear in the incident field amplitudes in each of the two incident polarization states.

Let us consider the effect of a linear dichroic polarizer. The word "dichroic" refers to the fact that in certain materials the absorption spectrum can vary greatly between the two polarization states, as is discussed in greater depth in Chapters 3 and 4. By illuminating such a material with a bandlimited signal whose center wavelength corresponds to a resonance for one polarization state

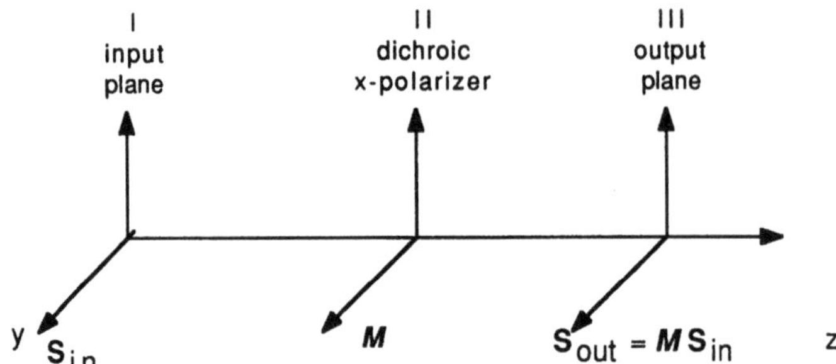

FIGURE 2.21.   Diagram indicating the location of a dichroic polarizer with respect to input and output planes aligned along the optic (z-axis). If one were to know the transformation matrix M of the polarizing element, one could determine $\mathbf{S}_{\text{out}}$ from any arbitrary $\mathbf{S}_{\text{in}}$.

but is well off resonance for the other, one can use the material as a polarizer, as a large quantity of the energy in one polarization state will be absorbed if the material is thick enough. Figure 2.21 illustrates the experimental setup, as well as indicating the analytical knowledge we would like to have about the setup. In essence, one can think of the polarizer in terms of the transformation M that it performs on the input Stokes vector $\mathbf{S}_{\text{in}}$ to give the Stokes vector $\mathbf{S}_{\text{out}}$, which would be measured at the output plane. This relation can therefore be expressed in the form

$$\mathbf{S}_{\text{out}} = M\mathbf{S}_{\text{in}} \tag{2-113}$$

As the dichroic x-polarizer removes the y-component of the wave, one immediately sees that, independent of the incident Stokes vector, the output vector will be

$$\mathbf{S}_{\text{out}} = (\langle a_x^2 \rangle_{\text{in}}, \langle a_x^2 \rangle_{\text{in}}, 0, 0) \tag{2-114}$$

which one can express in terms of linear combinations of the input Stokes vector components as

$$\mathbf{S}_{\text{out}} = \left(\left(\frac{S_0 + S_1}{2} \right)_{\text{in}}, \left(\frac{S_0 + S_1}{2} \right)_{\text{in}}, 0, 0 \right)^{\dagger} \tag{2-115}$$

and therefore the Mueller matrix of the linear x-polarizer can be written in the form

$$(M)_{\text{linear } x\text{-polarizer}} = \frac{1}{2} \begin{bmatrix} 1 & 1 & 0 & 0 \\ 1 & 1 & 0 & 0 \\ 0 & 0 & 0 & 0 \\ 0 & 0 & 0 & 0 \end{bmatrix} \tag{2-116}$$

Evidently, this Mueller matrix has the effect of moving any point on or in the Poincaré sphere to the intersection of the Greenwich Meridian and the equator, on a sphere of a new radius. It will be left as an exercise for the reader to show that the Mueller matrix for a polarizer oriented with transmission axis at ψ degrees from the x-axis is given by

$$(\boldsymbol{M})_{\text{linear }\psi \text{ from } x} = \frac{1}{2} \begin{bmatrix} 1 & \cos 2\psi & \sin 2\psi & 0 \\ \cos 2\psi & \cos^2 2\psi & \dfrac{\sin 4\psi}{2} & 0 \\ \sin 2\psi & \dfrac{\sin 4\psi}{2} & \sin^2 2\psi & 0 \\ 0 & 0 & 0 & 0 \end{bmatrix} \tag{2-117}$$

As a polarizer can operate, in general, on an unpolarized incident wave, it is hard to represent its true operation by a Jones matrix. On a completely polarized incident wave, it is easy to see that the Jones matrix for an x-polarizer would be

$$J = \begin{bmatrix} 1 & 0 \\ 0 & 0 \end{bmatrix} \tag{2-118}$$

To find the effect of a polarizer at another angle, we can use rotation matrices of the form of equation (2-60). The idea is that, if we had an initial state E_i and a final E_0, we can always rotate our coordinates to an x-polarized system by

$$R(\psi)\mathbf{E}_0 = JR(\psi)\mathbf{E}_i \tag{2-119}$$

which can be inverted to yield

$$\mathbf{E}_0 = R^\dagger(\psi)JR(\psi)\mathbf{E}_i \tag{2-120}$$

where J is the "simple" matrix. This simply shows that

$$J(\psi) = R^\dagger(\psi)J(\psi = 0)R(\psi) \tag{2-121}$$

that is, J need only be calculated for a "simple" case and transformed to more complex ones. One can use a similar trick with the Mueller matrices, but its efficacy is not nearly so great as for the Jones matrices.

Another interesting component of polarizing systems is the so-called quarter-wave plate. Here, we will always take the quarter-wave plate to be oriented with respect to the x-axis, and therefore its action will be taken to retard the y-polarized component of a plane wave by a quarter of a wavelength with respect to the x-component of the plane wave, thereby causing the angle φ to become $\varphi + \pi/2$ in the output wave. Obviously this device will be strongly wavelength-dependent and will require that the details of the optical spectrum be known before predicting its effect. Let us consider its effect, therefore, on strictly monochromatic waves and leave the polychromatic cases to a later paragraph. It is evident that the Mueller matrix for this device will be

$$(M)_{\lambda/4} = \begin{bmatrix} 1 & 0 & 0 & 0 \\ 0 & 1 & 0 & 0 \\ 0 & 0 & 0 & -1 \\ 0 & 0 & 1 & 0 \end{bmatrix} \qquad (2\text{-}122)$$

and that this matrix has the effect of causing a $\pi/2$ rotation in the S_2–S_3 plane on the Poincaré sphere. The Jones matrix for this case is simply

$$J = \begin{bmatrix} 1 & 0 \\ 0 & i \end{bmatrix} \qquad (2\text{-}123)$$

It is interesting to consider the working of (2-122) on various input vectors. For example, consider Table 2.1. Clearly (2-122) does not affect the pure x and y linear states. These states are eigenstates of the transformation. As these points lie on the equator at $\pm\pi/2$ longitude of the Poincaré sphere, it is clear that a $\pi/2$ S_2–S_3 plane rotation will leave them unaffected. However, in general, a linear state is transformed by (2-122). For example, a state polarized at 45° from the x-axis will be transformed into a left-hand circularly polarized state,

TABLE 2.1

Polarization State	(Stokes Vector)†
x-polarized	$(S_0, S_0, 0, 0)$
y-polarized	$(S_0 - S_0, 0, 0)$
θ from x-polarized	$(S_0, S_0 \cos 2\theta, S_0 \sin 2\theta, 0)$
L–H-polarized	$(S_0, 0, 0, -S_0)$
R–H-polarized	$(S_0, 0, 0, S_0)$
unpolarized	$(S_0, 0, 0, 0)$

and a right-hand circularly polarized state will be transformed into a state polarized at 45° from the x-axis. These observations hint to one that a circular polarizer could be constructed from a system such as that illustrated in Figure 2.22. Here an initial dichroic polarizer at 45° to the x-axis transforms an arbitrary initial state to a linearly polarized one, and the x-aligned quarter-wave plate transforms the linear state to a right-hand circular state. The composite transformation can be written as the product

$$M_{\text{circ pol}} = M_{\lambda/4} M_{\text{linear } 45°} \tag{2-123}$$

as the linear polarizer is first to operate on the incident Stokes vector. On the Poincaré sphere, the operation of the product is, again, just the product of transformations. The polarizer reduces any state to one at the Greenwich Meridian at the equator, and the quarter-wave plate then can rotate this state to the pole. From this, it is straightforward to show that

$$M_{\text{circ pol}} = \tfrac{1}{2} \begin{bmatrix} 1 & 0 & 1 & 0 \\ 0 & 0 & 0 & 0 \\ 0 & 0 & 0 & 0 \\ 1 & 0 & 1 & 0 \end{bmatrix} \tag{2-124}$$

With the aid of Table 2.1, it is easy to show that the operation of (2-124) will transform any of these states to a left-hand circular state. As was alluded to above, this is somewhat misleading when applied to an unpolarized wave, which by nature must be polychromatic. It can be shown that a quarter-wave plate designed for a design wavelength of λ_0 will perform a transformation on a wave-

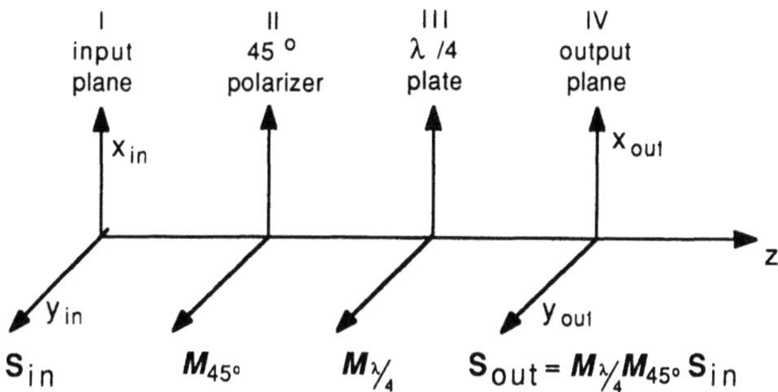

FIGURE 2.22.   Illustration of the placement of a 45° oriented dichroic polarizer and a quarter-wave plate in an optical system to effect a circular polarizer.

length λ of

$$M_{\lambda_0/4}(\lambda) = \begin{bmatrix} 1 & 0 & 0 & 0 \\ 0 & 1 & 0 & 0 \\ 0 & 0 & \cos\epsilon & -\sin\epsilon \\ 0 & 0 & \sin\epsilon & \cos\epsilon \end{bmatrix} \qquad (2\text{-}125)$$

where ϵ is defined by

$$\epsilon = \frac{\lambda_0}{\lambda}\frac{\pi}{2} \qquad (2\text{-}126)$$

For an intensity spectrum $I(\lambda)$ incident on a circular polarizer, therefore, the actual transformation would be given by

$$M_{\text{circ pol}} = \int M_{\lambda_0/4}(\lambda)M_{\text{linear } 45°}I(\lambda)\,d\lambda \qquad (2\text{-}127)$$

For a completely polarized monochromatic incident wave, the situation as handled by Jones matrices is similar to the Mueller matrix treatment. Indeed, using

$$J_{cp} = J_{\lambda/4}J_{45°} \qquad (2\text{-}128)$$

and the identifications

$$J_{45°} = \frac{1}{2}\begin{bmatrix} 1 & 1 \\ 1 & 1 \end{bmatrix} \qquad \text{(a)}$$

$$J_{\lambda/4} = \begin{bmatrix} 1 & 0 \\ 0 & i \end{bmatrix} \qquad \text{(b)}$$

$$(2\text{-}129)$$

one finds that

$$J_{cp} = \frac{1}{2}\begin{bmatrix} 1 & 1 \\ i & i \end{bmatrix} \qquad (2\text{-}130)$$

It should probably be noted here that there can be a drawback to applying Stokes matrices to find the polarization state of multiple beams of interfering monochromatic light. The Stokes vectors of monochromatic beams do not add, but rather the Stokes vectors of incoherent beams add. That is, the Stokes vectors represent intensities, and intensities of incoherent waves add. To add together monochromatic waves, one needs to add fields. The earlier mentioned Jones vectors represent fields and therefore can be used to add monochromatic beams together. Multiple beams of partially coherent light cannot easily be

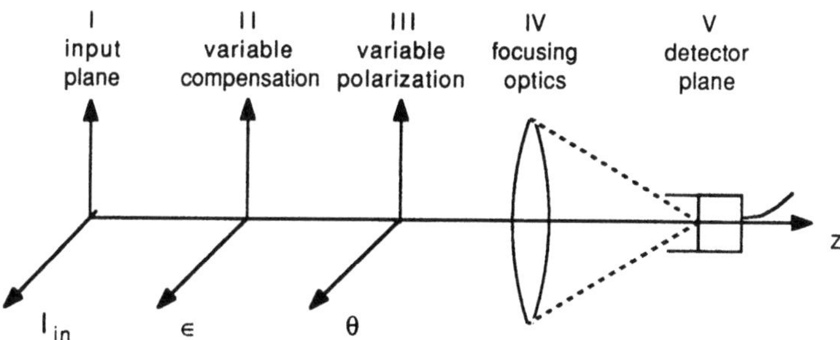

FIGURE 2.23.   Illustration of a system for generating an $I(\theta, \epsilon)$ that can be Fourier-analyzed to determine the complete set of Stokes parameters of the incident (input plane) wave.

treated by either method. In such a case, one must simply calculate the propagated fields and perform the appropriate time averages.

An important question, both practically and theoretically, is that of how one can experimentally determine the Stokes parameters of a given incident wave. A possible experimental arrangement is given in Figure 2.23. Here an incident wave is first acted upon by a compensating plate that gives a tunable phase delay ϵ to the y-polarized component of the wave, followed by an analyzer whose pass axis can be adjusted to any angle ψ with respect to the x-axis, and then the resulting signal is read out by an intensity detector. Such an operation is easily carried out with commercially available components such as a Babinet-Soleil compensator followed by a variable analyzer, lens, and pin detector.

Some straightforward analysis of the setup leads to the following expression for the received intensity:

$$I(\psi, \epsilon) = \tfrac{1}{2}[S_0 + S_1 \cos 2\psi + (S_2 \cos \epsilon - S_3 \sin \epsilon) \sin 2\psi] \quad (2\text{-}131)$$

But (2-131) is in the form of a double Fourier series of the Stokes parameters, and therefore if one takes enough measurements to perform integrals over θ and ϵ with some accuracy, one can find the Stokes parameters by

$$S_0 = \frac{2}{\pi} \int_{-\pi/2}^{\pi/2} I(\psi, \epsilon)\, d\psi \qquad (a)$$

$$S_1 = \frac{4}{\pi} \int_{-\pi/2}^{\pi/2} I(\psi, \epsilon) \cos 2\psi\, d\psi \qquad (b)$$

$$(2\text{-}132)$$

$$S_2 = \frac{4}{\pi^2} \int_{-\pi}^{\pi} \int_{-\pi/2}^{\pi/2} I(\psi, \epsilon) \sin 2\psi\, d\psi \cos \epsilon\, d\epsilon \qquad (c)$$

$$S_3 = -\frac{4}{\pi^2} \int_{-\pi}^{\pi} \int_{-\pi/2}^{\pi/2} I(\psi, \epsilon) \sin 2\psi\, d\psi \sin \epsilon\, d\epsilon \qquad (d)$$

Clearly, if $I(\psi, \epsilon)$ is found to be independent of ψ and ϵ, then it can be concluded that the wave is unpolarized.

Another interesting example that can be treated by Jones matrices is that of an electro-optic sampling head. The electro-optic sampling technique takes advantages of the electric field-dependent changes in the polarization of the transmitted light in an electro-optic material to measure electric field strength. In reflection mode sampling, the incident and reflected beams can be separated by manipulating their polarization as shown in the Figure (2.24). The system can be qualitatively described as follows. The $\lambda/4$-plate produces an elliptically polarized beam with the major axis oriented at $\varphi_{\lambda/4} = \pi/8$. The $\lambda/2$-plate rotates the major axis an additional $2(\varphi_{\lambda/2} - \varphi_{\lambda/4}) = \pi/8$ to $\varphi_s = \pi/4$ to orient it properly relative to the principal axis of the substrate. After the beam reflects off the groundplane, the $\lambda/2$-plate rotates the major axis an angle $2(\varphi_s - \varphi_{\lambda/2}) = \pi/8$ back to $\pi/8$. Hence, after propagating twice through the $\lambda/2$-plate, which is equivalent to one pass through a λ-plate, the polarization state is unchanged. Therefore, the effect of the two wavelengths is just equivalent to two passes through the $\lambda/4$-plate. Propagating through a $\lambda/4$-plate twice is equivalent to a $\lambda/2$-plate. Hence the polarization of the reflected beam at the polarizer is rotated $2\varphi_{\lambda/4} = \pi/4$ with respect to the incident beam. Quantitatively, the polarization state can be analyzed using Jones calculus. The Jones matrix of a waveplate with phase retardation Γ rotated an angle θ is

$$M_\Gamma(\theta) = R(-\theta)J_\Gamma(0)R(\theta)$$

$$= e^{i\Gamma/2} \begin{bmatrix} \cos^2\theta + e^{-i\Gamma}\sin^2\theta & (1 - e^{-i\Gamma})\sin\theta\cos\theta \\ (1 - e^{-i\Gamma})\sin\theta\cos\theta & \sin^2\theta + e^{i\Gamma}\cos^2\theta \end{bmatrix} \tag{2-133}$$

The total matrix of the sampling head can be expressed as

$$J = J_y J_{\pi/2}\left(\frac{\pi}{8}\right) J_\pi\left(\frac{3\pi}{16}\right) J_\pi\left(\frac{3\pi}{16}\right) J_{\pi/2}\left(\frac{\pi}{8}\right) J_x \tag{2-134}$$

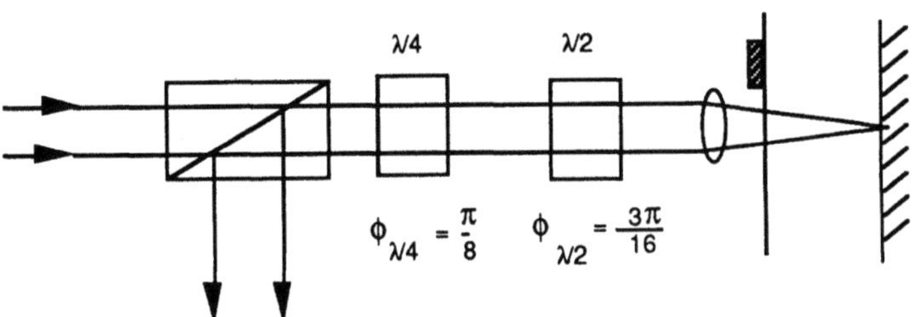

FIGURE 2.24.   A schematic depiction of an optical sampling head.

where J_x and J_y are the matrices for the x- and y-polarizer representing the two points of the beamsplitter. First, multiplying the two $\lambda/2$ matrices

$$J_\pi\left(\frac{3\pi}{16}\right) J_\pi\left(\frac{3\pi}{16}\right) = e^{i\pi/2} \begin{bmatrix} \cos\left(2\,\frac{3\pi}{16}\right) & \sin\left(2\,\frac{3\pi}{16}\right) \\ \sin\left(2\,\frac{3\pi}{16}\right) & -\cos\left(2\,\frac{3\pi}{16}\right) \end{bmatrix}$$

$$e^{i\pi/2} \begin{bmatrix} \cos\left(2\,\frac{3\pi}{16}\right) & \sin\left(2\,\frac{3\pi}{16}\right) \\ \sin\left(2\,\frac{3\pi}{16}\right) & -\cos\left(2\,\frac{3\pi}{16}\right) \end{bmatrix} \tag{2-135}$$

$$= e^{i\pi} \begin{bmatrix} 1 & 0 \\ 0 & 1 \end{bmatrix}$$

Thus:

$$J_{\pi/2}\left(\frac{\pi}{8}\right) J_\pi\left(\frac{3\pi}{16}\right) J_\pi\left(\frac{3\pi}{16}\right) J_{\pi/2}\left(\frac{\pi}{8}\right)$$

$$= e^{i\pi} e^{i\pi/4} e^{i\pi/4} \begin{bmatrix} \cos\left(2\,\frac{\pi}{8}\right) - i\sin^2\left(2\,\frac{\pi}{8}\right) & (1+i)\sin\left(\frac{\pi}{8}\right)\cos\left(\frac{\pi}{8}\right) \\ (1+i)\sin\left(\frac{\pi}{8}\right)\cos\left(\frac{\pi}{8}\right) & \sin\left(2\,\frac{\pi}{8}\right) - i\cos^2\left(2\,\frac{\pi}{8}\right) \end{bmatrix}$$

$$\begin{bmatrix} \cos\left(2\,\frac{\pi}{8}\right) - i\sin^2\left(2\,\frac{\pi}{8}\right) & (1+i)\sin\left(\frac{\pi}{8}\right)\cos\left(\frac{\pi}{8}\right) \\ (1+i)\sin\left(\frac{\pi}{8}\right)\cos\left(\frac{\pi}{8}\right) & \sin\left(2\,\frac{\pi}{8}\right) - i\cos^2\left(2\,\frac{\pi}{8}\right) \end{bmatrix}$$

$$= e^{i3\pi/2} \begin{bmatrix} \cos\left(2\,\frac{\pi}{8}\right) & \sin\left(\frac{2\pi}{8}\right) \\ \sin\left(\frac{2\pi}{8}\right) & \cos\left(\frac{2\pi}{8}\right) \end{bmatrix} = e^{i3\pi/2} \frac{1}{\sqrt{2}} \begin{bmatrix} 1 & 1 \\ 1 & 1 \end{bmatrix} \tag{2-136}$$

With x-polarized light incident on the first waveplate, the reflected beam has the Jones vector

$$\frac{1}{\sqrt{2}} e^{i(3\pi/2)} \begin{bmatrix} 1 \\ 1 \end{bmatrix}$$

that is, $\pi/4$-polarized. The total matrix can now be expressed as

$$
J = \begin{bmatrix} 0 & 0 \\ 0 & 1 \end{bmatrix} e^{i(3\pi/2)} \begin{bmatrix} \cos \pi/4 & \sin \pi/4 \\ \sin \pi/4 & \cos \pi/4 \end{bmatrix} \begin{bmatrix} 1 & 0 \\ 0 & 0 \end{bmatrix}
$$
$$
= e^{i(3\pi/2)} \begin{bmatrix} 0 & 0 \\ \sin \pi/4 & 0 \end{bmatrix} = \frac{1}{\sqrt{2}} e^{i(3\pi/2)} \begin{bmatrix} 0 & 0 \\ 1 & 0 \end{bmatrix} \tag{2-137}
$$

With x-polarized light to the input we find that

$$
\mathbf{E}_{\text{out}} = \frac{1}{\sqrt{2}} e^{i(3\pi/2)} \begin{bmatrix} 0 & 0 \\ 1 & 0 \end{bmatrix} \begin{bmatrix} 1 \\ 0 \end{bmatrix} = \frac{1}{\sqrt{2}} e^{i(3\pi/2)} \begin{bmatrix} 0 \\ 1 \end{bmatrix} \tag{2-138}
$$

that is, the output is y-polarized with an amplitude $1/\sqrt{2}$. This corresponds to an intensity transmission of $1/2$.

2.6 STRIATED MEDIA

The area of thin films has been important to optics since the discovery of Newton's rings in the seventeenth century (see, for example, Klein and Furtak 1986, Chapter 5). It is of great importance now because of the extensive use of dielectric multilayers for wavelength-sensitive reflectors and filters and will probably increase in importance in the future as the technology of thin films improves and various new insulating as well as semiconductor quantum well materials become available for optical use. This section will concentrate on formulation of a method of analysis useful for propagation through arbitrary multilayer structures, with emphasis placed on the design and analysis of reflecting filtering structures, this problem being considered archetypical of those encountered with multilayers. This section will begin with a quick review of some of the physical effects associated with the Fresnel problem. In this discussion, it will simply be assumed that propagation in a material with an index n is similar to freespace propagation. In the next chapter, a much more detailed study of the nature of the dielectric constant is given, which leads to some understanding of dispersive propagation, a subject not considered at present.

The Fresnel problem is the simplest of all electromagnetic boundary value problems, the problem of reflection and transmission of plane waves at a simple dielectric interface (see, for example, the treatment in Born and Wolf 1975, Chapter 2). As was argued earlier, the plane wave is a good representation of many propagating wave forms, and, in fact, arbitrary wave forms can be expanded in plane wave representations. (Unfortunately, in general, these representations require the use of evanescent as well as propagating plane waves, as can be seen from the treatment in Clemmow 1966. We will encounter evanescent waves in a following paragraph on total internal reflection). Therefore, despite the simplicity of the Fresnel problem, its simple analytical answers and physical insights are broadly applicable to the interpretation of optical phenomena.

Figure 2.25 illustrates the geometric relationships between the various quantities used to solve the problem. Because a plane wave is assumed as the incident wave, plane waves will also be the solutions for the transmitted and reflected waves. The solution to the problem of finding the amplitudes of these components therefore reduces to a linear algebraic problem of solving the boundary conditions, as the differential parts of Maxwell's equations have already been solved. From doing pillbox integrations of the curl equations across the dielectric surfaces, one can readily show that at a boundary the tangential components of the electric and magnetic vectors must be continuous across the interface in the absence of current at the interface. Similar integrals of the divergence equations in the absence of surface also show the normal components D_z and B_z are also continuous. The problem can be split into the two parts, as is shown in Figure 2.25, namely, the case in which the total electric field is tangent to the interface (perpendicular case) and the case in which the total magnetic field is tangential to the interface (parallel case). The equations governing these two cases can be derived more or less along the lines of the earlier derivation of the basic relations for a plane wave. For the perpendicular or transverse electric (TE) case, the only non-zero components of the field are H_x, E_y, and H_z in a coordinate system in which x would point upward (parallel to the interface). In Figure 2.25, y would be outwardly perpendicular to the plane of the paper in Figure 2.25, and z would be perpendicular to the interface but in the plane of the paper of Figure 2.25. The z-coordinate in this case is the

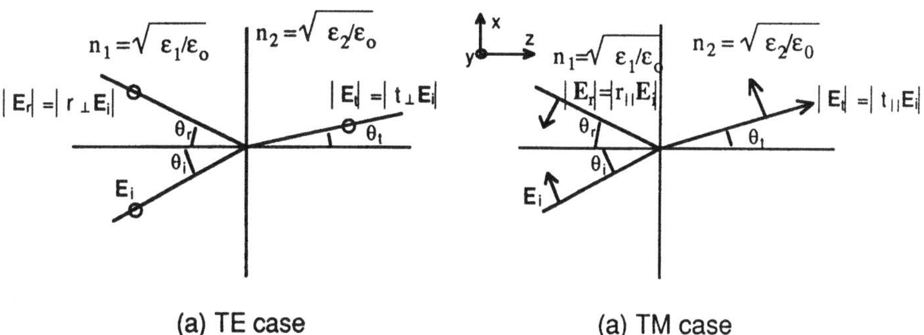

FIGURE 2.25.   Illustration of the geometric relationships between the field vectors and propagation directions in the Fresnel problem for (a) the TE case and (b) the TM case.

optical axis. Maxwell's equations in this case reduce to

$$\frac{\partial E_y}{\partial z} = -i\omega\mu_0 H_x \qquad \text{(a)}$$

$$\frac{\partial E_y}{\partial x} = i\omega\mu_0 H_z \qquad \text{(b)} \quad \text{(2-139)}$$

$$\frac{\partial H_z}{\partial x} - \frac{\partial H_x}{\partial z} = i\omega n^2 \epsilon_0 E_y \qquad \text{(c)}$$

where harmonic time dependence has been assumed, and the y-derivative has been taken as equal to zero. The materials have been assumed nonmagnetic, as essentially all materials are at optical frequencies, because of the inability of material charges to set up circulating currents at such high frequencies. (For a particle to traverse a 1 μm circular path in an optical period, its linear velocity would have to exceed the speed of light.) Eliminating H_x and H_z from (2-129), one finds that

$$\frac{\partial^2 E_y}{\partial x^2} + \frac{\partial^2 E_y}{\partial z^2} + k^2 E_y = 0 \qquad \text{(TE)} \qquad \text{(2-140)}$$

where $k^2 = \omega^2 \mu_0 \epsilon_0 n^2 = (\omega^2/c^2)n^2$, subject to the boundary conditions that E_y and $\partial E_y/\partial_z$ are continuous across the interface. Had one worked with the parallel or transverse magnetic (TM) case, one would have found that

$$\frac{\partial^2 H_y}{\partial x^2} + \frac{\partial^2 H_y}{\partial z^2} + k^2 H_y = 0 \qquad \text{(TM)} \qquad \text{(2-141)}$$

here subject to the boundary conditions that H_y is continuous across the interface, but $\partial H_y/\partial z$ must jump by an amount equal to the discontinuity in ϵ across the boundary. It is clear from (2-140) that fundamental solutions for the TE wave can be written in the form

$$E_y = E_{y_0} e^{+ikx \sin\theta} e^{\pm ikz \cos\theta} \qquad \text{(2-142)}$$

where the plus or minus signs give the two possible directions of propagation inside either of the media. One can see from Figure 2.25, together with the form of (2-142), that the angle of incidence θ_i, angle of reflection θ_r, and angle of transmission θ_t must be related by

$$\theta_i = \theta_r \qquad \text{(a)}$$

$$n_1 \sin \theta_i = n_2 \sin \theta_t \qquad \text{(b)}$$

$$(2\text{-}143)$$

independently of polarization, where n_1 and n_2 are the indices of refraction of the first medium and the second medium, respectively. The calculations for the reflection and transmission coefficients are straightforward and appear in many references, and are written as

$$t_\perp = \frac{2n_1 \cos \theta_i}{n_1 \cos \theta_i + n_2 \cos \theta_t} \qquad \text{(a)}$$

$$t_\parallel = \frac{2n_1 \cos \theta_i}{n_2 \cos \theta_i + n_1 \cos \theta_t} \qquad \text{(b)}$$

$$(2\text{-}144)$$

$$r_\perp = \frac{n_1 \cos \theta_i - n_2 \cos \theta_t}{n_1 \cos \theta_i + n_2 \cos \theta_t} \qquad \text{(c)}$$

$$r_\parallel = \frac{n_2 \cos \theta_i - n_1 \cos \theta_t}{n_2 \cos \theta_i + n_1 \cos \theta_t} \qquad \text{(d)}$$

where the amplitude transmittances $t_\perp$ and $t_\parallel$ and the reflectances $r_\perp$ and $r_\parallel$ are defined for the perpendicular and parallel incidence cases by the relations

$$\mathbf{E}_r(z = 0) = r_\perp \, \mathbf{E}_i(z = 0) \qquad \text{(a)}$$

$$\mathbf{E}_t(z = 0) = t_\perp \, \mathbf{E}_i(z = 0) \qquad \text{(b)}$$

$$\mathbf{H}_r(z = 0) = r_\parallel \mathbf{H}_i(z = 0) \qquad \text{(c)}$$

$$n_2 \mathbf{H}_t(z = 0) = t_\parallel n_1 \mathbf{H}_i(z = 0) \qquad \text{(d)}$$

$$(2\text{-}145)$$

and where intensity transmittances and reflectances can be defined by

$$R_\perp = |r_\perp|^2 \qquad \text{(a)}$$

$$T_\perp = |t_\perp|^2 \qquad \text{(b)}$$

$$R_\parallel = |r_\parallel|^2 \qquad \text{(c)}$$

$$T_\parallel = |t_\parallel|^2 \qquad \text{(d)}$$

$$(2\text{-}146)$$

Some insight into the physical content of these relations (2-144) can be obtained by looking at the case of normal incidence. From (2-143), one sees that, if $\theta_i = 0$, then $\theta_r = \theta_t = 0$, and from (2-144) one finds that

$$t_\perp = \frac{2n_1}{n_1 + n_2} = t_\parallel \qquad \text{(a)}$$

$$r_\perp = \frac{n_1 - n_2}{n_1 + n_2} = -r_\parallel \qquad \text{(b)}$$

(2-147)

Figure 2.26 can clarify the minus sign of (2-147)(b). As the figure shows, for normal incidence, the perpendicular and parallel cases look identical from symmetry, with the exception of a sign reversal that is simply due to an initial assumption concerning polarization direction. It is of some interest here to consider the energy flow to and from the interface. Generally, one would think that intensity reflectances and transmittances [i.e., (2-146)] should sum to the assumed incident intensity. However, from (2-147) above one notes that

$$|r_\perp|^2 + |t_\perp|^2 = \frac{5n_1^2 - 2n_1 n_2 + n_2^2}{(n_1 + n_2)^2} \qquad (2\text{-}148)$$

which is only equal to 1 if $n_1 = n_2$. It would seem, superficially, that energy is not conserved. The problem here is that one must recall the earlier discussion of Poynting's theorem. There, we saw that $\mathbf{S}$, the Poynting vector, which was something like energy flow, at least for a plane wave, could be written as

$$\mathbf{S}(t) = \frac{1}{2\eta} |E|^2 \hat{\mathbf{e}}_k \left(\begin{matrix} \text{plane} \\ \text{wave} \end{matrix} \right) \qquad (2\text{-}149)$$

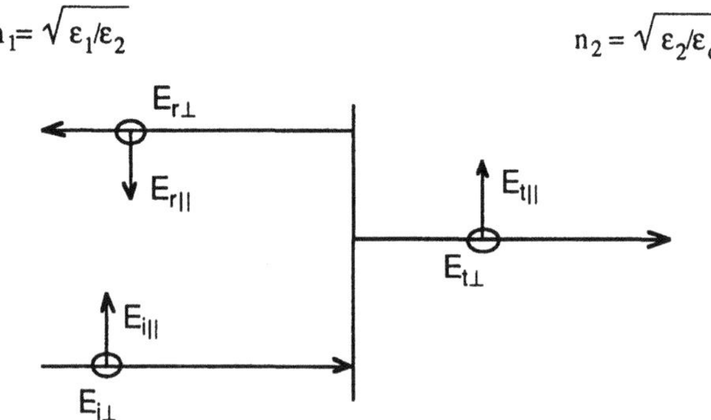

FIGURE 2.26.   Illustration like that of Figures 2-25(a) and (b), but here specialized to the case of normal incidence.

where we have taken the liberty of using ϵ of the medium instead of ϵ_0 in the expression for η. The impedance η on the two sides of the interface is not the same, and therefore, by equating the reflected plus transmitted Poynting vectors with the incident, one could surmise that

$$|r|^2 + \frac{n_2}{n_1} |t|^2 = 1 \quad \left(\begin{array}{c} \text{normal} \\ \text{incidence} \end{array}\right) \tag{2-150}$$

where $r(t)$ can represent either $r_\perp (t_\perp)$ or $r_\parallel (t_\parallel)$. It is not hard to show that, for arbitrary incidence, the intensity of reflectances and transmittances satisfies the relation

$$|r|^2 + \frac{n_2 \cos \theta_t}{n_1 \cos \theta_i} |t|^2 = 1 \tag{2-151}$$

Before leaving the case of normal incidence, we should probably touch on the four percent reflection rule. On an optical bench, the rule of thumb for optical power flow is that, at each air–glass (glass–air) interface, 4% of the incident power is reflected. If one takes the index of air to be close to 1 and that of glass to be close to 1.5 and plugs this into (2-147), one indeed finds that $R_\perp = R_\parallel \approx .04$. The rule looks all right at first glance. Unfortunately, one must look a little deeper to save oneself from some possibly serious problems. Figure 2.27 illustrates the multiple bounce problem. If one has a plane parallel glass plate, the incident plane wave will experience multiple bounces, and to find the reflected and transmitted intensities one must perform an infinite sum over the fields and square the results. This will be done in Chapter 6 when we consider the Fabry-Perot interferometer. Then we will notice that there is no

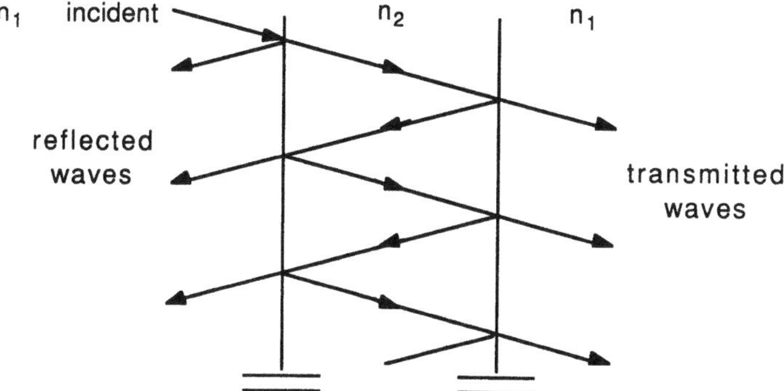

FIGURE 2.27.   Schematic depiction of the multiple bounce problem, wherein the plate of index n_2 is so perfect that phase coherence is preserved across the whole phase front.

concept of a 4% from each interface rule, and, in fact, if the incoming light is polychromatic, complementary rainbows will be transmitted and reflected. Indeed, this latter effect can be observed if one holds a $\lambda/20$ optical flat up to a light bulb. The trick is that the components on an optical bench are not all plane and parallel to a twentieth of a wavelength. Normal glass plates are rough on the order of a wavelength, and therefore the plane wave phase front is destroyed on reflection. Lenses are curved, and therefore successive reflections take off at different angles, even if the lens is highly polished. The 4% rule therefore "works" as long as nothing "funny" happens, such as unwanted, ignored reflections coming up in the center of a photographic plate. Unfortunately, such things can happen.

An interesting effect that is predicted by the Fresnel relations is one first observed by Brewster. Let us consider an interface where $n_2 > n_1$. In this case, we find that $\sin \theta_t < \sin \theta_i$ and $\cos \theta_t > \cos \theta_i$, and thus $n_2 \cos \theta_t > n_1 \cos \theta_i$, so that $r_\perp$ is negative for all incident angles. This is simply a statement that the reflected perpendicularly polarized field always must subtract from the incident in order to satisfy the boundary condition. The situation for the parallel field is more interesting. One could indeed find $n_1 \cos \theta_t$ equal to $n_2 \cos \theta_i$, and the incident angle for which this occurs is called the Brewster angle, θ_B, which is mathematically defined by

$$\tan \theta_B = \frac{n_2}{n_1} \tag{2-152}$$

At the Brewster angle $r_\parallel = 0$, and therefore all of the incident energy in the parallel polarization is transmitted. The effect is reasonably easy to interpret as follows: When θ_i is at the Brewster angle, it is straightforward to show that $\theta_t = \pi/2 - \theta_i$, a situation depicted in Figure 2.28. If one considers the polarizations of the waves as in Figure 2.28(a), one notes that the dipoles at the medium interface that give rise to the reflected and transmitted waves are polarized along the **k** vector of the reflected wave. As a dipole cannot radiate along its polarization direction, there can be no reflected wave at the Brewster angle.

The effect of Brewster can be used for various polarization applications. Two of these are illustrated in Figure 2.29. In Figure 2.29(a), a Brewster angle polarizer is illustrated. The device consists of a stack of plates alternately of indices n_1 and n_2. When the stack is illuminated at the proper angle, the angle at which the wave in n_1 strikes the n_2 interface at just the n_1–n_2 Brewster angle, then only the perpendicular component is reflected. At the other interfaces, the wave will be incident at the Brewster angle of the n_2–n_1 interface, and the roles of the incident and reflected angles are interchanged. Therefore, after traversal of a long-enough stack of plates, the transmitted wave should contain only the parallel component of polarization. The reflected wave will, of course, not be

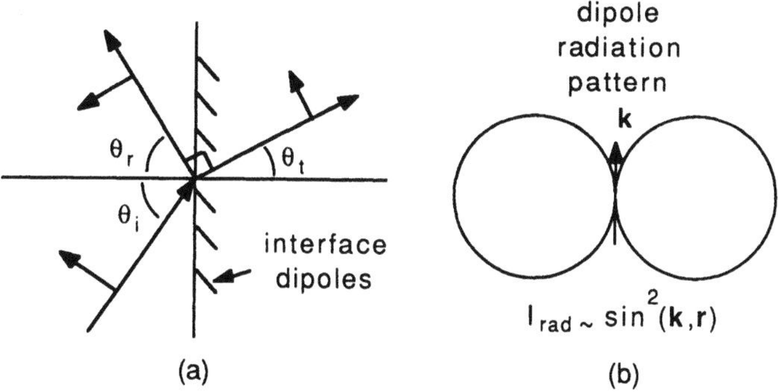

FIGURE 2.28.   Illustration of the physics of Brewster's effect. (a) There must exist a $\theta_i = \theta_B$ for which $\theta_t = \pi/2 - \theta_r$; (b) $\theta_i = \theta_B$, the dipoles at the material interface are directed along the reflected $\mathbf{k}$ vector, and therefore can radiate no intensity in this direction.

completely perpendicularly polarized. A problem with this arrangement is that, depending on the details of the structure, it could take many layers to achieve complete reflection of the perpendicular component. Further, for monochromatic illumination, the reflectivity can be a very complicated function of the stack parameters, that is, thickness and index. If the illumination is incoherent, and the layers are thick with somewhat rough interfaces, then the reflected intensities should sum, and the total transmitted intensity in the perpendicular polarization will vary according to $(n_1/n_2)^N$, where n_1 is the smaller index, and N is the total number of layers traversed. A practical problem, however, even with this arrangement is that the end layers must really be free space with an index of $n_0 = 1$. These two non-Brewster interfaces will cause coupling noise,

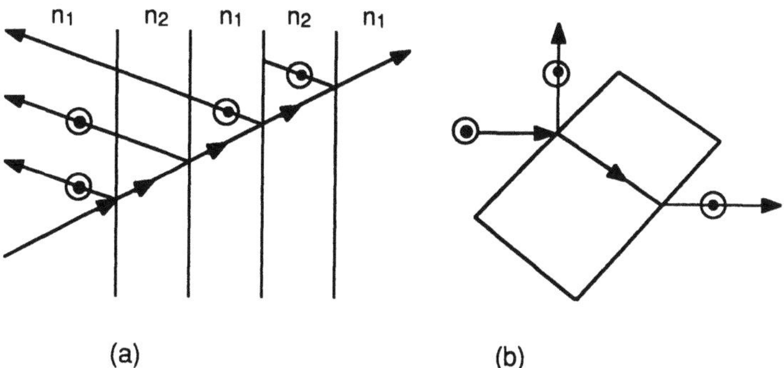

FIGURE 2.29.   Illustrations of the principles behind (a) a Brewster plate where an unpolarized incident wave, incident at the Brewster angle, will be perpendicularly polarized on reflection and (b) a Brewster plate placed in a laser cavity to cause higher losses for the perpendicular than the parallel component.

and might lead to a situation in which there is no real angle in air that can be chosen such that the incidence on the 1–2 interface will be at Brewster's angle, as will be further discussed in the next paragraph. This problem can be alleviated by careful choice of the indices n_1 and n_2.

A second and more widely used application of Brewster's effect is illustrated in Figure 2.29(b). Here, a Brewster plate, that is, a plate aligned to be at the Brewster angle for z-directed plane waves, is placed in a laser cavity aligned along the z-axis. The perpendicularly polarized component of the light in the cavity will experience a reflection loss from the plate and, therefore, a higher lasing threshold than the parallel component. In this way, single polarization lasing can be induced.

A second important effect that is predicted by Fresnel's equations is total internal reflection. It is clear that, if $n_1 > n_2$, there are incident angles for which equation (2-143) (b) has no solutions. For such angles, the $\sin \theta_t$ that is equal to $(n_1/n_2) \sin \theta_i$ is greater than 1. For such a case, one can write that

$$\cos \theta_t = i \sqrt{\sin^2 \theta_t - 1} \qquad \text{(a)}$$

$$r_\perp = \frac{n_1 \cos \theta_i - n_2 \cos \theta_t}{n_1 \cos \theta_i + n_2 \cos \theta_t} = \frac{A - iB}{A + iB} = e^{i\delta_\perp} \qquad \text{(b)} \qquad \text{(2-153)}$$

$$r_\parallel = e^{i\delta_\parallel} \qquad \text{(c)}$$

where $\delta_\parallel \neq \delta_\perp$. It is clear here that the incident wave is totally reflected as $|r_\perp| = |r_\parallel| = 1$. The form of the transmitted wave can be written, for $\sin \theta_t > 1$, as

$$\mathbf{E}_t = t e^{ikx \sin \theta_t} e^{-kz \sqrt{\sin \theta_t - 1}} \, \hat{\mathbf{e}}_t \qquad \text{(2-154)}$$

where $\hat{\mathbf{e}}_t$ is the unit vector in the polarization direction, and t is the magnitude of the transmissivity, which can be seen from (2-144) to be non-zero. It is left to the reader as an exercise to interpret the meaning of t in this case, in light of Poynting's theorem. This form of the transmitted wave is a prime example of an inhomogeneous plane wave, a type of wave that was mentioned earlier.

Total internal reflection has many applications. Perhaps the earliest was Fresnel's rhomb as is illustrated in Figure 2.30(a). Fresnel's rhomb is a device, generally made of glass, that uses two internal reflections to obtain a $\pi/2$ phase shift between the parallel and perpendicular components, thereby converting linear polarization to circular polarization and vice versa. A second possible application is that of optical guidance as is illustrated in Figure 2-30(b). Here, one sandwiches a higher-index medium between lower-index media and excites this inner medium with "plane waves" that meet the cladding layers at a glancing-enough angle such that the waves are continually and totally internally

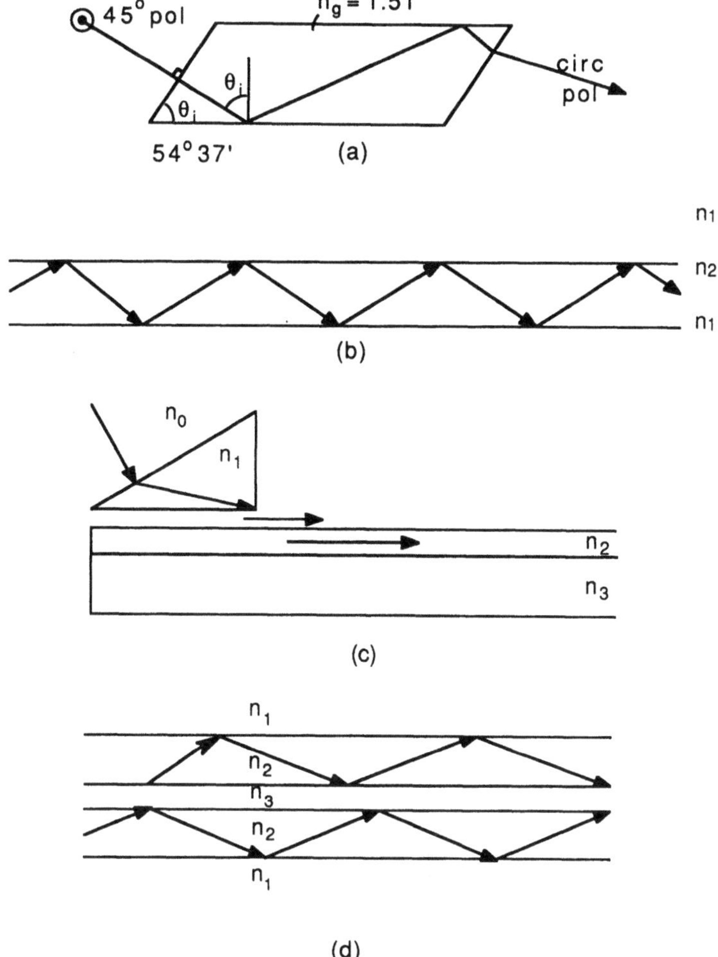

FIGURE 2.30. Schematic depiction of some applications of total internal reflection: (a) Fresnel's rhomb; (b) an optical waveguide; (c) a prism coupler; and (d) coupled optical waveguides.

reflected. This guidance principle is the same one that is basic to fiber and integrated optics. Another use of total internal reflection is for optical coupling as is illustrated in Figure 2.30(c). Here, a prism is illuminated, and the "plane wave" is totally internally reflected at the interface between the prism and the air. The air layer, however, is thin compared to a wavelength, and therefore the optical wave extends into the medium below the air layer and will couple energy to this medium. This technique is very useful for coupling energy to an integrated optic surface waveguide in which end fire coupling is not possible.

A last application that will be mentioned here is that of the optical directional coupler of Figure 2.30(d). Here, two waveguides are separated by a cladding that is sufficiently thin that the field from one waveguide has an appreciable "wing" amplitude in the other guide. As in the prism coupler, the guided waves from one guide will extend into the other, and the basic mode structures will be modified. These modifications cause the energy distribution to appear to oscillate back and forth between the two guides with propagation distance, much as an excited dipole oscillates between positions on either side of equilibrium with time. Much more could be said about these last three applications, but they basically lie within the realm of guided wave optics, and their detailed descriptions belong in another course (see, for example, Mickelson 1992).

The natural extension to the simple plane interface problem is the multiple plane interface, or striated medium, problem, which is a fundamental and important problem in optics for various historical and practical reasons. Perhaps the most important contemporary application is the dielectric mirror. In any application where one needs to minimize losses, the dielectric mirror is a natural design solution, as the dielectric absorption is much less than the metallic absorption associated with the finite skin depth of usual metals at optical frequencies. The dielectric mirror also exhibits a wavelength selective characteristic that is widely tunable by varying design parameters. It is such design problems that we will presently introduce by analyzing multilayer structures.

One could attack the multilayer problem by applying Fresnel's relations repeatedly at each interface. As was mentioned previously, however, even for a single slab, this technique leads to infinite sums of reflections. Such a case will be solved for a Fabry-Perot cavity in Chapter 6, but here we choose to present a more general formulation whereby one can analyze arbitrary numbers of layers. We consider structures such as that illustrated in Figure 2.31 with the associated illumination direction definitions. We will consider the problem as a boundary value problem. As in the Fresnel problem, the incident wave can be separated into perpendicular (transverse electric or TE) and parallel (transverse magnetic or TM) components, and these two cases can be considered separately. Here we will consider the TE case and leave the TM case to the concerned reader and the problem sets.

As was discussed along with the Fresnel problem, the fundamental solutions within each individual slab are already known, and one need not solve Maxwell's equations at all. Within the ith slab, the general solution will take the form (TE)

$$\mathbf{E}_y = A_i e^{ik_i x \sin \theta_i} e^{ik_i z \cos \theta_i} + B_i e^{ik_i x \sin \theta_i} e^{-ik_i z \cos \theta_i} \tag{2-155}$$

where k_i is the free space $k = \omega/c$ multiplied by the index of the ith slab, n_i, and θ_i is the propagation angle in the ith medium, which is related to the incident angle in the first medium by Snell's law that $n_1 \sin \theta_1 = n_i \sin \theta_i$. What is necessary to solve the propagation problem is to relate the A's and B's of each

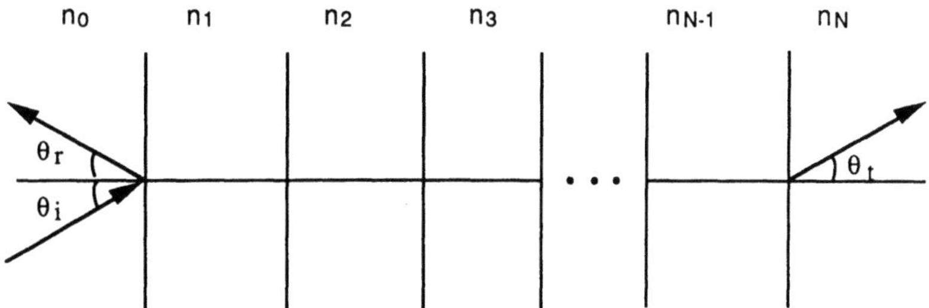

FIGURE 2.31.   A multilayered dielectric structure.

layer to those in the other layers. The equations we have to do this with are the boundary condition equations. For the TE waves (the TM are left for the exercises), these boundary conditions at the boundary between the ith and $(i + 1)$th layer take the form

$$e^{ik_i z_{i,i+1} \cos \theta_i} A_i + e^{-ik_{i+1} z_{i,i+1} \cos \theta_i} B_i \qquad (a)$$

$$= e^{ik_{i+1} z_{i,i+1} \cos \theta_{i+1}} A_{i+1} + e^{-ik_{i+1} z_{i,i+1} \cos \theta_{i+1}} B_{i+1}$$

$$ik_i \cos \theta_i e^{ik_i z_{i,i+1} \cos \theta_i} A_i - ik_i \cos \theta_i e^{-ik_i z_{i,i+1} \cos \theta_i} B_i \qquad (2\text{-}156)$$

$$= ik_{i+1} \cos \theta_{i+1} e^{ik_{i+1} z_{i,i+1} \cos \theta_{i+1}} A_{i+1} \qquad (b)$$

$$-ik_{i+1} \cos \theta_{i+1} e^{-ik_{i+1} z_{i,i+1} \cos \theta_{i+1}} B_{i+1}$$

where $z_{i,i+1}$ is the z coordinate at which the interface occurs, and where all the terms in x went out as they were identically canceled with the use of Snell's law. Note that (2-156) can be written in a matrix form:

$$R_i V_i = L_{i+1} V_{i+1} \qquad (2\text{-}157)$$

if the identifications

$$V_i = \begin{pmatrix} A_i \\ B_i \end{pmatrix} \qquad (a)$$

$$R_i = \begin{bmatrix} e^{ik_i z_{i,i+1} \cos \theta_i} & e^{-ik_i z_{i,i+1} \cos \theta_i} \\ n_i \cos \theta_i \, e^{ik_i z_{i,i+1} \cos \theta_i} & -n_i \cos \theta_i \, e^{-ik_i z_{i,i+1} \cos \theta_i} \end{bmatrix} \qquad (b) \quad (2\text{-}158)$$

$$L_{i+1} = \begin{bmatrix} e^{ik_{i+1} z_{i,i+1} \cos \theta_{i+1}} & e^{-ik_{i+1} z_{i,i+1} \cos \theta_{i+1}} \\ n_{i+1} \cos \theta_{i+1} \, e^{ik_{i+1} z_{i,i+1} \cos \theta_{i+1}} & -n_{i+1} \cos \theta_{i+1} \, e^{-ik_{i+1} z_{i,i+1} \cos \theta_{i+1}} \end{bmatrix}$$

$$(c)$$

where the $\mathbf{V}_i$'s are the wave coefficient vectors, $\mathbf{R}_i$ is the boundary matrix for the right-hand side of the ith layer, and $\mathbf{L}_{i+1}$ is the boundary matrix for the left-hand side of the $(i + 1)$th layer. Note that both sides of (2-156) were divided by $k_0 = 2\pi/\lambda$ to obtain (2-158).

Equation (2-157) greatly simplifies the process of formally obtaining a solution to the propagation problem. In general, the problem one wishes to solve is that in which a unity amplitude plane wave is incident upon the interface of medium 1 from medium 0 at an angle of θ_0. From this information, one wishes to calculate the reflection coefficient r and the transmission coefficient t, assuming that there is no wave incident on the $(N - 1)$th interface from the medium N. These boundary conditions define the quantities

$$A_0 = e^{-ik_{0z}z_{01}\cos\theta_i} \qquad \text{(a)}$$

$$B_0 = re^{ik_{0z}z_{01}\cos\theta_i} \qquad \text{(b)}$$

$$A_N = te^{-ik_{Nz}z_{N-1,N}\cos\theta_N} \qquad \text{(c)} \qquad \text{(2-159)}$$

$$B_N = 0 \qquad \text{(d)}$$

where (2-159) was obtained by setting

$$\mathbf{R}_0\mathbf{V}_0 = \begin{bmatrix} 1 + r \\ n_0\cos\theta_0(1 - r) \end{bmatrix} \qquad \text{(a)}$$

$$\qquad \qquad (2\text{-}160)$$

$$\mathbf{L}_N\mathbf{V}_N = \begin{bmatrix} t \\ n_N\cos\theta_N t \end{bmatrix} \qquad \text{(b)}$$

where the first component of the vector is simply the E-field at the surface, and the second component of the vector is a constant times the H-field at the surface. It is clear from (2-158) that once either r or t is known (i.e., once either $\mathbf{V}_0$ or $\mathbf{V}_n$ is known), then all the other A's and B's can be determined. For concreteness, let us consider a single slab. Here one can write that

$$L_2\mathbf{V}_2 = R_1\mathbf{V}_1 \qquad \text{(a)}$$

$$\qquad \qquad (2\text{-}161)$$

$$L_1\mathbf{V}_1 = R_0\mathbf{V}_0 \qquad \text{(b)}$$

One can eliminate $\mathbf{V}_1$ from (2-161) to obtain

$$R_0\mathbf{V}_0 = L_1R_1^{-1}L_2\mathbf{V}_2 \qquad \qquad (2\text{-}162)$$

The combination $L_1 R_1^{-1}$ is generally called the characteristic matrix of medium 1, M_1, and one can therefore write that

$$R_0 V_0 = M_1 L_2 V_2 \qquad (2\text{-}163)$$

From (2-163), it is straightforward to obtain

$$r = \frac{(m_{11} + n_2 \cos \theta_2 m_{12})n_0 \cos \theta_0 - (m_{21} + m_{22}n_2 \cos \theta_2)}{(m_{11} + n_2 \cos \theta_2 m_{12})n_0 \cos \theta_0 + (m_{21} + n_2 \cos \theta_2 m_{22})} \qquad \text{(a)}$$

$$t = \frac{2n_0 \cos \theta_0}{(m_{11} + n_2 \cos \theta_2 \, m_{12})n_0 \cos \theta_0 + (m_{21} + n_2 \cos \theta_2 m_{22})} \qquad \text{(b)}$$

$$(2\text{-}164)$$

where the m_{ij}'s are the elements of M_1, and the n's are the refractive indices of the layers. The results of (2-163) and (2-164) can be immediately generalized to the N-layer case, where

$$R_0 V_0 = M_1 M_2 \ldots M_{N-1} L_N V_N \qquad (2\text{-}165)$$

and equation (2-115) goes over unchanged if one simply makes the identification

$$M = M_1 M_2 \ldots M_{N-1} \qquad (2\text{-}166)$$

The characteristic matrix has some interesting salient features. Comparison of (2-162) with (2-163) shows that the characteristic matrix effectively translates an interface from one coordinate to another for the purposes of applying boundary conditions. In this sense, the matrix is a true mathematical representation of the propagation effects induced by the layer. Students of transmission line theory would say that this observation is trivial, as the multilayer is obviously just another set of transmission lines; but this material really belongs to another course (see the book by Kuester and Chang 1992). A second observation concerning the characteristic matrix can be made with reference to equation (2-166). The characteristic matrix need not represent a simple layer but could represent a quite complex refractive index. One could easily think of a limit in which each layer became infinitely thin, and therefore the matrix product of (2-166) became an infinite one (continuous limit of a staircase approximation). Here, the characteristic matrix could be used to represent a continuous index distribution. Actually, one need not take this continuous limit to obtain such an M. As is shown in Born and Wolf (1975), one can construct an M matrix for any medium in which one knows the fundamental solutions to the differential equation. In this sense, one could use another fundamental type of M matrix rather than that for a step to expand a problem, such as expanding a fiber profile in terms of thin parabolas. Here we will stick to staircase approximations.

Perhaps it is best to end this section with some specific examples of layered media. For a layer of index n_1 and thickness l_1, one finds a characteristic matrix of the form

$$
M_1 = \begin{bmatrix} \cos\,(kn_1 l_1 \cos\,\theta_1) & -\dfrac{i}{n_1 \cos\,\theta_1}\,\sin\,(kn_1 l_1 \cos\,\theta_1) \\[2ex] -in_1 \cos\,\theta_1 \sin\,(kn_1 l_1 \cos\,\theta_1) & \cos\,(kn_1 l_1 \cos\,\theta_1) \end{bmatrix}
$$

$$(2\text{-}167)$$

An interesting and important case is that of a beam splitter. The simplest form of a beam splitter is simply a flat plate that is illuminated at an angle so that the reflected beam is not collinear with the incident. The reflection and transmission coefficients can be found by using (2-176) in (1-164) (a) and (b), to find

$$
r = -i\,\frac{\left(n_1 \cos\,\theta_1 \cos^2\,\theta_0 - \dfrac{1}{n_1 \cos\,\theta_1}\right) \sin\,(kn_1 l_1 \cos\,\theta_1)}{2 \cos\,\theta_0 \cos\,(kn_1 l_1 \cos\,\theta_1) - i\,(n_1 \cos\,\theta_1 + (1/n_1 \cos\,\theta_1))\,\sin\,(kn_1 l_1 \cos\,\theta_1)}
$$

$$(a)$$

$$
t = \frac{2n_0 \cos\,\theta_0}{2 \cos\,\theta_0 \cos\,(kn_1 l_1 \cos\,\theta_1) - i\,(n_1 \cos\,\theta_1 + (1/n_1 \cos\,\theta_1))\,\sin\,(kn_1 l_1 \cos\,\theta_1)}
$$

$$(b)$$

$$(2\text{-}168)$$

where n_0 and $n_2 = 1$ and $\theta_0 = \theta_2$ have been used. The denominators of the two expressions are identical. A most noteworthy point is that the two terms are in quadrature, and will be in quadrature independent of any of the parameters of the problem. This is a common feature of all of the so-called directional coupling structures, of which the beam splitter is the most ubiquitous. To go further with the directional coupler analogy, one could easily think of exciting both sides of the beam splitter, as is illustrated in Figure 2.32. In this case, the reflected wave of one input will be coincident with the transmitted beam of the other. Perusual of equation (2-168) indicates that if the incident beams are of equal amplitude and zero phase, the exiting disturbances will have equal amplitude and zero phase. This says that the symmetric excitation is an eigenmode of the coupling structure. A similar argument can be used to show that equal excitations with a π phase shift between them will also be an eigenexcitation. These symmetric and antisymmetric eigenmodes are common features of all directional coupling structures. Perhaps the important point here is that one can use the simple picture of a beam splitter as a model for reasonably complicated switching structures, as well as a model for loss mechanisms. The beam splitter

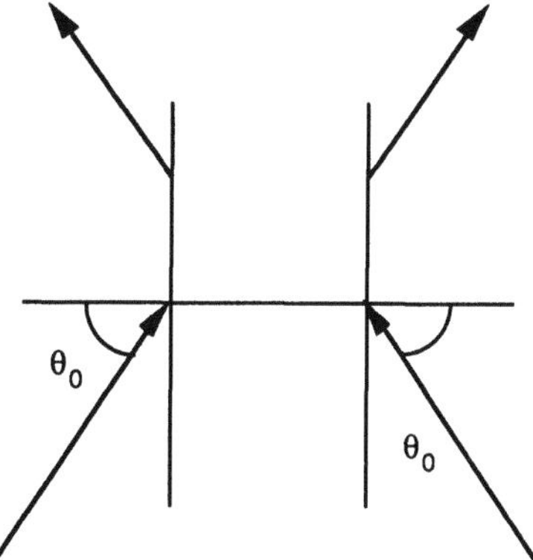

FIGURE 2.32.   Excitation of both input ports of the beam splitter.

also has some interesting quantum mechanical properties although a discussion of them would take us too far afield.

Another interesting example is that of a dielectric mirror. Say, for example, that l_1 is chosen to be a quarter wavelength, that is, $l_1 = \lambda/4n_1 \cos \theta_1$. Then (2-167) reduces to

$$M\left(\frac{\lambda}{4n_1 \cos \theta_1}\right) = \begin{bmatrix} 0 & -\dfrac{i}{n_1 \cos \theta_1} \\ -n_1 \cos \theta_1 & 0 \end{bmatrix} \tag{2-169}$$

and (2-164) (a) reduces to

$$r\left(\frac{\lambda}{4n_1 \cos \theta_1}\right) = i\,\frac{n_1 \cos \theta_1 - \dfrac{n_0 \cos \theta_0 n_2 \cos \theta_2}{n_1 \cos \theta_1}}{n_1 \cos \theta_1 + \dfrac{n_0 \cos \theta_0 n_2 \cos \theta_2}{n_1 \cos \theta_1}} \tag{2-170}$$

Indeed, one can see that if the index of medium 1 is chosen such that

$$n_1 \cos \theta_1 = \sqrt{n_0 n_2 \cos \theta_0 \cos \theta_2} \quad \text{(AR coating)} \tag{2-171}$$

then $r = 0$ and the $n_0 - n_2$ interface has been antireflection (AR)-coated. It is seen from (2-164) together with (2-167), however, that this AR coating is strongly dependent upon wavelength and angle, and, therefore, trying to AR coat an interface for all optical wavelengths at all angles is not an easy (if possible) task.

A second interesting limiting case of (2-167) would be that in which l_1 is chosen to be a half-wavelength. Here, (2-167) reduces to

$$M\left(\frac{\lambda}{2n_1 \cos \theta_1}\right) = \begin{bmatrix} 1 & 0 \\ 0 & 1 \end{bmatrix} \tag{2-172}$$

and therefore (2-164) (a) reduces to

$$r\left(\frac{\lambda}{2n_1 \cos \theta_1}\right) = \frac{n_0 \cos \theta_0 - n_2 \cos \theta_2}{n_0 \cos \theta_0 + n_2 \cos \theta_2} \tag{2-173}$$

Equation (2-173) is exactly the reflection coefficient one would obtain for an uncoated medium. This leads to the rather surprising result that the half-wavelength coating is effectively invisible.

Some very practically important cases are those of periodic slab media. In such a case, one repeats an arrangement of slabs many times, as is illustrated in Figure 2.33. For such a case where a given motif is repeated N times, it is clear that the M matrix will take the form

$$M_{\text{tot}} = M^N(l) = (M_1(l_1)M_2(l_2))^N \tag{2-174}$$

where $l = l_1 + l_2$, and l_1 is the thickness of layer 1 and l_2 the thickness of layer 2. For the case of a stack of quarter-wave plates illuminated at normal incidence, one finds that (2-174) will reduce to

$$r(N_{\lambda/4} \text{ layers}) = \frac{\left(\dfrac{n_2}{n_1}\right)^N - \left(\dfrac{n_1}{n_2}\right)^N}{\left(\dfrac{n_2}{n_1}\right)^N + \left(\dfrac{n_1}{n_2}\right)^N} \tag{2-175}$$

and the stack will become highly reflecting at normal incidence at the design wavelength. The real questions that arise out of such a case ask what are the

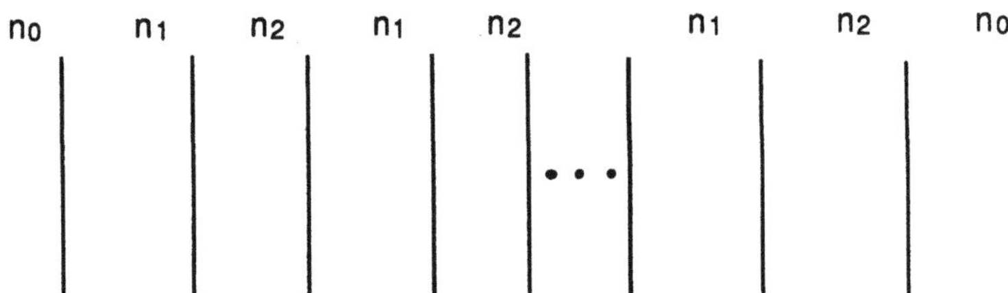

FIGURE 2.33.   A diagram depicting a periodic multilayer structure.

bandwidth and the acceptance angle of this dielectric mirror? In some applications, one might want a very small bandwidth for laser line control, whereas in another, one might want a wide bandwidth for power applications. Such questions will be addressed in the problems.

Although we have concentrated on studying the properties of solutions of Maxwell's equations that propagate normally to the layered structure, the same formulation can also be used to study the propagation of waves inside the layered structure along the layers. In such a case, the angles of propagation become complex in most of the layers, and the condition that the fields must die off to zero at infinite distance from the structure can be used to solve for the propagation constant of the bound wave. The basic idea is that of total internal reflection, as discussed earlier in this chapter. In one layer (or some subset of layers) the propagating wave is totally internally reflected on each contact with the boundary. The resulting waves in the other layers are inhomogeneous plane waves that die off toward infinity. Calculation of the Poynting vector flux would then show that the energy is propagating along the guiding layers. Such is the nature of a guided wave, but this material is in the domain of a separate course on guided wave optics (see, for example, Mickelson 1992).

References

Azzam, R. M. A. and N. M. Bashara, *Ellipsometry and Polarized Light*, North-Holland, New York (1989).

Born, M. and E. Wolf, *Principles of Optics*, Fifth edition, Pergamon Press, New York (1975).

Clemmow, P. C., *The Angular Plane Wave Representation of Electromagnetic Fields*, Pergamon, Oxford (1966).

Einstein, A., Die quanten theorie der Strahling, *Phys. Zeit. 18*, 121 (1917).

Haken, H. *Light*, Volume I, *Wave, Photons, Atoms*, North-Holland, New York (1981).

Hermann, A. *The Genesis of the Quantum Theory*, M.I.T. Press, Cambridge, MA (1971).

Hurwitz, H., Jr. and R. C. Jones, A new calculus for the treatment of optical systems II. Proof of three general equivalence theorems, *JOSA 31*, 493–499 (1941).

Jackson, J. D., *Classical Electrodynamics*, Second edition, John Wiley and Sons, New York (1975).

Johnk, C. T. A., *Engineering Electromagnetic Fields and Waves*, John Wiley and Sons, New York (1975).

Jones, R. C., A new calculus for the treatment of optical systems, I. Description and discussion of the calculus, *JOSA 31*, 488–493 (1941a).

Jones, R. C., A new calculus for the treatment of optical systems III. The Sohncke theory of optical activity, *JOSA 31*, 500–503 (1941b).

Jones, R. C. A new calculus for the treatment of optical systems IV, *JOSA 32*, 486–493 (1942).

Jones, R. C. A new calculus for the treatment of optical systems V. A more general formulation, and description of another calculus, *JOSA 37*, 107–110 (1947a).

Jones, R. C., A new calculus for the treatment of optical systems VI. Experimental determination of the matrix, *JOSA 37*, 110–112 (1947b).

Jones, R. C., A new calculus for the treatment of optical systems VII. Properties of the *N*-matrices, *JOSA 38*, 671–685 (1948).

Kittel, C. and H. Kroemer, *Thermal Physics*, W. H. Freeman, San Francisco (1980).

Klein, M. V. and T. E. Furtak, *Optics*, Second edition, John Wiley and Sons, New York (1986).

Kraus, J. D. *Electromagnetics*, Third edition, McGraw-Hill, New York (1984).

Kuester, E. F. and D. C. Chang, *Theory of Waveguides and Transmission Lines*, course Notes for ECEN5114, University of Colorado (to be published).

Maimom, T. H., Stimulated optical radiation in ruby masers, *Nature 187*, 493 (1960).

Mickelson, A. R., *Guided Wave Optics* (to be published by Van Nostrand Reinold in 1992).

Papas, C. H. *Theory of Electromagnetic Wave Propagation*, McGraw-Hill, New York, Chapter 5 (1965).

Schawlow, A. L. and C. H. Townes, Infrared and Optical Masers, *Phys. Rev. 112*, 1940–1949 (1958).

Shurcliff, W. A., *Polarized Light: Production and Uses*, Harvard University Press, Cambridge, MA (1962).

Shurcliff, W. A. and S. S. Ballard, *Polarized Light*, D. Van Nostrand, Princeton, NJ (1964).

Silverberg, E. C., Operating and performance of a lunar laser ranging station, *Appl. Opt. 13*, 565–574 (1974).

Trebes, J. E. et al., Demonstration of X-ray holography with an X-ray laser, *Science 238*, 517–518 (1987).

Yariv, A., *Quantum Electronics*, Third edition, John Wiley and Sons, New York (1989).

Problems

1. Use complex notation as introduced by equation (2-7) to derive the expression of equation (2-26) for the time-averaged Poynting vector as defined in equation (2-24).

2. Consider the overall Poynting vector for a superposition of forward- and backward-propagating plane waves. Assume that the plane waves are propagating in the positive and negative z-directions. Also assume that both plane waves have their electric fields polarized in the x-direction along with having equal amplitudes and optical frequencies.

   (a) Calculate the Poynting vector as defined in equation (2-17).

   (b) Now we place a "square law" detector of surface area A oriented perpendicular to the z-axis facing the negative z direction. Calculate the detector current as defined by the time-averaged Poynting vector as given in equation (2-24).

   (c) Calculate the optical intensity as defined in expression (2-28) of this field.

   (d) Explain in terms of this problem what one would "see" in an optical cavity containing these two plane waves, both with a detector and in terms of the optical intensity.

3. For the following plane wave fields, calculate the Poynting vector, the optical intensity from the flux crossing the plane $z = 0$ (i.e., $\langle \mathbf{S} \cdot \mathbf{e}_z \rangle$) at $x = 0$, $y = 0$, and the optical intensity from the rule $I(x, y) = \mathbf{S} \cdot \hat{\mathbf{e}}_k$.

(a) For an x-polarized, z-directed plane wave of amplitude E_0.
(b) For the sum of z- and minus z-directed plane waves, both of amplitude E_0.
(c) For a plane wave with E-field $E_0 \hat{e}_y e^{ikz\cos\theta} e^{ikx\sin\theta}$

4. Assume the following:

$$E_x \cong \hat{e}_x E_x e^{-r^2/2\omega_0^2} e^{ikz}, \ E_y = 0$$

and

$$H_y \cong \hat{e}_y H_y e^{-r^2/2\omega_0^2} e^{ikz}, \ H_x = 0$$

Using the fact that $\nabla \cdot \mathbf{E} = 0$, and the power, $\mathbf{P}$, is given by:

$$P \ \alpha \ \text{Re} \ (E \times H^*)$$

compute the power flowing in the $\hat{e}_x$-direction. How physically accurate is the result? What improvement in the original assumption would most improve the power calculation?

5. Consider a free electron at coordinate $z = 0$, directly in the path of an incident plane wave, whose electrical polarization vector is given by

$$\mathbf{E} = E_x \hat{e}_x + E_y \hat{e}_y$$

with

$$E_x = a_x \cos(\omega t - kz) \ \text{rect}\left(\frac{t - \tau/2}{\tau}\right)$$

$$E_y = a_y \cos(\omega t - kz + \varphi) \ \text{rect}\left(\frac{t - \tau/2}{\tau}\right)$$

where the rect function is defined by

$$\text{rect}(x) = \left\{ \begin{array}{ll} 1 & |x| < 1/2 \\ 0 & \text{otherwise} \end{array} \right\}$$

and where it is tacitly assumed that $\tau \gg 2\pi/\omega$. We wish to find the subsequent motion of the particle, given that its initial state is described by initial conditions

$$x(0) = x_0 \ \ y(0) = y_0$$

$$\frac{dx}{dt}(0) = \dot{x}_0 \ \ \frac{dy}{dt}(0) = \dot{y}_0$$

(a) Solve the Lorentz force law using Newton's equations (assuming nonrelativistic motion) for the particle's motion, taking $a_x = a_y = a$ and $x_0 = y_0 = \dot{y}_0 = 0$.

(b) Sketch the motion for several values of φ.

(c) and (d)   What are dx/dt $(t > \tau)$ and dy/dt $(t > \tau)$?

6. An electron at rest is illuminated by an elliptically polarized plane wave of frequency ω, starting at time $t = 0$.

(a) Find an expression for the electron's motion.

(b) Find the average velocity for the electron, assuming that the electron was surrounded by free space and that the plane wave is circularly polarized with an intensity of 1 μW/cm^2, and a wavelength of 0.5 mm. Is a relativistic correction (i.e., taking the **B** field into account) necessary?

(c) Write the equations of motion for the particle if the electron has an initial velocity $v_0 = v_x$ that is comparable to the speed of light.

7. An electron is traveling in a straight line at 1/10 the speed of light. Now imagine that the electron enters a region with uniform magnetic field of 0.3 T as depicted in Figure 2-34. The velocity vector of the entering electron makes an angle of 30° with respect to the direction of the magnetic field as shown in the figure. Describe the motion of the electron in the field.

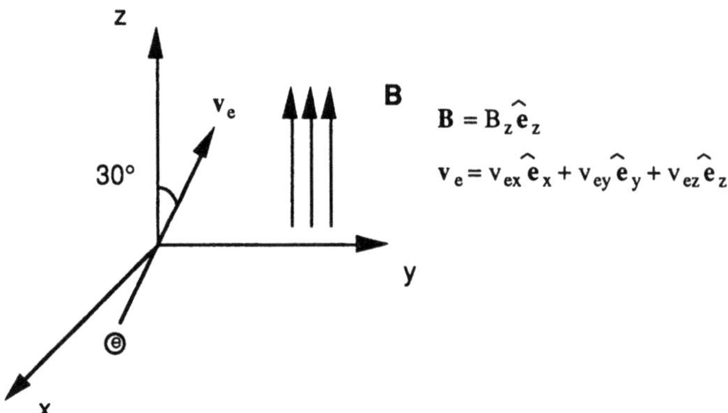

FIGURE 2.34.   Figure for problem 7.

8. The following questions relate to thermal radiation.

(a) What is the maximum temperature that a microwave cavity (say, at 1 GHz) could be kept at so that thermal noise would not be a dominant pumping mechanism for the quantum two-level system generating the microwave signal?

(b) What is the minimum temperature of a blackbody for it to have the peak of its spectrum in the X-ray (say, 1 keV) spectral region?

(c) At what temperature does the spontaneous rate equal the stimulated rate for optical frequencies (say, around 5×10^{14} Hz)?

(d) At what temperature would the occupation N_2 of the upper level of a two-level system be twice that of the occupation N_1 of the lower level?

9. This problem concerns the 3°K background radiation distribution that permeates the universe.

(a) Sketch the $\rho(\nu)$ for this distribution, labeling important points on the curve (i.e., maxima) as to their frequency.

   (b) Find the ratio of the stimulated to the spontaneous energy production for this distribution at a photon energy of 2eV.
   (c) Suppose we wish to detect this spectrum with a detector at $300°K$. What would the S/N ratio [that is, the ratio of $\rho(\nu)$ to the background $\rho(\nu)$] at $300°K$ be as a function of frequency?
   (d) Devise a better technique for measuring the background radiation spectrum than that proposed in (c).

10. We wish to construct an X-ray laser by externally pumping a box filled with two-level atoms that have a transition energy of 1 keV. Suppose that the box is at a temperature T.
   (a) If T is room temperature, what is the thermally induced ratio of the spontaneous to the stimulated emission rates?
   (b) If the density of excited atoms in the box is N_2, what would the density of injected photons at the transition frequency in the box have to be for the stimulated rate to exceed the spontaneous?
   (c) Suppose that we wish to force the stimulated rate to exceed the spontaneous by raising the temperature of the box. How high a T would be necessary? How could this be done?

11. A soft X-ray laser ($\lambda = 21$ nm) is to be built using neon-like selenium (that is, selenium that has been ionized 24 times).
   (a) At room temperature, what is the ratio of spontaneous emission to stimulated emission, assuming that the working medium is at thermal equilibrium with its surroundings?
   (b) Again, assuming thermal equilibrium, what would the temperature of the working medium need to be (i) for the stimulated emission rate to equal the spontaneous rate; (ii) for the simulated role to exceed the spontaneous rate by a factor of 16.4 (12.1 dB)?

12. What is the minimum temperature of a blackbody for it to have the peak of its wavelength in the gamma-ray (say, 0.03 Å)?

13. Two of the energy levels of a molecule X are $\epsilon_1 = 6.1 \times 10^{-21}$ J and $\epsilon_2 = 8.4 \times 10^{-21}$ J. What is the occupation ratio of the two energy levels in an assembly of molecules of X at (i) $300°K$ (room temperature); (ii) $3000°K$?

14. (a) What would the temperature have to be for a 1 mm source if the emission is not dominated by spontaneous emission?
   (b) Consider a two level system with a transition energy of 1.98×10^{-19} J. What is the temperature for an atomic-level ratio (N_2/N_1) of 0.5 at thermal equilibrium?
   (c) To achieve laser operation in a two-level system it is necessary to have an inversion, $N_2 > N_1$. If we *think* of the system as being at thermal equilibrium, what would the *effective* temperature be for $N_2/N_1 = 5.0$?
   (d) Sketch the effective temperature as a function of N_2/N_1. From this sketch, what can be said about the necessary effective temperature (at thermal equilibrium) to achieve laser operation in a two-level system?

15. The thermally stimulated emission rate equals the spontaneous emission rate for radiation that has a wavelength of $\lambda \approx 67$ μm. For a higher energy density one can make the stimulated and spontaneous emission rates equal for shorter wavelengths. Calculate the necessary intensity of a light beam in the visible region (λ

~ 0.6 μm, $h\nu$ ~ 2eV) to equalize the two emission rates. *Hint: $I = c/n \, \rho(\nu) \, d\nu$.* Assume $d\nu = 10$ GHz and $n = 1$.

16. The maser can be used as an extremely low-noise amplifier. The noise in the maser results from spontaneous transitions between energy levels. Use the relation between spontaneous and stimulated transition rates to calculate the noise factor of a maser amplifier with 10 dB gain. *Hint:* The noise factor F of a conventional amplifier is defined as

$$P_{\substack{\text{amplifier} \\ \text{noise}}} = (F - 1)G \, kT \, d\nu$$

Assume $N_2 \gg N_1$, beam cross-sectional area ~λ^2, $\nu = 10$ GHz.

17. We wish to evaluate the Wiener-Khintchine relation of equation (2-86) in the text for several different correlation functions $R(\tau)$. For each of the cases below, evaluate the spectral density function $W(\omega)$, and try to give some explanation of what kind of source the $R(\tau)$ might represent.
    (a) $R(\tau) = \delta(\tau)$.
    (b) $R(\tau) = \begin{cases} 1 - |\tau|/\tau_c & |\tau| < \tau_c \\ 0 & \text{otherwise} \end{cases}$.

    (c) $R(\tau) = e^{-c|\tau|} \cos \bar{\omega}\tau$.
    (d) $R(\tau) = e^{-c\tau^2} \cos \bar{\omega}\tau$.

18. Find the output Jones vector for the following cases where a general input is given:
    (a) A general polarizer that passes light oriented at an angle θ to the x-axis. Verify that linear polarization at an angle θ to the x-axis is unchanged.
    (b) An x-directed polarizer followed by a y-directed polarizer. What if a 45° polarizer is inserted between the two?
    (c) The configuration of (b) with a θ-directed polarizer between the x-directed and y-directed polarizers.

19. A Jones matrix for a mode converter (i.e., converts x-polarized to y-polarized and vice versa) may be constructed by rotating the axes by $-45°$, introducing a π phase shift, and then rotating the coordinate system back to its original configuration. This effectively couples the two orthogonal states of polarization.
    (a) Calculate the mode converter matrix. Take $\Delta\varphi_y - \Delta\varphi_x = kL$, and neglect extraneous common phase terms.
    (b) Show that any input polarization state can be converted to any arbitrary output state by replacing the π phase shift with an arbitrary phase shift.

20. Consider the optical system of Figure 2.35. Suppose that the input is unpolarized and of field strength $\mathbf{E}_{\text{in}}$. Find the a_x, a_y in the expression $\mathbf{E}_{\text{out}} = a_x\hat{\mathbf{e}}_x + a_y\hat{\mathbf{e}}_y e^{i\delta}$ where $\mathbf{E}_{\text{out}}$ is the field in the x_{out}, y_{out} plane for the following set of polarizers:
    (a) I and II have x-directed polarizers, and III is directed at an angle θ (with respect to x).
    (b) I is the x-directed, II is θ-directed and III is $(\theta + \pi/2)$-directed.
    (c) I is x-directed, II is a piece of glass, and III is y-directed.
    (d) Repeat (c), but replace the glass in II with a 45° directed polarizer.

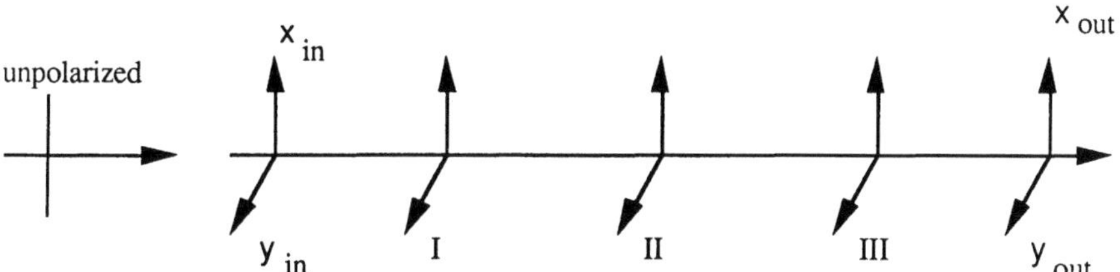

FIGURE 2.35.   Figure for problem 20.

21. Consider Figure 2.36 with light of an arbitrary state of polarization (SOP) incident at the input plane. Give expressions for the transmitted electric field vector for the following situations and places:
    (a) At the output of polarizer 1, where polarizer 1 is an x-directed (passes x) polarizer.
    (b) At the output plane where polarizer 3 is y-directed, polarizer 1 is as in (a), and there is no polarizer 2.
    (c) At the output plane where polarizers 1 and 3 are as in (b), but polarizer 2 is directed at $45°$ between x and y.
    (d) At the output plane if polarizers 2 and 3 of (c) are interchanged.

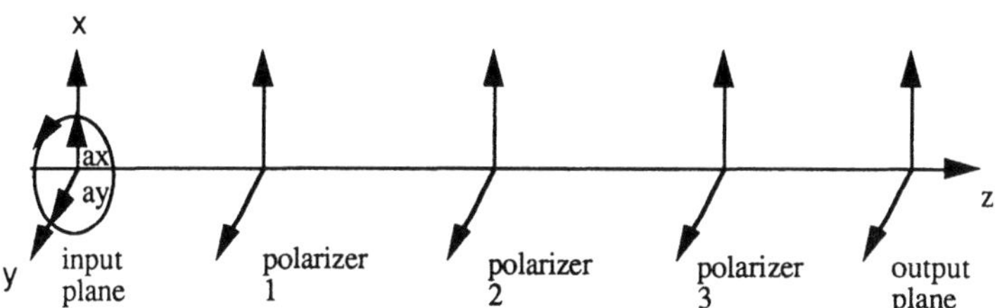

FIGURE 2.36.   Figure for problem 21.

22. A quarter-wave plate with polarization oriented in the direction making an angle θ with the vertical direction is placed in the path of a vertically polarized light wave. What is the output polarization?
    (a) What is the result if a half-wave plate is used instead of a quarter-wave plate?
    (b) How can you use a quarter-wave plate and a half-wave plate to transform the vertically polarized wave into any polarization state?
23. Mueller matrices were used in the text to describe polarization devices.
    (a) Find the Mueller matrix for a device that transforms amplitudes a_x, a_y of the incoming wave to $a_x' = t_x a_x$, $a_y' = t_y a_y$, where t_x and t_y are real. (*Hint:* Write the transformed Stokes parameters in terms of the initial a_x's and a_y's, and then write them as linear combinations of the incident Stokes parameters.)
    (b) Use the result of (a) to find the Mueller matrix for a dichroic polarizer at an arbitrary angle θ.

(c) Find the Mueller matrix for a device that retards the x-polarized component of the incident wave by an arbitrary phase α.

(d) Find the Mueller matrix for a set of three polarizers where the first is y-directed, the second is at angle $\theta(t) = \pi/2 \cos \omega t$ with respect to x, and the third is x-directed, where t is time.

24. Consider an x-polarized wave normally incident on a polarizer oriented at angle θ to x. Assume that the polarizer is perfectly AR-coated.

(a) Find the loss suffered by the wave on passing through the polarizer as a function of θ.

Consider a stack of N such above-described polarizers, each oriented at angle θ_0/N with respect to the preceding one, the first at angle θ_0/N with respect to the x-polarized incident wave.

(b) Find the loss suffered by the wave on traversing the stack of polarizers as a function of N, the number of polarizers.

25. Find the Stokes parameter and the degree of polarization for the wave at the output plane of Figure 2.37, given the following source spectra:

(a) $I(\lambda) = \delta(\lambda = .5\ \mu m)$.

(b) $I(\lambda) = \frac{1}{3} \delta(\lambda - .25\ \mu m) + \frac{1}{3} \delta(\lambda - .5\ \mu m) + \frac{1}{3} \delta(\lambda - .75\ \mu m)$

(c) $I(\lambda) = \dfrac{1}{.5\ \mu m} \operatorname{rect} \left(\dfrac{\lambda - .5\ \mu m}{.5\ \mu m} \right)$, where

$$\operatorname{rect}(x) = \begin{cases} 1 & |x| \le \frac{1}{2} \\ 0 & \text{otherwise} \end{cases}$$

(d) Repeat (a) through (c) for the case where the $x(\lambda_0/4)$ plate is followed by a $y\ (\lambda/4)$ plate.

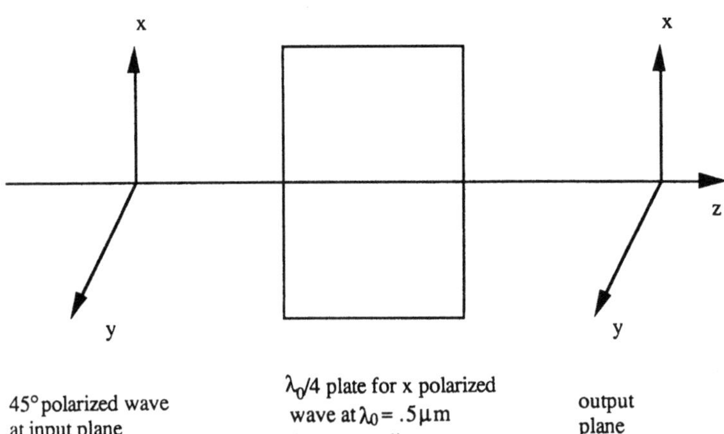

45° polarized wave
at input plane

$\lambda_0/4$ plate for x polarized
wave at $\lambda_0 = .5\ \mu m$
(i.e. x retarding)

output
plane

FIGURE 2.37.   Figure for problem 25.

26. Let us say that we have a stack of N polarizers, where N can be quite large. In (a) and (b) assume that each polarizer has its major axis oriented at $45°$ to the one in front of it.
   (a) What is the energy transmitted through the stack as a function of N if the input is unpolarized?
   (b) What is the transmitted polarization state as a function of N?
   (c) Let us say that the angle of the next polarizer is $45°/2^Q$, where Q can be large. Repeat (a) and (b).

27. The Stokes vector is an intensity quantity; therefore different Stokes vectors should add on an intensity basis. In particular, if we combine two mutually incoherent beams, their Stokes vectors should add. What is the resulting Stokes vector for the incoherent addition of
   (a) Equal amplitude x-polarized and y-polarized waves?
   (b) Equal amplitude unpolarized and right circularly polarized waves?
   Consider the setup of Figure 2.38, where the polarizing beam splitter separates the equal-amplitude x- and y-polarized components into two paths, one of which can be lengthened with respect to the other by a corner cube reflector. What Stokes vectors and degrees of polarization will be seen at the output plane, as a function of the offset from perfect path alignment, if
   (c) The source is monochromatic?
   (d) The source is incoherent with coherence length l_c?

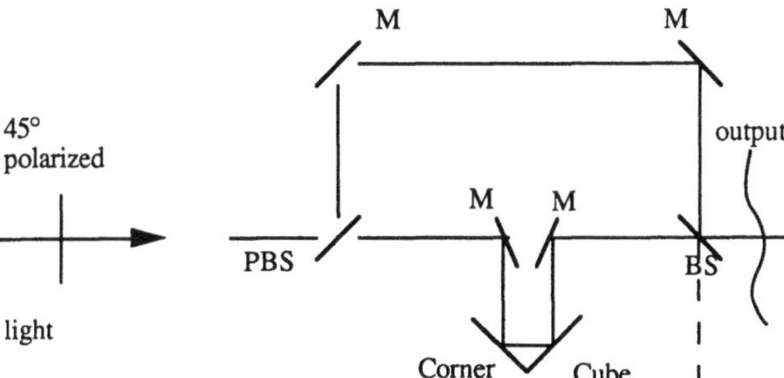

FIGURE 2.38.   Figure for problem 27.

28. As has been discussed, an arbitrary transformation of polarization can be decomposed into a phase shift between the x and y states (δ_1), followed by a rotation from a_x into a_y (θ_2), followed by a second phase of shift (δ_3).
   (a) Find the Mueller matrix for this general transformation.
   Then say that each element of the transformation system can be voltage-controlled, that the shifts are linear in the voltage, but that θ_2 requires a much higher voltage than the other two. Find the "least voltage" (i.e., minimize $V_1 + V_2 + V_3$) transformation to transform the following Stokes vectors to a linear x state:

$$\text{(b) } \mathbf{S}_{\text{in}} = \frac{S_0}{2} \begin{bmatrix} 1 \\ 0 \\ 0 \\ \pm 1 \end{bmatrix} \qquad \text{(c) } \mathbf{S}_{\text{in}} = \frac{S_0}{2} \begin{bmatrix} 1 \\ -1 \\ 0 \\ 0 \end{bmatrix} \qquad \text{(d) } \mathbf{S}_{\text{in}} = \frac{S_0}{2} \begin{bmatrix} 1 \\ 1/\sqrt{3} \\ 1/\sqrt{3} \\ 1/\sqrt{3} \end{bmatrix}$$

(e) What effect will the transformer have on the following state, and why?

$$\frac{S_0}{2} \begin{bmatrix} 1 \\ 0 \\ 0 \\ 0 \end{bmatrix}$$

29. Let us say we put a polarizer in the path of an unpolarized wave. Evidently the entropy carried by the wave must be diminished by the polarizer. Where did the entropy go? Can you use the definition of T, $\dfrac{1}{k_B T} = \dfrac{dS}{du}$, where k_B is Boltzmann's constant and u internal energy, to show that the entropy of the universe increases? Can you use the Helmholtz free energy $F = U - k_B TS$ and the chemical potential $\left(\mu = \dfrac{dF}{\delta N} \right)_{T,V}$, where N is the number of particles, to track the entropy flow?

30. Refer to Figure 2.25. Show that $E_y = E_{y_0} e^{ikx\sin\theta} e^{\pm ikz\cos\theta}$ (2-142) is indeed a solution to the TE wave equation

$$\left(\frac{\partial^2}{\partial x^2} + \frac{\partial^2}{\partial z^2} + k^2 \right) E_y = 0$$

31. Given an air–glass interface at $x = 0$, as illustrated in Figure 2.39, with a plane wave incident on the interface whose electric field is given by $\mathbf{E}_{\text{inc}} = a_y E_0 e^{-ikon_1(-x\cos\theta + z\sin\theta)} \hat{e}_y$:
    (a) Is this a TE or a TM polarized wave?
    (b) What is the magnetic field of this incident plane wave?
    (c) What is the electric field of the reflected plane wave?

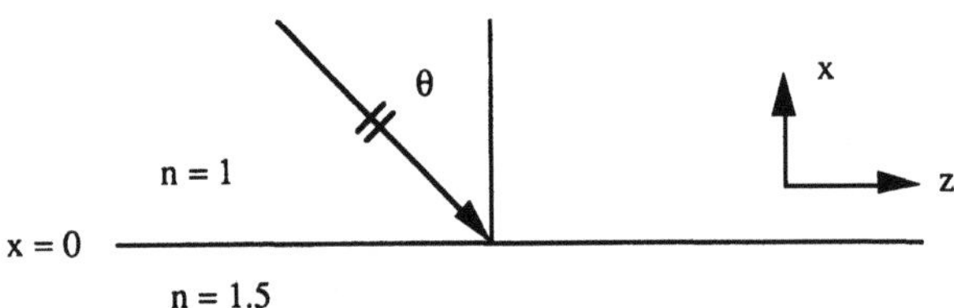

FIGURE 2.39.   Figure for problem 31.

32. Consider an air–glass interface with a wave incident at an angle of θ_i, as depicted in Figure 2.40:
    (a) Plot $t_\parallel$, $t_\perp$, $r_\parallel$, $r_\perp$ as functions of the incident angle θ_i.

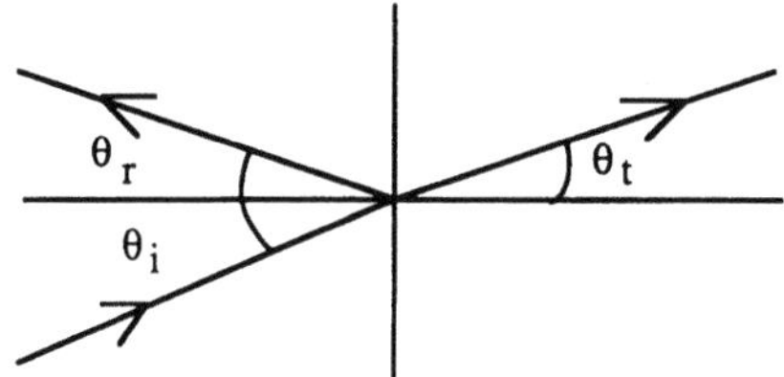

FIGURE 2.40.   Figure for problem 32.

(b) Take a_y as the perpendicular component of the incident polarization and a_x as the parallel component. Plot a_x and a_y as functions of the incident angle for the transmitted and reflected waves.

(c) Plot the Stokes parameters of the reflected and transmitted waves as a function of θ_i when the incident wave is unpolarized.

(d) A Mueller matrix is an operator that represents a process, in that when it is applied to an incident set of Stokes parameters, the result is the set of Stokes parameters that would result from the process. Find the Mueller matrix for reflection and transmission from an interface in terms of $t_{\parallel}$, $t_{\perp}$, $r_{\parallel}$, $r_{\perp}$.

33. For (a) and (b) consider a plane wave incident on an air–glass interface at an angle θ with respect to a surface normal, as is illustrated in Figure 2.39.

(a) Find the Mueller matrices that transform the incident Stokes parameters to the reflected and transmitted Stokes parameters.

(b) Consider the case of (a) where θ is the Brewster angle. What are the reflected and transmitted polarization states for unpolarized, x-polarized, y-polarized, right-hand-polarized, and left-hand-polarized incident states?

(c) Repeat (a) for the case of a glass–air interface.

(d) Consider the result of (c) and repeat (b), but for the wave incident at the total internal reflection angle.

34. (a) Show that when a plane wave is incident on a suitable dielectric boundary, at the Brewster angle, the **k** vectors of the transmitted and reflected waves are perpendicular.

(b) Consider a gas laser with two glass ($n = 1.5$) Brewster windows bounding the gas tube within the laser cavity, which is assumed to have an index of 1.0 (see Figure 2.41).

(i) Calculate the Brewster angle for the air–glass interface.

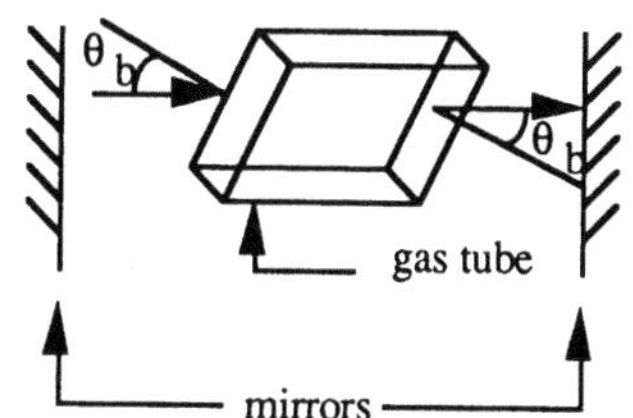

FIGURE 2.41.   Figure for problem 34.

(ii) Assuming that the windows are not optically flat (i.e., multiple reflections do not add coherently), calculate the fraction of the power transmitted by a light pulse traversing the cavity with both perpendicular and parallel polarization.

35. Consider a multilayer, such as in Figure 2.42, that is so imperfect that one can consider intensity transmission at each interface.

(a) Find the Mueller matrix for the transmitted wave at the 0-1 interface using a coordinate system in which TE corresponds to x and TM to y.

(b) Repeat (a) for the interfaces 1-2 and 2-1.

(c) Suppose that the incident angle is such that the wave is incident on the 1-2 interface at the Brewster angle. What Mueller matrix describes the transition from the first to the third layer?

(d) Repeat (c) for N layers. How large must N be to polarize the incident light if $n_0 = 1$, $n_1 = 1.4$, and $n_2 = 1.6$?

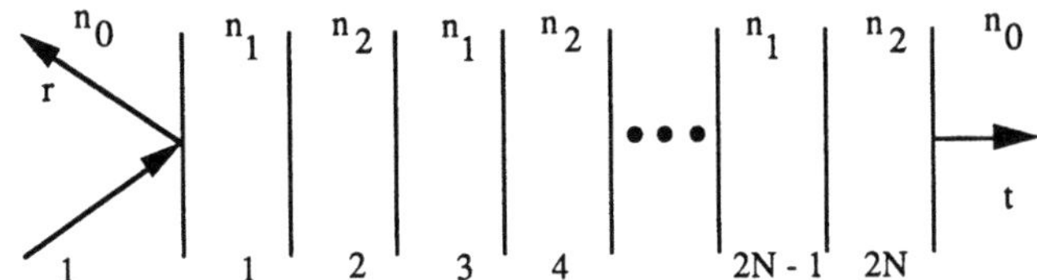

FIGURE 2.42.   Figure for problem 35.

36. Consider the structure of Figure 2.43.

(a) For an incident angle, θ_i, equal to $65°$, what is the value of θ_t?

(b) Write an expression for the fraction of power, P_f, transmitted along the z-axis in region 3. Assume that we can ignore multiple reflections, and that $\theta_i > \theta_c$,

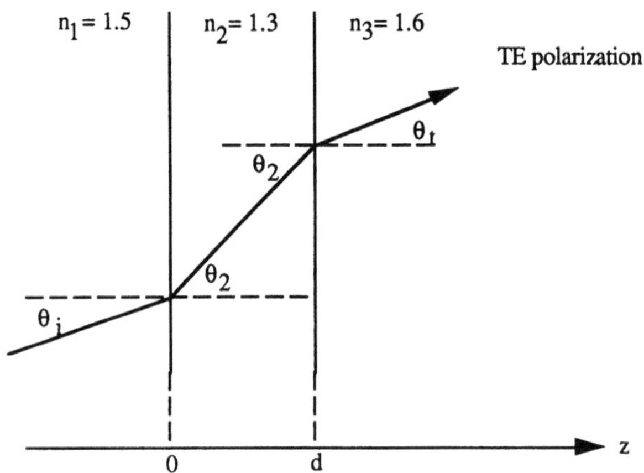

FIGURE 2.43.   Figure for problem 36.

where θ_c is the critical angle. P_f should be a function of d and θ_i only. Evaluate P_f for $I = 1$ μm, $\theta_i = 65°$, and $d = 3$ μm.

(c) Why does the simple expression for P_f fail for small values of d?

(d) For $\lambda = 1$ μm, $d = 3$ μm, plot P_f as a function of θ_i for $\theta_c < \theta_i \leq 1.14$ rad.

37. Consider a plate of index n_2 immersed in a medium of index n_1 $(n_2 < n_1)$, as illustrated in Figure 2.44, for TM polarization. Write down the boundary condition equations at the two interfaces for the cases:

(a) $\theta_i < \theta_{\text{critical}}$.

(b) $\theta_i > \theta_{\text{critical}}$.

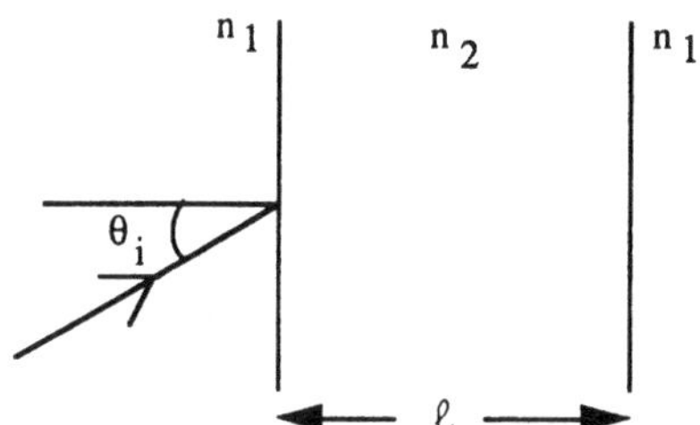

FIGURE 2.44.   Figure for problem 37.

38. Use the **R** and **L** matrices discussed previously in the text to solve the problem illustrated in Figure 2.45 for r and t, and thereby derive the form given in the text for the reflection from a glass plate. What happens to the matrices in the limit where $n_1 > n_0$ and $\sin \theta_1 > 1$?

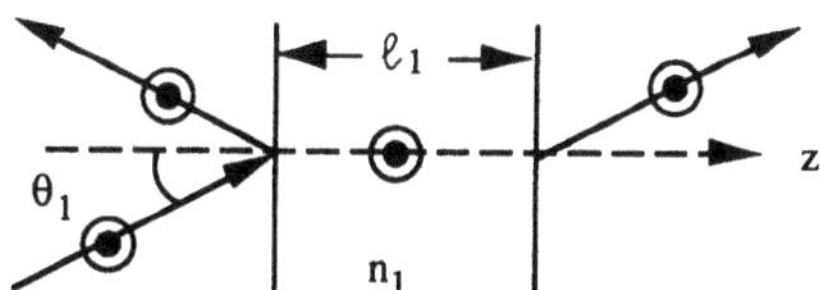

FIGURE 2.45.   Figure for problem 38.

39. Consider a circularly polarized wave incident on an n_1/n_2 interface. Find the polarization states of the transmitted and reflected waves for:

(a) Normal incidence with $n_1 < n_2$.

(b) Normal incidence with $n_1 > n_2$.

(c) Incidence at Brewster's angle.

(d) Incidence above the total internal reflection angle for $n_2 < n_1$.

40. Solve the TM Fresnel problem, and in particular, the following aspects thereof:

(a) Write out the curl equations in detail.

(b) Derive the resulting wave equation.

(c) Derive the boundary conditions using Gauss and Stokes theorems.

(d) Find the $r_{\parallel}$ and $t_{\parallel}$ from the linear system of boundary equations.

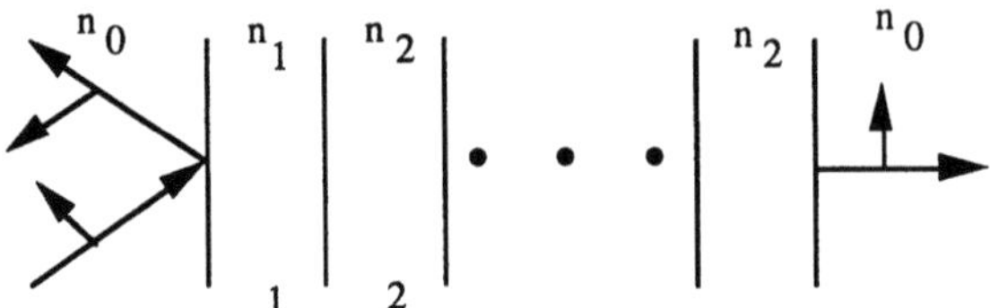

FIGURE 2.46.   Figure for problem 41.

41. Here we wish to consider the multilayer problem, as illustrated in Figure 2.46, for TM polarization.
    (a) Write down the form of the solutions in each region i.
    (b) Write out a set of boundary condition equations at one of the interfaces.
    (c) Define boundary condition matrices L and R as was done in the text.
    (d) Solve the Fresnel problem using the matrices of (c).
    (e) Given that we had found the characteristic matrix M for the multilayer, find expressions for r and t in terms of M.

42. Consider a stack of $2N$ half-wave layers of alternating indices n_1 and n_2. We wish to investigate the sensitivity of the transmissivity of this stack to changes in input angle and wavelength. Suppose that the design angle is zero, and the design wavelength is λ_0.
    (a) Find the characteristic matrix of this medium for a plane wave incident at θ with the wavelength λ.
    (b) Assume $\theta \ll 1$ and expand M and M^N to first order in θ. What does this tell you about the angular sensitivity of the stack relative to that of a single layer?
    (c) Repeat (b) but for first order changes in λ, that is, for $\lambda = \lambda_0(1 + \delta\lambda)$ where $\delta\lambda \ll 1$.
    (d) How do r and t depend on $\delta\lambda$ if the stack is placed between two media with n_0? Recall that for $x \ll 1$ one can use

$$\cos x \sim 1 - \frac{x^2}{2} \sim 1$$

$$\sin x \sim x$$

$$\frac{1}{1 - x} \cong 1 + x$$

etc.

43. Consider the multilayer problem as illustrated in Figure 2.47, where all the layers 1, 2, and 3 are a quarter-wavelength (in the material) thick, and $n_4 > n_0$.
    (a) Find the characteristic matrix of the composite material made up of 1, 2, and 3.
    (b) Find the $r_\perp$ and $t_\perp$ of this material.
    (c) What is the condition that must be satisfied by n_1, n_2, n_3, θ_1, θ_2, θ_3, l_1, l_2, and l_3 for $r = 0$?

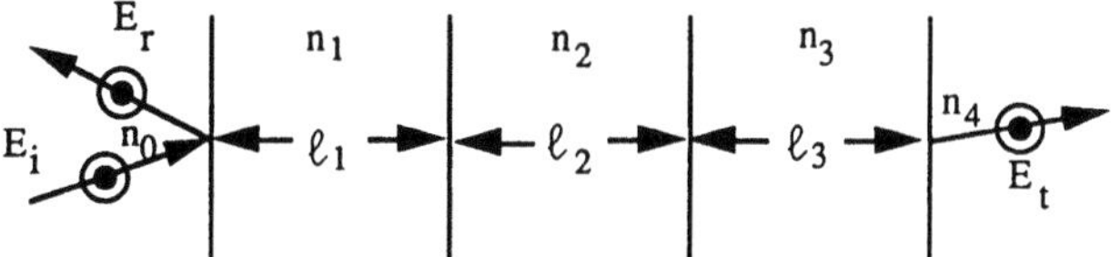

FIGURE 2.47.   Figure for problem 43.

(d) If one designs the layer to be an AR coating at normal incidence, how should one choose n_1, n_2, and n_3 for maximum wavelength insensitivity? For maximal angular insensitivity?

3

Material Polarization
and Dispersion

3.1 INTRODUCTION

The present chapter is designed to serve as a bridge between the polarization discussion of Chapter 2 and the anisotropic propagation theme of Chapter 4, as well as to stand on its own as a description of the relationship between the microscopic and the macroscopic pictures of the interaction of light and matter.

In Chapter 2, the discussion of polarization included a discussion of the field-induced motion of free particles in electromagnetic fields. It was seen that the polarization direction of a plane wave indicates a true direction of force on a charge. In that discussion, allusion was made to the fact that polarization forces on bound charges sap energy from the forward-going plane wave and store it in the electron–nucleon bonds. The displacement of the valence electrons from their equilibrium position, however, led to the generation of a real dipole moment, and, correspondingly, electromagnetic fields. For a time-varying incident field, these dipole fields would correspondingly be time-varying and, therefore, involve a radiative portion that would add to the incident field. It was then stated, in Chapter 2, that this was the basis of the refractive index. In the present chapter, this fact will be explicitly demonstrated through careful discussion of the relation between microscopic and macroscopic field quantities as well as through a semiclassical model for the field–matter interaction. This semiclassical model will be shown to reduce to a classical spring model of matter in a typically classical limit. This spring model simplifies the calculation of refractive indices from a calculation using atomic parameters to one based on phenomenological parameters that can be determined from atomic parameters.

The microscopic/macroscopic relations derived in this chapter will further serve as a foundation for the discussion of anisotropic propagation in Chapter

4. For purposes of the present chapter, it will be assumed that wave functions exhibit strict symmetry and that, therefore, the only direction that can enter into a given calculation is the direction of electric field polarization. Real solid materials, however, essentially always exhibit some degree of anisotropy, that is, have preferential directions built into their wave functions. The arguments in this chapter can, however, be seen to serve as a microscopic basis for the macroscopic component of the dielectric constant. It should also be mentioned that the dipole response equation derived in this chapter is general enough to describe anisotropic nonlinear optical effects, as will be pointed out. The weak perturbation limit of this equation, however, will be the one considered here, as the field of nonlinear optics is rich enough to require a whole text on its own.

To carry out the ambitious purpose of this chapter, discussion will begin with the most simple-minded picture imaginable, that of propagation by emission and absorption. This discussion leads directly to the concepts of real and virtual processes, which lead to a discussion of the nature of the refractive index.

After these introductory comments are completed, attention turns to a detailed derivation of the Lorentz-Lorenz relation and discussion of its consequences. The section begins with some qualitative discussion of the characteristics of the microscopic world, the world that exists on a dimensional scale in which an atomic radius is large compared to the standard yardstick. Microscopic Maxwell's equations, a constitutive relation-free yet source-dependent form, then are introduced. The averaging procedure necessary to couple the macro and micro worlds is discussed in terms of the physical processes that define its parameters. Averaging brackets are applied to Maxwell's microscopic equations to obtain a multipole expansion parameterized form of the macroscopic equations. Use of the dipole approximation then allows a definition of the usual macroscopic polarization. The last piece in the puzzle is to derive the dielectric constant in terms of a microscopic material constant. This is done through actually evaluating the polarization vector in terms of the microscopic dipole moments for a model material in an external field, and leads to the desired Lorentz-Lorenz equations. Some discussion is given at the close of the section as to how the Lorentz-Lorenz relation could be generalized to other than the model medium.

In the fourth section of the chapter, discussion turns to finding the microscopic material constant, α, the molecular polarizability, in terms of atomic constants of the material, and thereby determining the frequency response of the dielectric constant. The presentation starts with a qualitative discussion of phonons and resonances before beginning the quantitative semiclassical derivation. The derivation begins with a short review of the most salient features of quantum mechanics in general and of the quantum theory of the atom in particular. The discussion then turns to semiclassical time-dependent perturbation theory, in which the electromagnetic field is considered classically but the bound electron as a quantum mechanical wave function. The evolution of the wave function under a harmonic electric field perturbation is derived. This solution is used to calculate the evolution of the microscopic dipole moment, as expressed as a quantum mechanical ensemble average. In a weak field (weak compared to the electronic binding energy) limit, and with the use of quantum classical correspondence, it is shown that the evolution of the dipole moment can be described by a spring model of the motion of the classical electron coordinate. This classical spring model can be used to drive an expression for the frequency dependence of the index of refraction in terms of atomic parameters.

With expressions for the material dispersion in hand, discussion naturally turns to wave propagation in dispersive media, in the fifth section of the chapter. An electromagnetic derivation of the evolution of the envelope of a quasi-monochromatic wave packet is used to introduce the concept of group velocity. Following this, the Poynting vector is revisited, but taking into account the nonmonochromatic nature of information signals. This argument is extended to the discussion of energy propagation. It is seen that the group velocity and energy velocity are identical only in lossless media, and that pulse clipping will occur in lossy resonant materials.

The last section of the chapter discusses methods for generalizing the classical spring equations to take into account more exotic effects than the one that occurs in an isotropic model medium. This discussion can be used as a preface to the more general anisotropic medium discussion of Chapter 4.

3.2 COMPLEXITY IN THE MICROSCOPIC WORLD

Perhaps one of the most surprising things about matter is how simply its electromagnetic properties can be described. Quite generally, one writes that

$$\mathbf{D} = \epsilon \mathbf{E} \qquad (3\text{-}1)$$

in matter, and one readily finds that this relation predicts that the velocity of the phase propagation, v_p, inside of the medium is given by

$$v_p = \frac{c}{n} \qquad (3\text{-}2)$$

where (3-2) can be deduced by the form that the wave equation takes in the material medium. If however, one, thinks of a microscopic picture of matter relations, equations (3-1) and (3-2), which are so well established experimen-

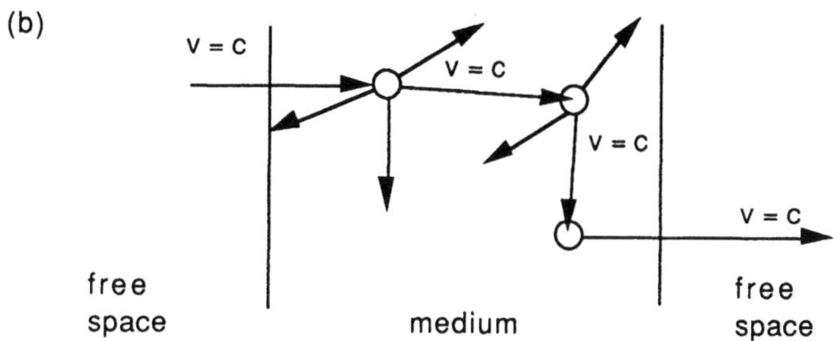

FIGURE 3.1. Comparison of wave propagation (a) in a macroscopic picture and (b) in a microscopic picture.

tally, become rather hard to understand. Figure 3.1 tries to illustrate the classical versus the microscopic situation. In the macroscopic picture, all effects of the medium can be described in terms of the slowing down of the phase front and the change in relation between **E** and **H** during propagation in the medium. In quantum mechanics, however, we are taught that interactions take place only when discrete quanta are interchanged. In the microscopic picture, therefore, one would think that one would need to sum over a multitude of scattering (absorption/remission) events that a photon experiences while propagating in the medium. It is moderately unclear why this sum should take such a simple form as the macroscopic picture illustrates that it does.

Part of the confusion arises from the fact that not all quantum-mechanical interactions need be actual. When an incident photon has a frequency that corresponds to a transition energy of the bound electron of an atom, then the photon can be absorbed, causing a shift in the state of the electron. As energy is conserved, the $\Delta E \Delta t$ form of the uncertainty principle, where ΔE is the absolute difference between the level spacing $E = E_b - E_a$ and the photon energy $h\nu$ and Δt is the lifetime for the transition, tells us that this state of affairs can last for a long time. In fact, it generally lasts so long as to have a decay mode into lattice phonons that heat the crystal, at least in cases where the incident photon stream is weak enough that the stimulated gain is negligible. However, interactions that are far from resonant also can take place, as long as they are very short-lived. The photon can, therefore, be arrested for a very short period and then sent on its way again. For a lossless material, that is, one whose resonances lie very far away from the incident photon frequency, the time is so short that the electron has little or no time to move to new position during the interaction. As we will see in subsequent sections, this is equivalent to saying that the amplitude coefficient of the excited state never takes on an appreciable value although, indeed, it takes on some non-zero value. Because the interactions are short, even for a weak incident photon stream, many interactions can take place if there are many atoms, and, therefore, each atom, on average, can have a non-zero excited state coefficient. In a relatively dense material, such as gallium arsenide (GaAs), there is a density of roughly 10^{23} atoms$/$cm^3, which amounts to as many as 10^{11} atoms$/\mu$m^3, or 0.1 atoms$/$Å^3. In a rarefied medium such as a molecular gas, one might have 5×10^{20} atoms$/$cm^3, which corresponds to 5×10^8 atoms$/\mu$m^3. In both cases, the number of atoms within an optical wavelength is very large and thus phase effects between scattering events are important only in an averaged sense. Therefore, one can treat many properties of the usual dielectric media in a continuum limit. That is what we will do in the present section, before giving a particular picture of dispersion in a later section where it will be necessary to give a full quantum-mechanical treatment of the atom in order to calculate the effects of the virtual transitions. There it will be seen that the interactions can be expressed in terms of a simple classical spring model in an important weak field limit.

3.3 A DERIVATION OF THE LORENTZ-LORENZ RELATION

In the present section, macroscopic quantities will be denoted by capital letters and microscopic quantities by small letters. Therefore, **E** and **B** will denote the macroscopic electric field vector and magnetic induction vector, respectively, and **e** and **b** the microscopic counterparts. A further distinction must be made between average macroscopic quantities in a medium and the macroscopic quantities measured outside the medium. The inside quantities will be subscripted l for local.

Perhaps now is an appropriate time to discuss what is meant by microscopic and local macroscopic fields. For example, inside a neutral chunk of material with no external field applied, one would expect there to be no macroscopic field, that is, a field measurable by a macroscopic apparatus such as a voltmeter. Such will not be the case for the local microscopic fields inside the lattice. For example, as is illustrated in Figure 3.2, fields near nuclei can become quite large and, in fact, the microscopic fields can easily take on values of 10^9 V$/$cm.

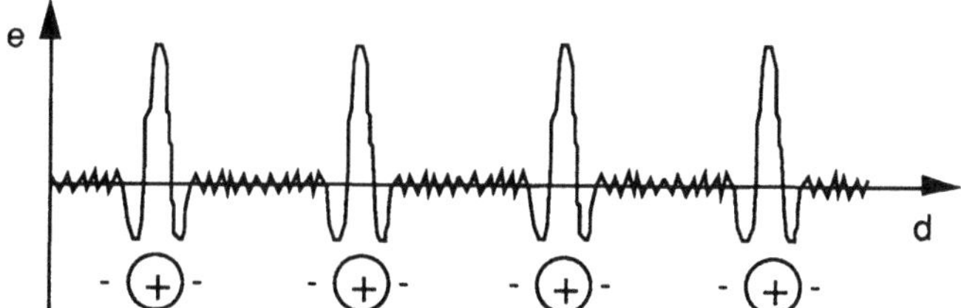

FIGURE 3.2.   The microscopic electric field inside a crystal lattice versus the distance (one-dimensional projection) along a lattice row, which is schematically depicted below the plot.

Further, these microscopic fields vary with time. The thermal velocity v_{th} of a particle in the lattice can be approximated to take on the free space value of

$$v_{th} = \sqrt{\frac{k_B T}{m}} \tag{3-3}$$

where k is Boltzmann's constant and m is the particle's mass. At room temperature, this thermal velocity is

$$v_{th}\bigg|_{\substack{\text{room} \\ \text{temperature}}} = \frac{1.2 \times 10^5}{(m/m_p)^{1/2}} \frac{\text{cm}}{\text{sec}} = \frac{1.2 \times 10^{12}}{\sqrt{m/m_p}} \frac{10 \text{ Å}}{\text{sec}} \tag{3-4}$$

where m_p is the proton mass. Equation (3-4) indicates that particles can move appreciably with respect to interatomic distances (which are on the order of 5 Å) in times on the order of 10^{-13} sec. Further, as electronic orbits have binding energies on the order of electron volts, these orbits can vibrate with periods ranging from 10^{14} to 10^{17} Hz. The thermal and electronic time constants, indeed, make the propagation of optical waves seem complex, as the material seethes with activity right at rates corresponding to the optical period. However, if one were to consider a lattice with no applied field, the above-discussed motions should be random, with correlation lengths on the order of the interatomic spacing. As this correlation length is on the order of a few Å's, yet the optical wavelength is on the order of a micron, these motions will average out in a spatial average, as long as the averaging regime is on the order of 0.01 wavelength, which is roughly 20 atomic spacings, or greater. One therefore sees that, in order to obtain the local, internal macroscopic fields, one must perform an averaging over the microscopic fields. As the time averaging is effectively carried out by the spatial averaging (spatial sampling generates an ensemble with time-stationary statistics), only the spatial averaging is necessary.

The microscopic fields inside a medium can be expressed in the following form (Jackson 1975, 227):

$$\nabla \times \mathbf{e} = -\mu_0 \frac{\partial \mathbf{b}}{dt} \qquad \text{(a)}$$

$$\nabla \times \mathbf{b} = \mu_0 \mathbf{j}_\mu + \frac{1}{c^2} \frac{\partial \mathbf{e}}{dt} \qquad \text{(b)} \qquad \text{(3-5)}$$

$$\nabla \cdot \mathbf{b} = 0 \qquad \text{(c)}$$

$$\nabla \cdot \mathbf{e} = \rho_\mu / \epsilon_0 \qquad \text{(d)}$$

where it has been assumed that $\mathbf{D}$ and $\mathbf{H}$ are macroscopic fields to be derived from the microscopic fields and their time averages and that j_μ and ρ_μ are the microscopic current and charge density due to the μth particle. The form of the averaging must be

$$\mathbf{E}(\mathbf{r}, t) = \frac{1}{\delta V} \int_0^{dr} d^3 r' \mathbf{e}(r - r', t) = \langle \mathbf{e}(r, t) \rangle \qquad \text{(a)}$$

$$\text{(3-6)}$$

$$\mathbf{B}(\mathbf{r}, t) = \frac{1}{\delta V} \int_0^{dr} d^3 r' \mathbf{b}(r - r', t) = \langle \mathbf{b}(r, t) \rangle \qquad \text{(b)}$$

where the volume δV must be large compared to a lattice constant but small compared to a wavelength of the radiation under consideration. The angular bracket to the far right in (3-6) can be thought of as being defined by the equation. As optical wavelengths are on the order of 1000 lattice spacings, one can perform averages over tens of lattice points to find the macroscopic fields, and, in fact, this allows one even to get by with certain static field calculations applied to the optical domain. One must bear this in mind when finding macroscopic polarization from microscopic polarizabilities. Clearly, for the quasistatic situation we discuss, the j_μ can be taken to be zero. It still remains to be seen how the $\langle \rho_\mu \rangle$ can be related to macroscopic quantities.

The most direct, physically motivated technique for calculating the effect of a material polarizability on wave propagation is to include a polarization term in the $\mathbf{E}$, $\mathbf{D}$ relation as (see, for example, Born and Wolf 1975, Chapter 2):

$$\mathbf{D} = \epsilon_0 \mathbf{E} + \mathbf{P} \qquad \text{(3-7)}$$

Using (3-7) in Maxwell's equations leads one rapidly to the expression that

$$\nabla^2 \mathbf{E} - \frac{1}{c^2} \frac{\partial^2}{\partial t^2} \mathbf{E} = \mu_0 \frac{\partial^2 \mathbf{P}}{\partial t^2} - \frac{1}{\epsilon_0} \nabla(\nabla \cdot \mathbf{P}) \qquad \text{(3-8)}$$

If one assumes that **P** is proportional to **E**, then it is clear that the first term on the right-hand side of (3-8) serves only to change the effective light velocity c. If **P** is proportional to **E**, it also seems reasonable to assume that plane wave-like solutions (or at least solutions that are locally plane wave-like, as will be discussed in the next chapter) are possible. If this is so, the second term on the right-hand side is negligible, and the total effect of the polarizable material is only to modify the effective light velocity. Results of Section 3.4 will also tell us how to quantitatively evaluate this expression when this proportionality is valid. For the present section, we will assume the plausability of the result.

In the present picture, we choose to ignore the effects of free charges. Here, dipole moments are due to the displacement of charged matter from its equilibrium position within the assumed electrically neutral material. The effect of the displacement can be mathematically modeled by a multipole expansion of the atomic charge distribution centered about its equilibrium point. With this, it is possible to expand the average of ρ_μ in the form (Jackson 1975, Section 6.7):

$$\langle \rho_\mu \rangle_{\text{atom}} = \langle q_a \delta(x - x_a) \rangle_a - \nabla \cdot \langle \mathbf{p}_a \delta(x - x_a) \rangle_a$$

$$+ \tfrac{1}{6} \nabla \nabla : \langle \boldsymbol{q}_a \delta(x - x_a) \rangle + \cdots \tag{3-9}$$

where the : denotes a double contraction of the second rank tensor $\boldsymbol{q}_a$ with the two divergence operators. Now, keeping only the dipole term in (3-8), one can average (3-5) (d) and compare it with the form of the equation obtained by substituting (3-7) in Maxwell's divergence **E** equation to obtain

$$\mathbf{P} = \left\langle \sum_{\text{atoms}} \mathbf{p}_N \delta(x - x_N) \right\rangle = \frac{1}{V} \sum_N \mathbf{p}_N \tag{3-10}$$

Equation (3-10) will be used in later considerations.

In the next section, we discuss in considerably more detail what the form of the relation between the microscopic dipole moment and the local field $\mathbf{E}_l$ would be as a function of material structure and field strength. We do know, however, that if the dipole moment is driven by a time-varying field, then if the field is periodic (or monochromatic), it stands to reason that the dipole moment will have the same period. As a crystal is, in general, anisotropic, that is, the dipole will not be in the direction of the field, the relation between the ith component, p_i, of the microscopic dipole moment, **p**, and components of the local field $\mathbf{E}_l$ is of the form

$$\frac{p_i}{\epsilon_0} = \alpha_{ij} E_{\ell j} + \beta_{ijk} E_{\ell j} E_{\ell k} + \gamma_{ijkl} E_{\ell j} E_{\ell k} E_{\ell 1} + \cdots \tag{3-11}$$

where the α's, β's, and γ's are the various susceptibility tensors of the medium, and the summation convention is assumed to hold. In the case where the medium is assumed to be (approximately) isotropic, one could write that some component of **p**, called p_i, could be expressed as a power series

$$\frac{p_i}{\epsilon_0} = \alpha E_{\ell i} + \beta E_{\ell i}^2 + \gamma E_{\ell i}^3 + \cdots \quad \text{(isotropic, noncentrosymmetric)} \quad (3\text{-}12)$$

but more discussion of this will appear in the next section. Here we wish to take the simplest relation possible, that is, that

$$\mathbf{p} = \epsilon_0 \alpha \mathbf{E}_l \quad \text{(linear, isotropic)} \quad (3\text{-}13)$$

What we wish to do now is to find the local internal macroscopic field when an external field is applied across the medium. The situation is depicted in Figure 3.3 (Born and Wolf 1975, Chapter 3; Kittel 1971, Chapter 13). Assuming charge neutrality and some nominal symmetry (or lack thereof) that does not allow for the type of order that leads to built-in dipole moments, the local internal electric field will average to zero in the absence of an external field. However, as was discussed above, these internal fields can be very large with respect to macroscopic externally applied fields, and, therefore, the small readjustments to internal structure that result from the external field may have a profound effect on the field setup internally. In any case, one would not expect the macroscopic local field to match the externally applied field except in the case of a near-vacuum gas, as we will see to be the case below.

One way to split up $\mathbf{E}_l$ for purposes of calculation is that indicated in Figure 3.3, where $\mathbf{E}_l$ is taken to be the sum

$$\mathbf{E}_l = \mathbf{E}_{\text{ext}} + \mathbf{E}_1 + \mathbf{E}_2 + \mathbf{E}_3 \quad (3\text{-}14)$$

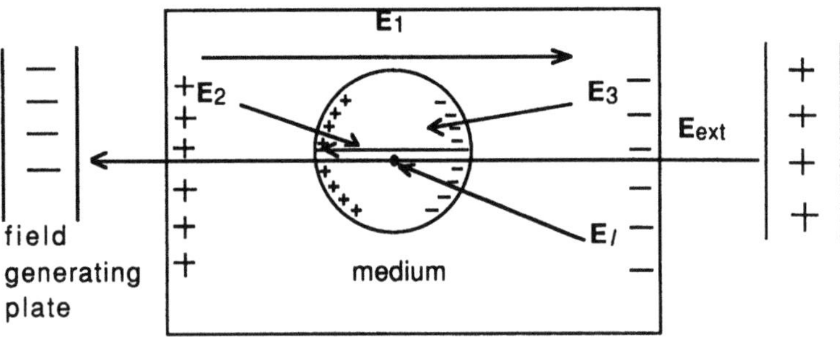

FIGURE 3.3.   A possible way to break up the local macroscopic field $\mathbf{E}_l$ into constituent parts, as indicated in equation (3-14). The field $\mathbf{E}_1$ is due to the dielectric surface charge, the $\mathbf{E}_3$ is due to volume dipoles in the sphere of integration, and the $\mathbf{E}_2$ is due to the induced dipole distribution on the surface of the sphere of integration.

where $\mathbf{E}_{ext}$ is the externally applied field, $\mathbf{E}_1$ is the polarization field setup on the surfaces of the dielectric sample due to the index discontinuity in the presence of the external field, $\mathbf{E}_2$ is the contribution of the surface charge on the surface of the averaging sphere, and $\mathbf{E}_3$ is the contribution to the local field due to the volume charge in the averaging sphere. We will proceed to carry out the calculation as if all these fields were static fields, despite the fact that we will subsequently apply the results to the optical domain. This approach is consistent with our averaging assumptions. As the present argument ignores resonant absorption, this argument is applicable at each time point during the optical period, and the static calculation is indeed applicable. To make the calculation, we begin by considering the sum of $\mathbf{E}_{ext}$ and E_1. The $\mathbf{E}_1$ field opposes the $\mathbf{E}_{ext}$ such that anywhere inside of the dielectric, the measured field would be the sum of the two. Therefore, if one were to write $\mathbf{D} = \epsilon\mathbf{E}$ inside the material, the $\mathbf{E}$ in this expression would be the sum of the $\mathbf{E}_1$ and the $\mathbf{E}_{ext}$. This field is still different from $\mathbf{E}_l$, however, as $\mathbf{E}_l$ indicates the field seen by a charge in the material. This charge, however, cannot act on itself. The field it sees is therefore the field one would find at its coordinate were it removed from the material without altering the position of any other charges. If it were really removed from the material, however, the other charges would realign because they are creating a field to counteract the field of our "test" charge. It is this field created by the local charges to the "test" charge that augments the field seen by the test charge and therefore causes $\mathbf{E}_l$ to differ from $\mathbf{E}$. Further, in order to determine the field that an electric generator applies, one monitors voltages and fields rather than coulombs of charge on the plates, and one notes that the actual measurably applied free space field $\mathbf{E}$ is also actually the sum of the fields due to the coulombs of charges on the plate and the reaction fields. One can therefore write that

$$\mathbf{E}_l = \mathbf{E} + \mathbf{E}_2 + \mathbf{E}_3 \qquad (a)$$
$$\mathbf{E} = \mathbf{E}_{ext} + \mathbf{E}_1 \qquad (b)$$

$$(3\text{-}15)$$

To find $\mathbf{E}_3$, one should recall that the field of a static dipole (or one that is slowly varying with respect to averaging time) can be written in the form (see, for example, Jackson 1975, Chapter 4):

$$\mathbf{E}_{\text{dipole}}(r) = \frac{3(\mathbf{p} \cdot \mathbf{r})\mathbf{r} - r^2\mathbf{p}}{4\pi\epsilon_0 r^5} \qquad (3\text{-}16)$$

where the field configuration can be sketched as in Figure 3.4. Because of the applied field direction, one could always limit one's consideration, in calculating the contribution of the volume dipole distribution, to configurations such as that depicted in Figure 3.5. It is clear from symmetry (i.e., Figures 3.4 and 3.5

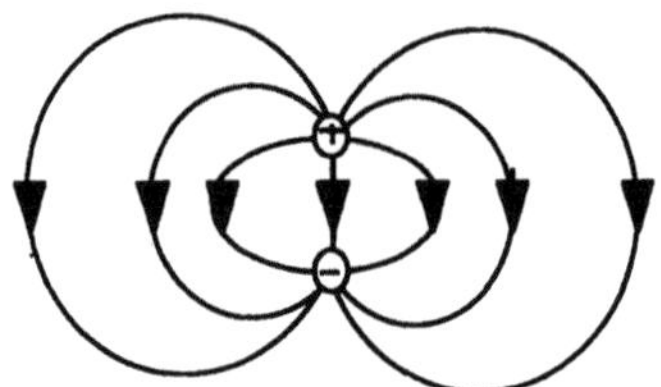

FIGURE 3.4.   The field of an elementary static dipole.

together) that the contribution to E_z must vanish. For those who do not believe in symmetry, one could assume a coordinate system as is indicated in Figure 3.5, and use equation (3-16) to make the following calculation, which involves summing over all the dipole contributions within the sphere. Starting with

$$E_3 = \sum_{\substack{\text{atomic} \\ \text{locations} \\ i}} E_{\text{dipole}} \tag{3-17}$$

and denoting $r_i = x_i\hat{e}_x + y_i\hat{e}_y + z_i\hat{e}_z$ as the ith dipole location and $\mathbf{p} = p\hat{e}_z$ as the dipole moment of each dipole, one sees that

$$\begin{aligned}
E_3 &= p \sum_i \frac{3z_i\mathbf{r}_i - r_i^2\hat{e}_z}{4\pi\epsilon r_i^5} \\
&= p \sum_i \frac{(2z_i^2 - x_i^2 - y_i^2)\hat{e}_z}{4\pi\epsilon r_i^5} + 3p \sum_i \frac{(x_i\hat{e}_x + y_i\hat{e}_y)}{4\pi\epsilon r_i^5} z_i \\
&= \frac{p}{4\pi\epsilon}\left[\left(2\left\langle\frac{z_i^2}{r_i^5}\right\rangle - \left\langle\frac{x_i^2}{r_i^5}\right\rangle - \left\langle\frac{y_i^2}{r_i^5}\right\rangle\right)\hat{e}_z \right. \\
&\quad \left. + \left\langle\frac{x_iz_i}{r_i^5}\right\rangle\hat{e}_x + \left\langle\frac{y_iz_i}{r_i^5}\right\rangle\hat{e}_y\right] = 0
\end{aligned} \tag{3-18}$$

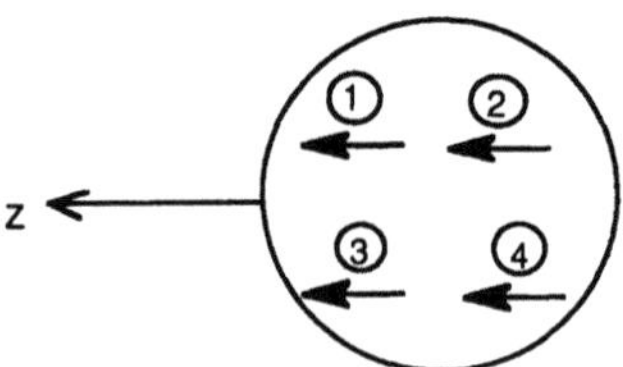

FIGURE 3.5.   All configurations of elemental dipoles possible in the sphere can be built up from such configurations.

where the angular brackets denote the average value over the atomic cell. The last line of equation (3-18) follows the fact that the angular brackets of each of the second-order moments must be equal to each other from symmetry, and the brackets of the cross-product moments must be identically zero from the fact that averages of x, y, and z are independent, and that the averages of first-order quantities must be zero. Do note, though, that this whole argument rests on the spherical shape of the averaging regime. Real materials have more complicated shapes. For further discussion one might want to consult a good freshman-level physics text such as Feynman, Leighton, and Sands (1964, Volume 2, Chapter 32). With the above considerations, we can write that

$$\mathbf{E}_l = \mathbf{E} + \mathbf{E}_2 \tag{3-19}$$

To calculate the contribution of $\mathbf{E}_2$ to the average local field, one must perform an integral over the charge distribution on the integrating surface, that is, the surface surrounding the dipole that we have "removed" from the material. Figure 3.6 gives a schematic depiction of this charge distribution. If one were to denote θ as a polar angle measured downward from the z-axis (dipole direction) and φ as an azimuthal angle measured around the dipole direction, one could express the charge element dq derived from $dq = (\mathbf{P} \cdot \mathbf{r})/r \, dS$ in the form

$$dq = P \cos \theta \, a \sin \theta d \, \theta a \, d\varphi \tag{3-20}$$
$$= -P \cos \theta \, 2\pi a^2 d \, (\cos \theta)$$

where a is the radius of the averaging sphere, and the last line is obtained from the symmetry of the problem with respect to φ. Now, if one writes Coulomb's law in the form

$$d\mathbf{E}_2 \cdot \hat{\mathbf{e}}_z = \frac{dq \cos \theta}{4\pi\epsilon_0 r^2} \tag{3-21}$$

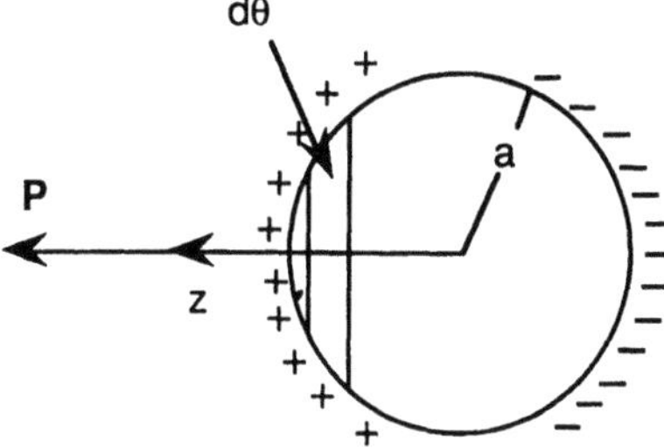

FIGURE 3.6.   Schematic depiction of the charge distribution on the surface of the sphere of integration.

one can carry out the indicated integration in (3-21) to obtain

$$4\pi\epsilon_0 a^2 \mathbf{E}_2 \cdot \mathbf{e}_z = -2\pi a^2 P \int_{-1}^{+1} \cos^2 \theta \, d\,(\cos \theta) \qquad (3\text{-}22)$$

with the result that

$$\mathbf{E}_2 = \frac{\mathbf{P}}{3\epsilon_0} \qquad (3\text{-}23)$$

One can then plug this back into equation (3-19) to obtain the Lorentz-Lorenz result that

$$\mathbf{E}_l = \mathbf{E} + \frac{\mathbf{P}}{3\epsilon_0} \qquad (3\text{-}24)$$

Now, as we have seen previously, the microscopic fields are quite complex. However, in a medium without a permanent dipole moment, the dipole fields are not nearly so complex. In fact, in the absence of an applied external field, there are no microscopic dipole moments or local macroscopic fields. It is clear that the two must be closely related, and here we assume that they are related by the mean atomic polarizability, α, as

$$\mathbf{p} = \alpha\epsilon_0 \mathbf{E}_l \qquad (3\text{-}25)$$

Now assuming that one is operating at optical frequencies or below (many dipoles within an averaging sphere) and that the medium is homogeneous down to the microscopic scale, it is reasonable to combine (3-10) with the above logic to obtain

$$\mathbf{P} = N\mathbf{p} \qquad (3\text{-}26)$$

where N is the average density of microscopic dipoles. Now using (3-24) and (3-25) in (3-26), one finds that

$$\mathbf{P} = N\alpha\epsilon_0 \left(\mathbf{E} + \frac{\mathbf{P}}{3\epsilon_0} \right) \qquad (3\text{-}27)$$

Now $\mathbf{P}$ is the polarization that is a response to the field $\mathbf{E}$, and therefore, in line with previous assumptions, one should be able to write that

$$\mathbf{P} = \chi \mathbf{E}\epsilon_0 \qquad (3\text{-}28)$$

and thus that

$$\epsilon = \epsilon_0(1 + \chi) \tag{3-29}$$

Solving equations (3-27) and (3-28) for χ, one finds

$$\chi = \frac{N\alpha}{1 - \dfrac{N\alpha}{3}} \tag{3-30}$$

Then substituting in equation (3-29), one finds

$$\epsilon = \epsilon_0 \left(\frac{1 + \dfrac{2N\alpha}{3}}{1 - \dfrac{N\alpha}{3}} \right) \tag{3-31}$$

or the reverse relation that

$$\alpha = \frac{3}{N} \frac{n^2 - 1}{n^2 + 2} \tag{3-32}$$

It can be interesting and instructive to determine how much error would be incurred by assuming that the internal field $\mathbf{E}_l$ is equal to the applied field $\mathbf{E}$. This approximation amounts to assuming that

$$n = \left[\frac{1 + \dfrac{2N\alpha}{3}}{1 - \dfrac{N\alpha}{3}} \right]^{1/2} \sim \sqrt{1 + N\alpha} \approx 1 + \frac{N\alpha}{2} \tag{3-33}$$

One can write that the dimensionless quantity

$$N\alpha = 3 \frac{n^2 - 1}{n^2 + 2} \tag{3-34}$$

For air, with an index of 1.0003, one calculates that $N\alpha$ is roughly 0.02, and, therefore, the error in the approximation of equation (3-32) is roughly 1%. In glass, where the index is roughly 1.5, one finds that $N\alpha$ is roughly 0.88, and therefore the error in n is roughly 4%. In LiNbO$_3$, with an index of roughly

2.2, one finds an $N\alpha$ of roughly 1.97, giving an index error of nearly 10%. In GaAs, with an index of roughly 3.5, the index error would be roughly 50%, as the $N\alpha$ is roughly 2.37.

3.4 THE SPRING MODEL OF MATTER

In this section, we wish to derive a simple classical model for matter that can predict, in particular, the details of the wavelength dependence of the index of refraction. To do this, we will first present some discussion of the resonance phenomenon in matter, followed by a derivation of a quantum dipole model. This model will describe the response of an electron charge cloud to the application of the local electric field, which we have related to both the applied electric field and the average electric field in the material in the last section. It then will be shown that a simple classical oscillator (spring) model combines the most important salient features of a much more complicated quantum interaction model. The frequency dependence of the refractive index will then be calculated by using the spring model as a basis.

Figure 3.7 shows a magnification of a portion of a lattice. The plus charges are the lattice points, and the minus charges indicate bound electrons that are free enough to perform some oscillation about the lattice point. Figure 3.8 illustrates three types of resonances that can take place in such a lattice. In part (a) of the figure, a rotational resonance, as might be excited by a millimeter wave absorption, is shown. In (b), a vibrational resonance, as might be excited by infrared absorption, is depicted. In (c), an electronic resonance, as might be excited by optical absorption, is shown. In each case of resonance, an absorption event excites the resonance, which subsequently decays. The decay process can occur in two ways, radiatively or nonradiatively, or some combination of

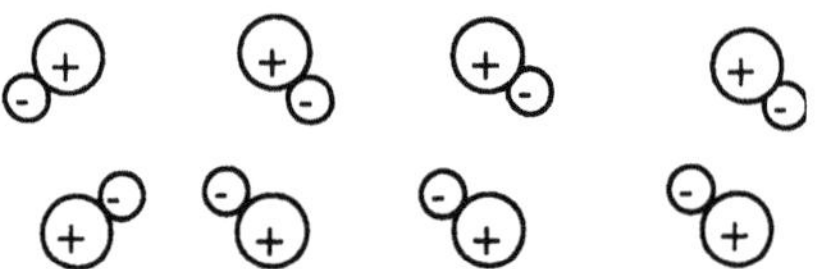

FIGURE 3.7.   A schematic depiction of a portion of a lattice.

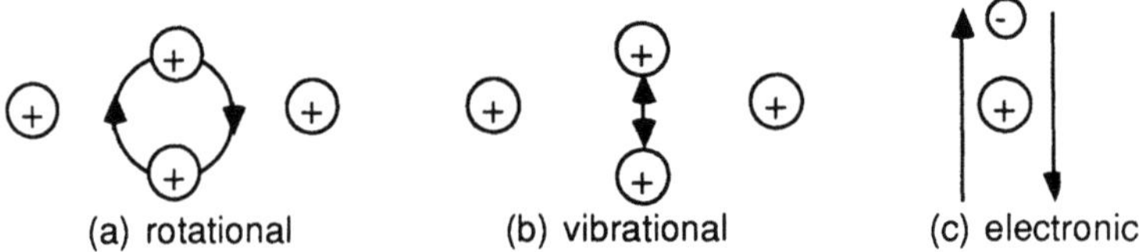

FIGURE 3.8.   Depiction of three types of material resonances.

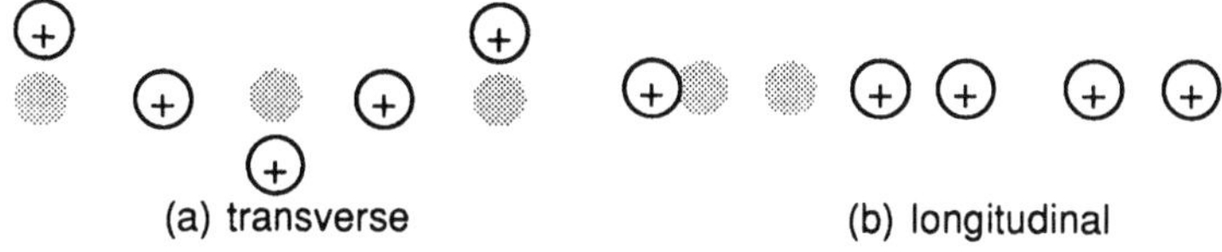

FIGURE 3.9.   Illustration of transverse and longitudinal phonons.

the two. If the decay is completely radiative, a photon just like the one that was incident is emitted, and the lattice returns to its quiescent state. In a totally nonradiative decay, all of the resonant energy is converted to lattice heating through phonon production. Figure 3.9 illustrates the two types of phonons that can be excited in such a lattice. In general, a decay will be some combination of these two types, in that a photon of lower energy than the incident one is emitted, and the remainder of the energy is converted to thermal phonons. In general, therefore, a resonance will dissipate a certain amount of energy and is associated with propagation loss.

Figure 3.10 illustrates a loss curve for fused silica, the material used for making optical fibers. As silicon dioxide is a very wideband material, we do not see the electronic resonances on this scale, as they lie in the UV, which would consist of wavelengths less than 0.3 μm. It is clear from the sketch that the OH^- vibrational resonances cause some considerable loss. Past 1.6 μm, the vibrational resonances become so numerous that the curve goes off scale. If one were to ignore the two OH^- resonances, one would note that the curve has a dependence of λ^{-4} for wavelengths shorter than about 1.6 μm. This is easily explainable in terms of an effect known as Rayleigh scattering. Glass is amorphous; that is, it has no long-range structure. Therefore, if one were to consider each atom as a little antenna, these antennas would be randomly located and so, in some sense, independent. Each antenna, therefore, has a probability of

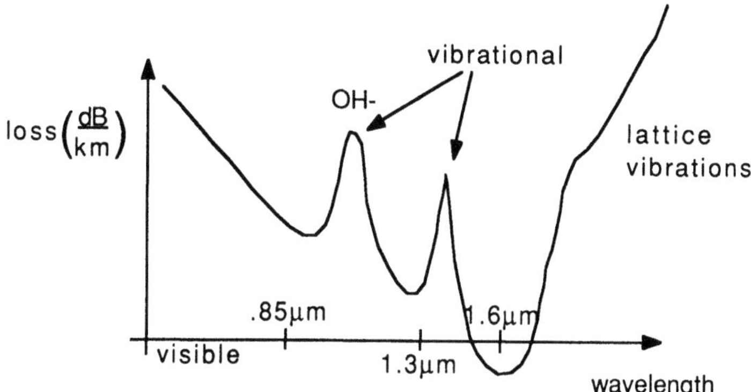

FIGURE 3.10.   Sketch of the loss versus wavelength curve for pure fused silica glass.

absorbing incident photon and a probability of reemitting that photon having absorbed it; and as we are assuming that the antennas are independent, we need not worry about the effect of one antenna loading the next. It is well known that a "small" antenna has an amplitude proportional to k to absorb a photon and an amplitude proportional to k to emit a photon, where small means that the antenna must have an extent much smaller than the wavelength of the incident radiation. As an atom has the extent of a few angstroms and the wavelength of light is nearly a micron, clearly an atom is a small optical antenna. Therefore the amplitude for the atom scattering the light is proportional to k^2, and the probability (or cross section) for the event is proportional to k^4. As $k = 2\pi/\lambda$, $k^4 \sim \lambda^{-4}$. It was Lord Rayleigh who first determined this dependence through the use of dimensional analysis. As a small dipole antenna radiates into a pattern $\sin\theta$, where θ is the incident angle, the scattering is primarily at right angles to the propagation direction and thus shows up directly as propagation loss, as it did in Figure 3.10. (The effects of Rayleigh scattering are not only important in glass, but play a big part in our everyday lives. In Colorado, for example, the sky is often blue. This is a manifestation of Rayleigh scattering. As the scattering is proportional to λ^{-4}, much more blue light than red light is scattered at right angles to the propagation path. On a clear day, therefore, if we look away from the sun we will see primarily Rayleigh-scattered light, and it is indeed primarily blue.)

To correctly describe resonance phenomena, it is necessary to use quantum theory. Here we wish to consider only the electronic resonances, as they are the ones most important optically. We will assume that we can model the atoms as a two-level system, as depicted in Figure 3.11. [Treatments of the time-dependent perturbation problem that we are about to study are also given in Sargent, Scully, and Lamb (1974, Chapters 2 and 3), Haken (1985, Chapter 5), and Sakurai (1985, Chapter 5).]

We will assume that the two wave functions are solutions of the Schroedinger equation

$$i\hbar\,\frac{\partial\Psi}{\partial t} = H_0\Psi \tag{3-35}$$

where h is Planck's constant and $\hbar = h/2\pi$. H_0 is a time-independent Hamiltonian, and ψ is the wave function. The time harmonic, or time-independent

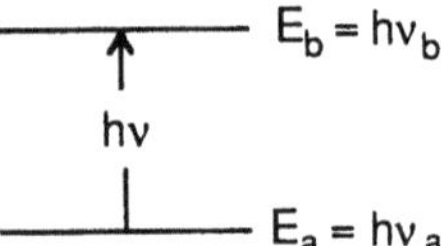

FIGURE 3.11.   Depiction of a two-level system.

(as they are sometimes called because $|\psi|^2$ is independent of time) solutions to (3-35) are

$$\psi_a(r,\ t) = u_a(r)e^{-i\omega_a t} \qquad \text{(a)}$$
$$\psi_b(r,\ t) = u_b(r)e^{-i\omega_b t} \qquad \text{(b)}$$

(3-36)

where u_a and u_b are chosen to be orthonormal, and where an arbitrary wave function should be expandable in the form

$$\psi(r,\ t) = C_a u_a(r)e^{-i\omega_a t} + C_b u_b(r)e^{-i\omega_b t} \qquad (3\text{-}37)$$

Of course, (3-37) cannot represent any wave function, as the set ψ_a, ψ_b cannot be complete. However, we are assuming that they are the only bound states, and thus any other terms in an expansion of the form of (3-37) would have to include radiation modes, which would radiate away in time. We can therefore say that "steady-state" wave functions must be of the form of (3-37), at least in some sense.

Now let us say that a perturbation, such as a time-varying $\mathbf{E}$ field, is applied to the system, leading to the Schroedinger equation

$$i\hbar\,\frac{\partial \psi}{\partial t} = [H_0 + V(r,\ t)]\psi \qquad (3\text{-}38)$$

where $V(r,\ t)$ is the perturbation. Now we have argued that (3-37) somehow represents any solution of (3-35). But, in the sense that ψ_a and ψ_b form a complete set, if one were to make C_a and C_b functions of time, one could use (3-37) to represent any solution of (3-38). Plugging (3-37) into (3-38) and using (3-35) for ψ_a and ψ_b, one finds that

$$i\hbar\left[\psi_a(r,\ t)\frac{dC_a}{\partial t} + \psi_b(r,\ t)\frac{dC_b}{\partial t}\right] = V(r,\ t)[C_a\psi_a(r,\ t) + C_b\psi_b(r,\ t)]$$

(3-39)

We wish to make a couple of simplifying assumptions before continuing. First, let us say that the perturbation is due to an applied electric field that is roughly constant in space (at least in the space surrounding the two-level atom). With this, we can write that

$$V(r,\ t) = -e\mathbf{r} \cdot \mathbf{E}_0(t) \qquad (3\text{-}40)$$

where $-e$ is the electron charge, $\mathbf{r}$ is a radius vector outward from the nucleus to a point in the electron cloud, and $\mathbf{E}_0(t)$ is the applied external electric field.

Now, equation (3-40) explicitly has assumed that $\mathbf{E}_0(t)$ is only a function of time and not of coordinates. In reality, the electric field will be a propagating entity. However, the wavelengths that correspond to atomic transitions generally lie in the thousands of angstroms range, whereas atomic dimensions tend to be in the several-angstrom range. Therefore, the optical phase front over the atomic coordinates (extent of the wave function) will tend to appear quite plain. The function $V(r, t)$, therefore, separates into a function of coordinates times a function of time. The function of coordinates further has odd symmetry about the origin. Without loss of generality, we will further assume that u_a and u_b are real functions. This is not a necessary assumption, but will reduce the amount of writing necessary to find the final form of the equations. Making these assumptions, then multiplying (3-39) by u_a and integrating, and then multiplying (3-39) by u_b and integrating, one finds

$$\frac{\partial C_a}{\partial t} = \frac{1}{i\hbar} e^{-i\omega t} C_b V_{ab} \qquad \text{(a)}$$

$$\frac{\partial C_b}{\partial t} = \frac{1}{i\hbar} e^{i\omega t} C_a V_{ab} \qquad \text{(b)}$$

$$\text{(3-41)}$$

where $\omega = \omega_b - \omega_a$, and V_{ab} is defined by

$$V_{ab} = \int u_a V(r, t) u_b d^3 r = \mathbf{E}(t) \cdot \wp \qquad \text{(3-42)}$$

where the terms V_{aa} and V_{bb} have identically disappeared because of symmetry, that is, the fact that the squared wave functions are symmetric, the radius vector $\mathbf{r}$ is antisymmetric, and $\wp$, the intrinsic dipole moment, can be taken as defined by (3-42). If we additionally say that the perturbation is in some sense weak, and that the system is initially in its ground state ψ_a, then it is easy to see that the solution to (3-41) is given approximately by

$$C_a(t) \sim 1 \qquad \text{(a)}$$

$$C_b(t) = \frac{1}{i\hbar} \int_0^t e^{i\omega t'} V_{ab} dt' \qquad \text{(b)}$$

$$\text{(3-43)}$$

It now is appropriate to look at some illuminating examples of how solutions of (3-43) vary. We could take, for example, a case in which the electric field is an impulse along the z-axis, expressible as

$$V(r, t) = -ezA\delta(t) \qquad \text{(3-44)}$$

and thereby find that

$$C_a(t) \sim 1 \tag{a}$$

$$C_b(t) = \frac{A\mathcal{P}_z}{i\hbar} \tag{b}$$

(3-45)

where the $\mathcal{P}_z$ is defined as the intrinsic dipole moment in the z direction

$$\mathcal{P}_z = -\int u_a e z u_b \, d^3 r \tag{3-46}$$

Plugging the result of (3-45) back into (3-37), we find that

$$\Psi(r, t) = e^{-i\omega_a t}\left[u_a(r) + \frac{A\mathcal{P}_z}{i\hbar} u_b(r) e^{-i\omega t}\right] \tag{3-47}$$

It can be quite instructive to look at how such a function varies in time, and in Figure 3.12 we do that. As is seen from the figure, the wave function of the two-level system just oscillates back and forth about the equilibrium point with

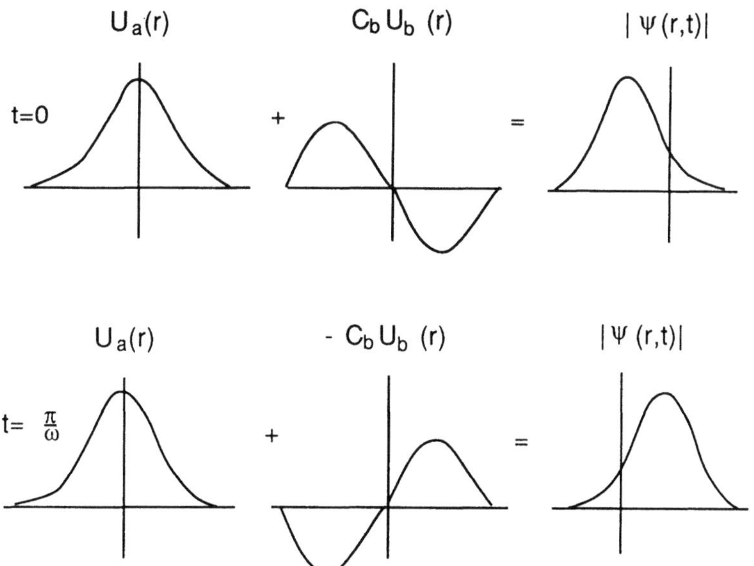

FIGURE 3.12. The time evolution of a quantum dipole after being subjected to an impulse of electric field at $t = 0$.

frequency ω, the natural frequency of the system. The behavior of the system is exactly analogous to the plucking of a guitar string and the string's subsequent oscillation. The only problem with the quantum solution is that it does not damp out. We could correct this problem by trying to couple the oscillator to the phonon heat bath of the lattice, but this is a problem in second quantization that is best left for a more advanced course, and, we will soon see that we can find a much simpler solution.

We have seen that our lossless quantum dipole oscillates like a guitar string or a spring with a hanging mass, at the resonant frequency of the string or spring, under the influence of an impulse. We might also be interested in the converse case, of seeing the system's steady-state response to a sinusoidal driving force in which the $\mathbf{E}_0(t)$ is given by a cosine of frequency ω_0 and is polarized in the z direction. As before, we will assume that $C_a(t)$ remains large and thus use (3-43) (b) to calculate the evolution of $C_b(t)$. Plugging the cosine into (3-43) (b) and expanding it into complex exponentials, one obtains two terms from the integral. One of these is unimportant, however, as it contains a $\omega + \omega_0$ in the denominator, while the other is "synchronous," containing $\omega - \omega_0$. The idea here is that, by assuming that the system is two-level, we have tacitly assumed that $\omega_0 \sim \omega$, lest other transitions become important. Keeping only the important term in (3-43) (b), we rapidly find that

$$C_b(t) \approx \frac{1}{2} \frac{\mathcal{P}_z E_0}{i\hbar} e^{i(\omega - \omega_0)t/2} \frac{\sin (\omega - \omega_0)t/2}{(\omega - \omega_0)/2} \tag{3-48}$$

We see immediately that C_b is more strongly driven the closer ω_0 lies to the resonance frequency. Further, if we expand (3-48) in complex exponentials and ignore the non-time-varying term, which is due to the artificial turn on transient, we find from (3-37) that

$$\psi(r, t) \sim e^{-i\omega_a t} \left[u_a(r) + \frac{\mathcal{P}_z E_0}{i\hbar(\omega - \omega_0)} u_b(r) e^{-i\omega_0 t} \right] \tag{3-49}$$

and we therefore see that the incident field with frequency ω_0 induces a dipole moment in the atomic medium that is identical in frequency to the driving force and has a "resonant" effect about the natural frequency ω of the transition. This behavior, however, is identical to that which would be exhibited by a classical spring loaded with a mass, driven by a sinusoidal force. We have now seen in two examples—those two examples that are most important to an electromagnetic model of matter—that the results of the analysis of the quantum dipole reduce to the results of analyzing a classical spring. In this sense, we can say that matter is made of springs and carry out our analysis classically. Presently, we will see why this is so in the cases we have investigated, and will find a quantitative validation of a classical oscillator model for a large number of practical problems.

At this point we would like to try to find a form as close to classical as we can find for the quantum-mechanical equations embodied in (3-41). To do this, we first note that the classical value, A, taken on by a quantum-mechanical operator, $\hat{A}$, is given by the expectation value

$$A = \langle \hat{A} \rangle = \int \psi^*(\mathbf{r}, t)\hat{A}\Psi(\mathbf{r}, t)d^3\mathbf{r} \tag{3-50}$$

over the wave function $\psi(\mathbf{r}, t)$. Now, to find the form of a quantum-mechanical operator, one generally uses classical correspondence; that is, one uses classical equations to define a classical quantity whose quantum-mechanical equivalent one wants, until one finds a form for which one knows all of the quantum-mechanical equivalents of the variables on the right-hand side of the classical equation. Now, if one denotes the radius of a negative charge from a positive nuclei, classically the microscopic dipole movement is given by

$$\mathbf{p} = -e\mathbf{r} \tag{3-51}$$

Using classical correspondence, one finds that the dipole operator would be given by

$$\hat{\mathbf{p}} = -e\hat{\mathbf{r}} \tag{3-52}$$

where it is well known that the operator $\hat{\mathbf{r}}$ is evaluated as if it were a classical variable. Now using the form (3-37) for $\Psi(\mathbf{r}, t)$ in (3-50), the form for the expectation value, one finds that

$$\mathbf{p} = -e\left[\int C_a^* C_b u_a u_b \mathbf{r} d^3 r e^{-i(\omega_b - \omega_a)t} + \int C_a C_b^* u_a u_b \mathbf{r} d^3 r e^{i(\omega_b - \omega_a)t}\right] \tag{3-53}$$

which, in turn, can be written in the form

$$\mathbf{p} = \mathcal{P}[C_a^* C_b e^{-i\omega t} + C_a C_b^* e^{i\omega t}] \tag{3-54}$$

where $\mathcal{P}$ and ω are as previously defined.

To try to find the relation between (3-54) and a harmonic oscillator model, we take derivatives of $\mathbf{p}$, substituting from relations (3-41) and (3-42), to find

$$\frac{\partial \mathbf{p}}{\partial t} = -i\omega(C_a^* C_b e^{-i\omega t} - C_a C_b^* e^{i\omega t})\mathcal{P} \tag{3-55}$$

and

$$\frac{\partial^2 \mathbf{p}}{\partial t^2} = -\omega^2 \mathbf{p} + \frac{\omega}{\hbar} d(t) \mathbf{E}(t) \cdot \wp \wp \tag{3-56}$$

where $d(t)$ is the inversion parameter defined by

$$d(t) = |C_b(t)|^2 - |C_a(t)|^2 \tag{3-57}$$

Now, equation (3-56) is getting quite close to the form of a harmonic oscillator equation for $\mathbf{p}$. In fact, if we were to note that we have everywhere ignored damping on the oscillator due to phonons and other lattice disturbances, and therefore include a damping term $\gamma \, \partial \mathbf{p}/\partial t$, we would have the form

$$\frac{\partial^2 \mathbf{p}(t)}{\partial t^2} + \frac{\gamma \partial \mathbf{p}}{\partial t} + \omega^2 \mathbf{p}(t) = -\frac{\omega}{\hbar} d(t) \mathbf{E}(t) \cdot \wp \wp \tag{3-58}$$

The left-hand side of (3-58) is in the form of an oscillator equation. The problem is just that the variables making up $\mathbf{p}$ also appear on the right-hand side. However, in the classical regime, the regime where fields are weak in the sense that atoms act collectively, the inversion parameter $d(t)$ can never be significantly different from -1, as, on the average, very few atoms can be excited to an upper state. Using this limit, one sees that (3-58) reduces to

$$\frac{\partial^2 \mathbf{p}}{\partial t^2} + \gamma \frac{\partial \mathbf{p}}{\partial t} + \omega^2 \mathbf{p}(t) = \frac{\omega}{\hbar} \mathbf{E}(t) \cdot \wp \wp \tag{3-59}$$

Now, to this point, no assumption has been made about the symmetry of the atom or, therefore, the wave functions. Discussion of nonsymmetric atoms will again be taken up from the very beginning of Chapter 4. For the present, however, we will assume for simplicity that the atom possesses no polar axis and that, therefore, the direction of $\wp$ must coincide with that of $\mathbf{E}(t)$. Using this ansatz in (3-59) and renaming constants, we find that (3-59) can be written in this limit as

$$\frac{\partial^2 \mathbf{p}}{\partial t^2} + \gamma \frac{\partial \mathbf{p}}{\partial t} + \omega^2 \mathbf{p} = \frac{e^2}{m} \mathbf{E}(t) \tag{3-60}$$

where e^2 and m can be thought of as effective charges and masses. The final transformation of this equation to the form we wish is to revise the relation for the classical microscopic dipole moment, equation (3-51), to find the equation

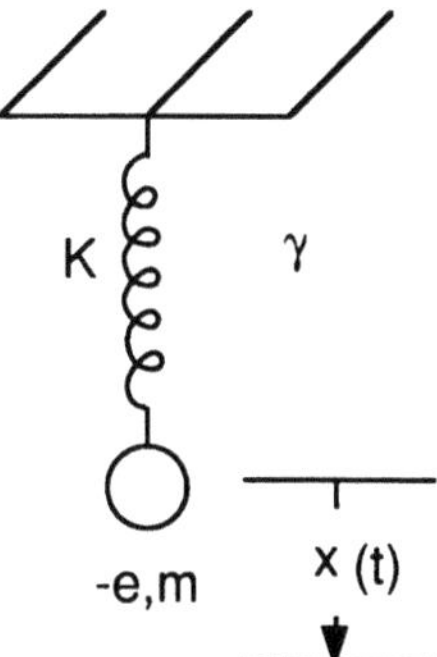

FIGURE 3.13.   A one-dimensional classical spring with an attached mass m with charge $-e$, a spring constant K, and a damping constant γm whose scalar displacement from equilibrium is denoted by $x(t)$.

for the change of radius of an electron from its nucleus to be of the form

$$\frac{\partial^2 \mathbf{r}}{\partial t^2} + \gamma \frac{\partial \mathbf{r}}{\partial t} + \omega^2 \mathbf{r} = -\frac{e}{m}\mathbf{E}(t) \tag{3-61}$$

Let us consider a classical spring, illustrated in Figure 3.13, as is treated in various texts such as Feynman, Leighton, and Sands (1964, Vol. 2 Chapter 32), Sargent, Scully, and Lamb (1974, Chapters 2 and 3), Haken (1985, Section 10.4), Sakurai (1985, Chapter 5), Yariv (1985, Chapter 5), Siegman (1986, Chapter 2), or Johnk (1988, Chapter 3). The three-dimensional vector equation of motion for a mass in a three-dimensional spring system would be identical to that of (3-61) with the exceptions that ω is the natural frequency of the spring defined by $\omega^2 = K/m$, and the **B**-field term in the Lorentz force law has been ignored here under the assumption that we are considering nominally equilibrium charges and therefore need not consider the possibility of relativistic velocities. Figure 3.13 illustrates only one dimension of the system defined by (3-61) because of the complexity of drawing a continuous spring system. It can be enlightening at this point to study examples such as those we considered earlier for the quantum dipole to see how our present model compares.

Example. Impulse where $\mathbf{E} = \mathbf{E}_0 \delta(t)$

Here our equation becomes

$$\ddot{\mathbf{r}} + \gamma \dot{\mathbf{r}} + \omega^2 \mathbf{r} = -\frac{e}{m}\mathbf{E}_0 \delta(t) \tag{3-62}$$

where the dots denote time derivatives. Taking the Fourier transform of both sides and solving for $\mathbf{r}(\omega_0)$, we find

$$\mathbf{r}(\omega_0) = \frac{1}{m}\frac{-e\mathbf{E}_0}{(\omega^2 - \omega_0^2 - i\gamma\omega_0)} \tag{3-63}$$

where ω_0 is the transform variable. Taking the inverse transform, one finds that

$$\mathbf{r}(t) = \begin{cases} -\dfrac{e}{m}\mathbf{E}_0 e^{-\gamma t}\cos\omega t, & t > 0 \\[2mm] 0, & t < 0 \end{cases} \tag{3-64}$$

and we see that the result is simply a damped exponential, as we would have expected from combining the quantum result with the loss due to phonon absorption.

Example. Harmonic excitation with $\mathbf{E} = \mathbf{E}_0\cos\omega_0 t$

Here, the easiest method of solution is to use complex steady-state notation such that

$$\mathbf{r}(t) = \mathrm{Re}\,[\mathbf{r}(\omega_0)e^{-i\omega_0 t}] \tag{3-65}$$

to find that $\mathbf{r}(\omega_0)$ is given by

$$\mathbf{r}(\omega_0) = -\frac{e}{m}\frac{\mathbf{E}_0}{(\omega^2 - \omega_0^2 - i\gamma\omega_0)} \tag{3-66}$$

Therefore, $\mathbf{r}(t)$ can be written in the form

$$\mathbf{r}(t) = \mathrm{Re}\,[\mathbf{r}(\omega_0)e^{-i\omega_0 t}]$$

$$= -\frac{e}{m}\mathbf{E}_0\left[\frac{\omega^2 - \omega_0^2}{(\omega^2 - \omega_0^2)^2 + \gamma^2\omega_0^2}\cos\omega_0 t \right. \tag{3-67}$$

$$\left. + \frac{\gamma\omega_0}{(\omega^2 - \omega_0^2)^2 + \gamma^2\omega_0^2}\sin\omega_0 t\right]$$

Perhaps the evolution of the particles' motion is best illustrated by plots such as those in Figure 3.14. Note that in the figure the vector natures of $\mathbf{r}(t)$ and $\mathbf{E}(t)$ have been ignored, and only a one-dimensional model of (3-67) has therefore been assumed where, in particular, we have assumed that the x direction

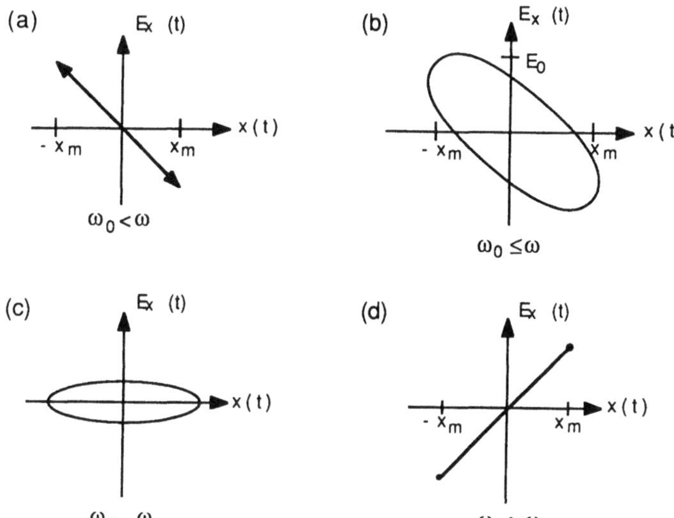

FIGURE 3.14.   Phase plots showing relative phase between the driving field $E_x(t)$ and the response $x(t)$ for various relations between the driving frequency ω_0 and the resonant frequency ω.

is the important one. This is consistent with our assumption about the collinearity of $\mathcal{P}$ and $\mathbf{E}(t)$. As can be clearly seen from the figure, for driving frequencies well below resonance, the response of the medium is in phase with the driving field and, further, little or no power is dissipated, as $\gamma\omega_0$, the loss term, is small compared to $\omega^2 - \omega_0^2$. The situation changes as the driving frequency increases. The sine component of the motion becomes relatively more important until, on resonance, it is the only one excited. This quadrature of phase between drive and response indicates the strongest absorption regime. As the frequency is increased still further, the response gets 180° out of phase with the incident radiation, and, further, the response becomes smaller and smaller with increasing frequency.

We are now ready to find the molecular polarizability $\alpha(\omega)$ that we had earlier discussed and as was defined, in the isotropic material case, by equations (3-25) and (3-32). Now, for a simple elemental dipole, one writes in general that the microscopic dipole moment can be expressed as it was in equation (3-52). Comparing (3-52) with (3-32) and (3-66), assuming that $\mathbf{E}_0$ and $\mathbf{E}_\ell$ are identical, one finds that

$$\alpha(\omega_0) = \frac{e^2}{\epsilon_0 m} \frac{\omega^2 - \omega_0^2 + i\gamma\omega_0}{(\omega^2 - \omega_0^2)^2 + \gamma^2\omega_0^2}$$

$$= \alpha_r + i\alpha_i$$

$$(3\text{-}68)$$

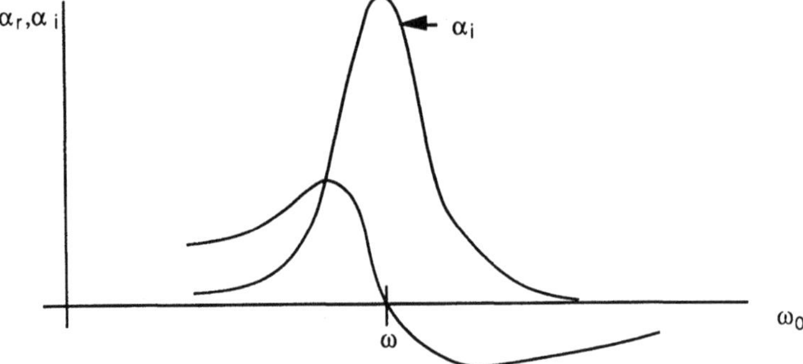

FIGURE 3.15.   The variation of the real and imaginary parts of the molecular polarizabilities as a function of frequency.

The dependencies of α_r and α_i on ω are sketched in Figure 3.15. To get a better understanding of the real meaning of α_R and α_i, we can recall (3-33) to obtain

$$n^2 = \frac{1 + \frac{2}{3} N(\alpha_r + i\alpha_i)}{1 - \frac{1}{3} N(\alpha_r + i\alpha_i)} \tag{3-69}$$

which shows us that having an imaginary part of α leads to an imaginary part of n, which indicates that the α_i can be associated with optical absorption and, therefore, damped propagation. At the same time, it is also clear that the resonance phenomenon leads to a frequency dependence of the real part of n, as is schematically depicted in Figure 3.16. Here we see that various resonances

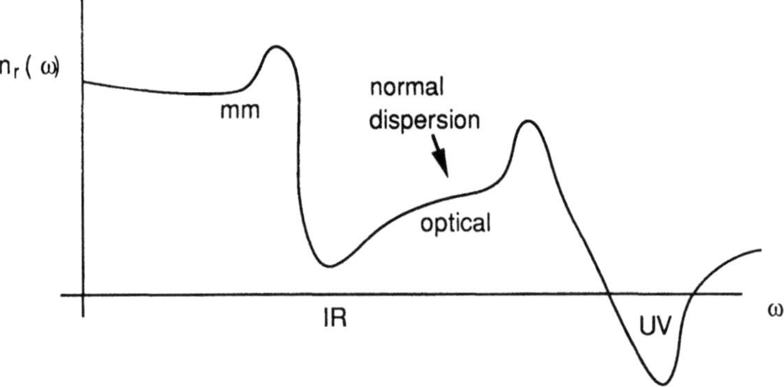

FIGURE 3.16.   Rough sketch of what a dispersion plot might look like for glass or water.

affect the dependence. The index is high in the millimeter wave part of the spectrum, drops precipitously in the IR, and rises gradually through the optical before experiencing its last electronic resonances in the UV. Indeed, an index of refraction must be written as a sum over resonances of the form

$$\frac{n^2 - 1}{n^2 + 2} = \frac{N}{3\epsilon_0} \frac{e^2}{m} \sum_{k=1}^{N} \frac{f_k}{\omega_k^2 - \omega_0^2 + i\gamma_k \omega_0} \tag{3-70}$$

where ω_k is the kth resonance frequency and γ_k is the kth damping constant, and f_k is a dimensionless constant often called the oscillator strength. The sum is due to the fact that transitions could occur between what we took to be the occupied state and any of the states above this state. In most atoms there are many such levels before one reaches the ionization energy of the electron. Each of these transitions adds to the index.

An interesting question is, how could one experimentally fix enough of the constants in (3-70) to use this equation as a predictive tool? Consider a possible measured curve, as illustrated in Figure 3.17. The index as a function of frequency could be measured by ellipsometry with a variable frequency source, while the absorption spectra could be measured by transmission spectroscopy. From the width and the height of the absorption resonances, it should be possible to determine the γ_k's and the f_k's relative to the D.C. index. (There can be a problem here if our material has resonances in the millimeter wave region, as variable sources can be hard to obtain.) The D.C. index can be measured in such an apparatus, as pictured in Figure 3.18. Here the idea is to measure the resonant frequency of the circuit ω_0, which is given in terms of the inductance L and capacitances C_s, and C_t

$$\omega_0 = \left[\frac{1}{L(C_s + C_t)} \right]^{1/2} \tag{3-71}$$

and to determine the low frequency dielectric constant from the relation

$$C_t \text{ (without dielectric)} = \frac{C_t \text{(with dielectric)}}{\epsilon(\omega \to 0)/\epsilon_0} \tag{3-72}$$

From (3-72) together with

$$\epsilon = \epsilon_0 \frac{1 + \dfrac{2N\alpha}{3}}{1 - \dfrac{N\alpha}{3}} \tag{3-73}$$

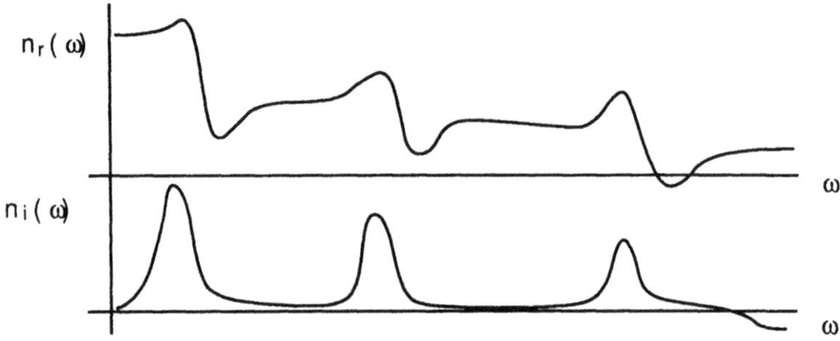

FIGURE 3.17.   A possible realization of a measurement of the index and absorption of a material medium.

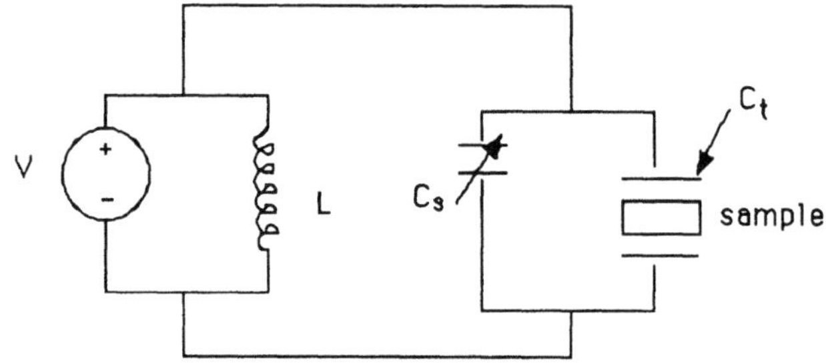

FIGURE 3.18.   Apparatus for measuring the D.C. index of a material sample.

one can determine $N\alpha(\omega \to 0)$, and, combining this knowledge with (3-70) and the resonance measurements, one can determine all of the independent constants in (3-70).

3.5 WAVE PROPAGATION IN DISPERSIVE MEDIA

Another interesting behavior of (3-70) is its very high frequency behavior. From perusing Figures 3.15 and 3.16 together with equation (3-70) and our knowledge that electronic ionization occurs in the UV, it is clear that, at some point, the real part of the index of refraction will become less than one. (See, for example, a table of metal indices in Born and Wolf 1975, 621.) If we recall that the magnitude of the **k** vector of a plane wave is given by

$$\mathbf{k} = \frac{n\omega}{c} \tag{3-74}$$

and the phase velocity of a plane wave is given by

$$v_p = \frac{\omega}{k} = \frac{c}{n} \tag{3-75}$$

we note that there can be a regime (see Figure 3.15) where n is approximately real yet less than 1, and therefore the wave's phase velocity exceeds that of light. This is confusing enough without considering the fact that the $\mathbf{k}$ and, therefore, the v_p become imaginary in the neighborhood of resonances. Understanding these and other propagation phenomena presently will require a thorough discussion of the effect known as dispersion.

The first situation we will consider is that in which we attempt to transmit a signal that contains information in the form of carrier modulation (perhaps this is the only form information can take). Figure 3.19 illustrates the appearance of a modulated carrier, where the envelope around the carrier is the one that propagates at the group velocity, and the carrier itself propagates at the phase velocity. Now to mathematically describe such a packet and how it propagates, one can use a formulation such as that of Birger and Vainshtein (1974) or Vainshtein (1976). Consider that we have a z-directed, y-polarized plane wave packet propagating from some initial plane $z = 0$ into a linear but possibly dispersive medium. The situation is as depicted in Figure 3.20 and described by the following expressions:

$$E_y(z, t) = f(z, t)e^{ik(\omega_0)z}e^{-i\omega_0 t} \tag{a}$$

$$E_y(0, t) = f(0, t)e^{-i\omega_0 t} \tag{b}$$

$$\tag{3-76}$$

Now we wish to relate $f(z, t)$ to $f(0, t)$. To do this, we note that one can write that

$$E_y(z, t) = \int_{-\infty}^{\infty} E_y(\omega)e^{ik(\omega)z}e^{-i\omega t}\,d\omega \tag{3-77}$$

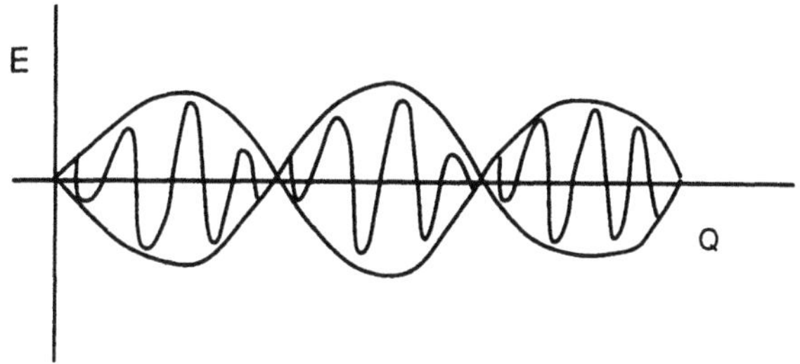

FIGURE 3.19.   Illustration of a modulated carrier.

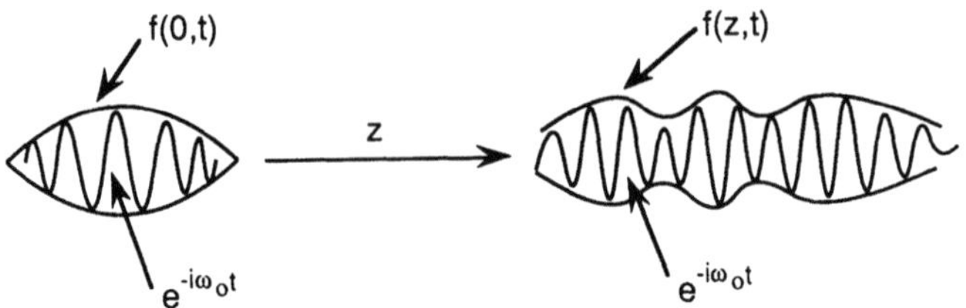

FIGURE 3.20.   Schematic illustration of the dispersive distortion suffered by a quasi-mono-chromatic plane wave disturbance propagating through a dispersive medium.

where $E_y(\omega)$ can be defined from

$$E_y(\omega) = \frac{1}{2\pi} \int_{-\infty}^{\infty} E(0,\, t')e^{i\omega t'}\, dt' \qquad (3\text{-}78)$$

If there were significant dispersive absorption, it should be taken care of by using the complex index when defining the $k(\omega) = k_0 n(\omega)$ in (3-77). This would imply, however, that the spectrum of the wave would be modified by the propagation. Now, substituting (3-76) (a) into (3-77) and rearranging terms, one obtains

$$f(z,\, t) = \int_{-\infty}^{\infty} d\omega\, E_y(\omega)e^{-i(\omega - \omega_0)[t - (k(\omega) - k(\omega_0))/(\omega - \omega_0)z]} \qquad (3\text{-}79)$$

which, together with (3-78), forms the desired relationship between $f(z,\, t)$ and $f(0,\, t)$.

To best determine the content of (3-79), it is useful to expand the dispersion relation $k(\omega)$ about the point ω_0, to obtain

$$k(\omega) = k(\omega_0) + (\omega - \omega_0) \left.\frac{dk}{d\omega}\right|_{\omega_0} + \frac{(\omega - \omega_0)^2}{2} \left.\frac{d^2 k}{d\omega^2}\right|_{\omega_0}$$

$$+ \frac{(\omega - \omega_0)^3}{6} \left.\frac{d^3 k}{d\omega^3}\right|_{\omega_0} + \cdots \qquad (3\text{-}80)$$

Now if one limits oneself to just the first two terms of (3-80), then (3-79) takes on the especially simple form

$$f(z,\, t) = f\left(0,\, t - \frac{dk}{d\omega}z\right) \qquad (3\text{-}81)$$

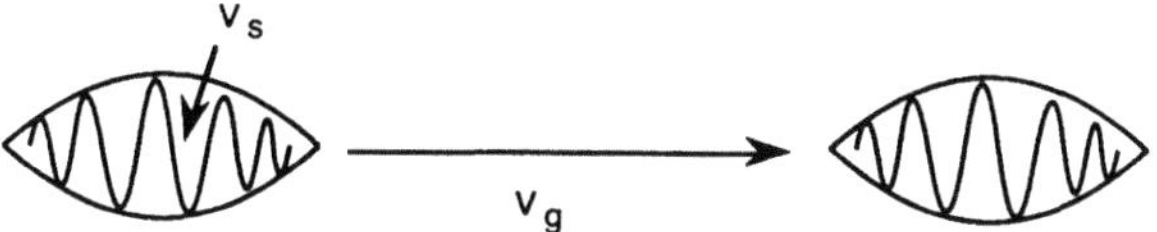

FIGURE 3.21.   Schematic depiction of the dispersionless propagation of a wave packet through a medium where second- and higher-order derivatives of the dispersion relation are negligible.

where (3-78) and (3-76) were used in the simplification. The situation is as illustrated in Figure 3.21. Clearly (3-81) indicates that we can define a group velocity, v_g, by

$$v_g = \frac{1}{\dfrac{dk}{d\omega}} \tag{3-82}$$

where this v_g represents the velocity of propagation of the wave packet. Clearly in the case where higher than first-order derivatives of k are negligible, one has dispersionless propagation, as (3-81) clearly shows that the functional form of the wave packet remains invariant under propagation. Clearly such is not the case if higher-order terms must be taken, as is evident from using the second-order expansion of (3-80) in (3-79), which appears as

$$f(z, t) = \int_{-\infty}^{\infty} d\omega\, E_y(\omega) e^{-i(\omega - \omega_0)[t - (z/V_g)(\omega - \omega_0)/2)(d^2k/d\omega^2)]} \tag{3-83}$$

which can be rewritten as

$$f(z, t) = \int dt' f\left(0, t - t' - \frac{z}{V_g}\right) \int d\omega'\, e^{-i(\omega - \omega_0)(t - (z/V_g))}$$

$$\cdot\, e^{i((\omega' - \omega_0)^2/2)(d^2k/d\omega^2)}$$

$$= f\left(0, t - \frac{z}{V_g}\right) * \mathscr{F}^{-1}\left[e^{[-i((\omega - \omega_0)^2 d^2k)/2]/d\omega^2}\right]\left(t - \frac{z}{V_g}\right) \tag{3-84}$$

where the asterisk denotes the operation of convolution, defined by

$$g(\omega) * h(\omega) = \int_{-\infty}^{\infty} g(\omega - \omega') h(\omega') d\omega' \tag{3-85}$$

and where the $\mathcal{F}^{-1}$ is the inverse Fourier transform operator, which can be defined by

$$\mathcal{F}^{-1}[f(\omega)](t - t') = \int d\omega\, e^{-i\omega(t - t')} f(\omega) \qquad (3\text{-}86)$$

The point of (3-84) is that the functional form of $f(z, t)$ is no longer the same as the input form. This change in form is generally referred to as dispersion and in general leads to such effects as pulse broadening. In fiber optic systems where one wants to propagate high-speed signals as far as possible without having to repeat them, this dispersion effect can be the limiting one on repeater spacing. The problem is an important enough one that the telecommunications industry switched its operating wavelengths from 0.83 μm to 1.3 μm, where optical fibers have a dispersion minimum, that is, where $d^2 k/d\omega^2$ has a zero, and the last term of the expansion of (3-80) is the limiting one. The changeover from 0.83 μm to 1.3 μm is an especially hard one, as radiation at 1.3 μm is especially invisible and cannot be seen with inexpensive IR scopes or television cameras as the slightly nonvisible radiation of 0.83 μm can be. However, the economic advantages of greater repeater spacing were deemed to greatly outweigh the inconvenience of operating at 1.3 μm.

The point here is that v_p is the velocity of the oscillations underneath the wave packet. As a carrier by itself cannot carry information, it can be assumed that the information velocity is the group velocity. As relativity only requires that information not be propagated faster than the speed of light, it is acceptable that the phase velocity exceed the speed of light as long as the group velocity does not. Writing the group velocity in the form

$$v_g = \left(\frac{dk}{d\omega}\right)^{-1} = \frac{v_p}{1 + \dfrac{\omega}{n}\dfrac{dn}{d\omega}} \qquad (3\text{-}87)$$

together with Figure 3.15, indicates that, in the regime where v_p exceeds c, v_g is clearly less than v_p and therefore less than c. However, near the center of the resonance where the absorption is greatest, v_g clearly becomes negative. This is so because $dn/d\omega$ becomes more negative than n/ω. Any region in which $dn/d\omega$ is negative is called a region of anomalous dispersion, to differentiate it from a loss-free region such as that exhibited by glass in the optical (Figure 3.16), where the index increases with frequency. (Blue bends more than red.) In this anomalous dispersion regime, however, it becomes hard to associate too much meaning with (3-87), as n and therefore k are complex. It is not clear how to interpret a complex velocity although much effort in the literature has

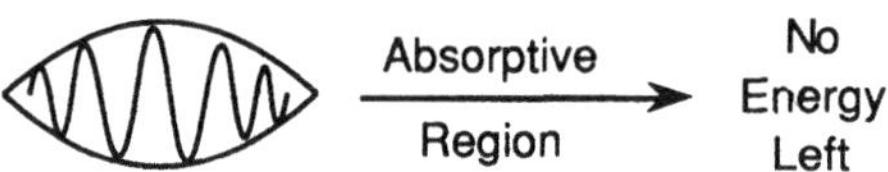

FIGURE 3.22.   Illustration of the dilemma of trying to define propagation velocity in a strongly absorbing regime.

been given to this, as in the works by Birger and Vainshtein (1974), Vainshtein (1976), Schulz-DuBois (1969), and Brillouin (1960). The problem is somewhat as illustrated in Figure 3.22. If all the energy in a wave form is rapidly absorbed, it is rather hard to describe information transport.

One way to try to describe what is happening in a strongly absorbing medium is to define an energy velocity v_E, which must be given by

$$v_E = \frac{\mathbf{S}_{av}}{\overline{W}} \tag{3-88}$$

where $\mathbf{S}_{av}$ is the time-averaged Poynting vector, and $\overline{W}$ is the average energy carried by the wave, which will be defined in (3-96) in terms of the W of Poynting's theorem.

Now, to find a relation between v_E and v_g, one can first define the centroid of a z-propagating wave packet by

$$t_0 = \frac{\displaystyle\int_{-\infty}^{\infty} t\, S_z\, dt}{\displaystyle\int_{-\infty}^{\infty} S_z\, dt} \tag{3-89}$$

Clearly, then, the energy velocity should be given by

$$\frac{1}{v_g} = \frac{dt_0}{dz} \tag{3-90}$$

We now wish to write Poynting's theorem in the form

$$\frac{dS}{dz} + \frac{dW}{dt} + P = 0 \tag{3-91}$$

where W is the electromagnetic energy, S is the magnitude of the Poynting vector, and P is the loss which is generally given by $\mathbf{J} \cdot \mathbf{E}$. One will also need to define the centroid of the lossy power flow by

$$t_1 = \frac{\int tP\,dt}{\int P\,dt} \tag{3-92}$$

Now, substituting (3-89) into (3-90), one finds that

$$\frac{1}{v_g} = \frac{\int t\,\dfrac{dS_g}{dz}\,dt}{\int S\,dt} - \frac{\int \dfrac{dS_z}{dz}\,dt\,\int tS_z\,dt}{\left(\int S\,dt\right)^2} \tag{3-93}$$

Using (3-91) in (3-93), one obtains

$$\frac{1}{v_g} = \frac{\int t\,\dfrac{dW}{dt}\,dt}{\int S\,dt} - \frac{\int t\,P\,dt}{\int S\,dt} + \frac{\int t\,S\,dt}{\left(\int S\,dt\right)^2}\left\{\int \frac{dW}{dt}\,dt + \int P\,dt\right\} \tag{3-94}$$

Using integration by parts and the definitions in (3-89) and (3-92), one finds that

$$\frac{1}{v_g} = \frac{\overline{W}}{S_{av}}\left[1 - \frac{\overline{P}t_1}{\overline{W}} + \frac{t_0\overline{P}}{\overline{W}}\right] \tag{3-95}$$

where

$$S_{av} = \int S_z\,dt \tag{a}$$

$$\overline{P} = \int P\,dt \tag{b} \qquad (3\text{-}96)$$

$$\overline{W} = \int W\,dt \tag{c}$$

which gives the final result that

$$v_g = \frac{v_E}{1 + \alpha(t_0 - t_1)}$$

(3-97)

where the α is given by

$$\alpha = \frac{\overline{P}}{\overline{\overline{W}}}$$

(3-98)

indicating that the group and energy velocity become identical in the limit where the loss becomes small. In cases of high loss, the packet shape can change so rapidly that one must be careful, about which questions one asks about energy propagation.

3.6 MACROSCOPIC MODELS OF MORE EXOTIC EFFECTS

A final point in this chapter concerns the origin of some of the more interesting optical propagation effects such as anisotropy, the electro-optic effect, the second harmonic coefficient, optical activity, etc. We have already generated all of the necessary machinery to take up this topic. Essentially, all of these propagation effects can be analyzed by the spring model (if one is willing to accept the uncertainties involved with the application of the Claussius-Mossetti relation, derived for spherically symmetric media, to nonspherically symmetric media). In the simple model we presented in (3-61), we assumed that the ω and γ were independent of direction in the lattice. In general, a particle's mass, the damping its motion experiences, and its resonances due to lattice interaction are all tensor quantities, and therefore must be represented by second-rank tensors, which appear just like matrices. In this general case, the equation of motion for a (bound) particle's coordinate could be expressed as

$$m\ddot{\mathbf{r}} + m\gamma\dot{\mathbf{r}} + k\mathbf{r} = -e\mathbf{E}_0 e^{-i\omega_0 t}$$

(3-99)

where γ is the damping tensor, and the microscopic dipole moment would be given by, as before,

$$\mathbf{p} = -e\mathbf{r}$$

(3-100)

Solving (3-89) for $\mathbf{r}(\omega_0)$, one finds that

$$\mathbf{r}(\omega_0) = -c\mathbf{\Lambda}\mathbf{E}_0$$

(3-101)

where

$$\Lambda = (-\omega_0^2 I - i\omega_0\gamma + m^{-1}K)^{-1}m^{-1} \tag{3-102}$$

where I is the 3 × 3 identity matrix, and the minus one superscripts refer to matrix inverses. If one assumes that the material polarizability can be given by a tensor α in the expression

$$\mathbf{P}(\omega_0) = \epsilon_0\alpha\mathbf{E}_0 \tag{3-103}$$

then one can find α in the form

$$\alpha = \frac{e^2}{\epsilon_0}\Lambda \tag{3-104}$$

Now, actually finding the relation between material polarization and the local electric field $\mathbf{E}_l$ is a very complicated task in a general anisotropic medium, as the electric field and polarization relation involves integrals over the material structure. It seems somewhat safe to assume, though, that the local field should be given by some linear combination of $\mathbf{E}$ and $\mathbf{P}$, such as

$$\mathbf{E}_l = \mathbf{E} + a\mathbf{P}/\epsilon_0 \tag{3-105}$$

where the a involves the material considerations. Using equations (3-25) and (3-26), one can therefore express

$$\mathbf{P} = N\alpha\epsilon_0\left(\mathbf{E} + a\frac{\mathbf{P}}{\epsilon_0}\right) \tag{3-106}$$

Assuming that the macroscopic relation in an anisotropic medium takes the form

$$\mathbf{P} = \epsilon_0\chi\mathbf{E} \tag{3-107}$$

where χ is the susceptibility tensor, which is related to the index of refraction tensor $n^2(\omega)$ by

$$n^2(\omega) = (I + \chi) \tag{3-108}$$

would thus give a generalized Claussius-Mossetti relation of the form

$$n^2(\omega) = (I - aN\alpha)^{-1}(I + (1 - a)N\alpha) \tag{3-109}$$

Although it is good to have a general form such as (3-109), it is rather hard to deduce all of its implication without resorting to a process of bootstrapping from simpler examples. In the simplest case of anisotropy, one would assume that the values of γ and ω were different along three principal axes, and would write a set of equations that would look like

$$\ddot{x} + \gamma_x\dot{x} + \omega_x^2 x = \frac{-eE_x}{m} \qquad\text{(a)}$$

$$\ddot{y} + \gamma_y\dot{y} + \omega_y^2 y = \frac{-eE_y}{m} \qquad\text{(b)} \quad\text{(3-110)}$$

$$\ddot{z} + \gamma_z\dot{z} + \omega_z^2 z = \frac{-eE_z}{m} \qquad\text{(c)}$$

By deriving the index tensor and resonance structure from (3-110), one can derive essentially all of the polarization properties of such an anisotropic medium, including polarization rotation and dichroic polarizing.

Electro-optic effects could also be explained in terms of spatial dependence of the ω's, but perhaps it is easier, for purposes of calculation, simply to make ω a function of the applied field. For example, say that a field along the z-direction induces a change in the index for the x-polarized wave such that $\omega_x^2(E_z) = \omega^2 x_0 + \Delta\omega_x E_z$. Then, one can express (3-50) in the form

$$\ddot{x} + \gamma_x\dot{x} + \omega_x^2(E_z)x = \frac{-eE_x}{m} \qquad\text{(3-111)}$$

where E_z is the (assumed) D.C. field, and E_x is the optical field. The same kind of analysis can be applied to nonlinear propagation, except that here the ω_x will become a function of the applied optical field such that

$$\omega_x^2 = \omega_{x0}^2 + \omega_{x1}^2 E_x + \omega_{x2}^2 E_x^2 + \cdots \qquad\text{(3-112)}$$

and therefore the equation corresponding to (3-111) becomes a nonlinear equation. Such an equation cannot generally be solved, but a perturbative solution can give important relations between the microscopic and the macroscopic parameters.

A very interesting case is that of an optically active medium. Here the molecules must have microscopic chiral symmetry, that is, handedness. That is, the molecules could all be attached to right-handed springs. Because the medium has a handedness, it couples to circular rather than linear polarization states. Here our linear springs are not quite the right thing to expand in. However, one could use linear combinations of the springs that would be driven by right- and left-hand circularly polarized waves. In such a manner, it is also possible to relate microscopic and macroscopic parameters for optically active media, but we will leave these interesting cases to the problems and the references (see, for example, Jaggard, Mickelson, and Papas 1979).

References

Birger, E. S. and L. A. Vainshtein, Propagation of high-frequency perturbations in absorbing and active media, *Sov. Phys. Tech. Phys.* *18*, 1405–1411 (1974).

Born, M. and E. Wolf, *Principles of Optics*, Fifth edition, Pergamon Press, New York (1975).

Brillouin, L., *Wave Propagation and Group Velocity*, Academic Press, New York (1960).

Feynman, R. P., R. B. Leighton, and M. Sands, *The Feynman Lectures on Physics*, Addison-Wesley, Reading, MA (1964).

Haken, H., *Light*, Volume 2 of *Laser Light Dynamics*, North-Holland, New York (1985).

Jackson, J. D., *Classical Electrodynamics*, Second edition, John Wiley and Sons, New York (1975).

Jaggard, D. L., A. R. Mickelson, and C. H. Papas, On electromagnetic waves in chiral media, Springer Verlag, *Appl. Phys.* *18*, 211–216 (1979).

Johnk, C. T. A., *Engineering Electromagnetic Fields and Waves*, John Wiley and Sons, New York (1988).

Kittel, C., *Introduction to Solid State Physics*, John Wiley and Sons, New York (1971).

Sakurai, J. J., *Modern Quantum Mechanics*, Benjamin/Cummings, Reading, MA (1985).

Sargent, M., M. O. Scully, and W. E. Lamb, *Laser Physics*, Addison-Wesley, Reading, MA (1974).

Schulz-DuBois, E. O., Energy transport velocity of electromagnetic propagation in dispersive media, *Proc. IEEE 57*, 1748–1757 (1969).

Siegman, A. E., *Lasers*, University Science Books, Mill Valley (1986).

Vainshtein, L. A., Propagation of pulses, *Sov. Phys. Usp 19*, 189–205 (1976).

Yariv, A., *Optical Electronics*, Holt, Rinehart and Winston, New York (1985).

Problems

1. In the optical time domain, one often writes constitutive relations in the form

$$\mathbf{J} = 0 \tag{a}$$

$$\mathbf{D}(\mathbf{r}, t) = \epsilon_0 \mathbf{E}(\mathbf{r}, t) + \mathbf{P}(\mathbf{r}, t) \tag{b} \quad (1)$$

$$\mathbf{B}(\mathbf{r}, t) = \mu_0 \mathbf{H}(\mathbf{r}, t) \tag{c}$$

(a) Derive a time-dependent wave equation using the above constitutive relations.

(b) Transform the relations of (a) and your wave equation to the frequency domain.

An alternative to the transformed relations of (1) are relations of the form

$$\mathbf{J}_\omega(\mathbf{r}) = \sigma_\omega \mathbf{E}_\omega(\mathbf{r}) \tag{a}$$

$$\mathbf{D}_\omega(\mathbf{r}) = \epsilon_\omega \mathbf{E}_\omega(\mathbf{r}) \tag{b} \quad (2)$$

$$\mathbf{B}_\omega(\mathbf{r}) = \mu_0 \mathbf{H}_\omega(\mathbf{r}) \tag{c}$$

(c) Use the relations of (2) to derive a frequency domain wave equation.

(d) Comparing your wave equations of (b) and (c), find an expression for $\mathbf{P}_\omega(\mathbf{r})$ in terms of σ_ω, ϵ_ω, and $\mathbf{E}_\omega(\mathbf{r})$.

2. Consider Maxwell's equations in a homogeneous space defined by permeability μ_0 and permittivity ϵ. In (a) and (b) assume that ϵ is constant, but there is a current $\mathbf{J}$ and charge density ρ.

(a) Derive wave equations of the form

$$\nabla^2 \mathbf{E} - \mu_0 \epsilon \frac{\partial^2 \mathbf{E}}{\partial t^2} = \text{electric something}$$

$$\nabla^2 \mathbf{H} - \mu_0 \epsilon \frac{\partial^2 \mathbf{H}}{\partial t^2} = \text{magnetic something}$$

and thereby derive the somethings.

(b) If $\mathbf{J}(\mathbf{r}, t) = \nabla(f(\mathbf{r}, t))$, are electromagnetic waves radiated? Explain your answer.

In (c) and (d), assume that ρ and $\mathbf{J}$ are zero, but that ϵ varies with coordinates but not time.

(c) Derive wave equations as in (a).

(d) Try to express the somethings in (a) as effective currents and charges.

3. For a material with no microscopic free charge ($\rho_{\mu\,\text{free}} = 0$), and no microscopic currents ($\mathbf{j}_\mu = 0$), assume that there are only microscopic bound charges ($\rho_{\mu\,\text{bound}}$). The bound charges (molecules) are randomly distributed with a distribution function $\mathbf{f}(\mathbf{r})$.

(a) Give the general form of the macroscopic fields $\mathbf{E}(\mathbf{r}, t)$ and $\mathbf{H}(\mathbf{r}, t)$ in terms of the microscopic fields $\mathbf{e}(\mathbf{r}, t)$ and $\mathbf{h}(\mathbf{r}, t)$.

(b) Using the expression derived in (a), show that the spatial averages of Maxwell's microscopic relations yield Maxwell's macroscopic equations.

4. Recall that in free space, outside of a monochromatic source the field is given by

$$\mathbf{E} = \frac{i}{k} \nabla \times \mathbf{B}$$

and the $\mathbf{B}$ field can be given in terms of a vector potential $\mathbf{A}$ by

$$\mathbf{B} = \nabla \times \mathbf{A}$$

For an infinitesimal radiating electric dipole, the vector potential **A** can be expressed as

$$\mathbf{A} = -ik\mathbf{p}\,\frac{e^{ikr}}{r}$$

wi.ere r is the distance from the dipole to the point of evaluation of the field, and **p** is the vector along the dipole direction. For a z-directed dipole located at the origin:

(a) Find and sketch the field lines for the near field $1/r^3$ part of the pattern, and verify that it is identical to the electrostatic dipole field.

(b) Repeat (a) for the intermediate $1/r^2$ part of the field.

(c) Repeat (a) for the far field $1/r$.

(d) Find the Poynting vectors for each of the parts of the dipole field and explain why the $1/r$ part is called the radiating field.

5. Photons incident on a crystal can cause periodic mechanical strain. This strain modulates the elastic properties of the crystal and produces a phonon. Consider the scattering of a photon by a phonon (known as Brillouin scattering), as illustrated in Figure 3.23. If the velocity of sound is constant, we have $\Omega = v_s K$. Also, because K is comparable to k and $v_s \ll c$, then $k \approx k'$.

(a) Using these assumptions and conservation of momentum, write an expression for the angular frequency Ω.

(b) Using a vacuum photon wavelength of 4000 Å, $v_s = 5 \times 10^5$ cm/sec, and $n = 1.5$, calculate the maximum frequency of the scattered phonon.

(c) What is the fractional change of the frequency of the photon?

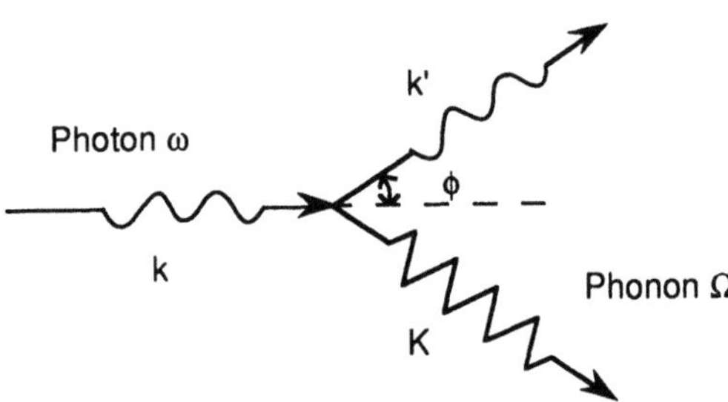

FIGURE 3.23.   Schematic depiction of photon phonon scattering from problem 5.

6. A charge per length of q is applied to parallel plates as shown in Figure 3.24.

(a) Calculate the electric field between the plates, neglecting fringing.

(b) If a uniform dielectric rod of length ℓ is inserted between the plates, as shown in Figure 3.25, calculate the resulting macroscopic electric field, again neglecting fringing. Assume that $\mathbf{P} = \chi\epsilon_0\mathbf{E}$.

(c) What is the local field in the center of the dielectric cylinder?

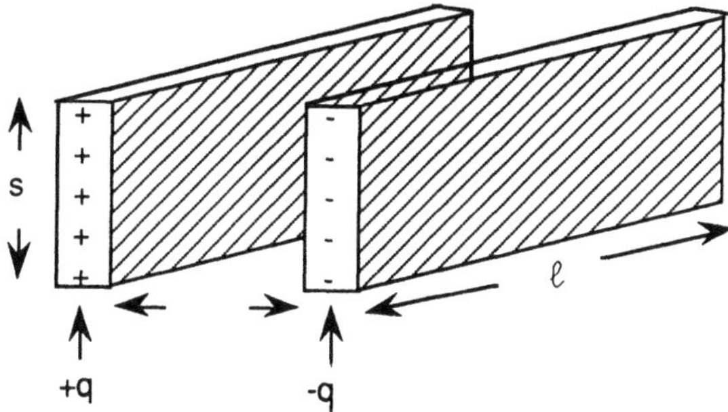

FIGURE 3.24.   Parallel charged plates as in problem 6.

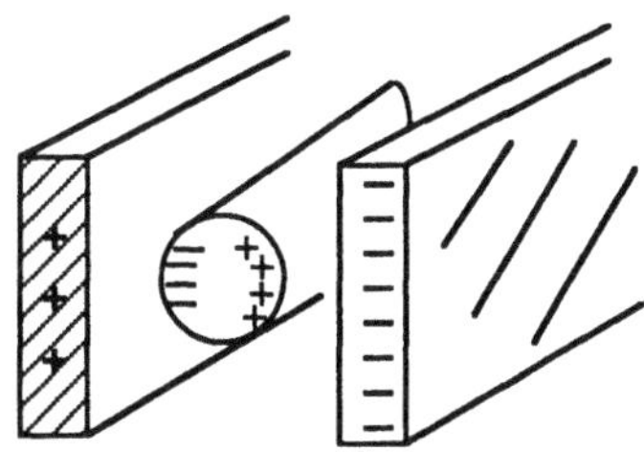

FIGURE 3.25.   Parallel charged plates loaded with a dielectric rod as in problem 6.

7. Consider a spring as depicted in Figure 3.26 with a restoring force that goes like $\kappa = m\omega_0^2 x + m\alpha x^2$, where m is the mass hanging from the spring. If the mass has a charge $-e$ and is acted on by a time-varying $\mathbf{E}$ field, $\mathbf{E}_0 e^{-i\omega t}$, then,
   (a) Find a recursion relation for the a_n's, the coefficients in the Fourier series expansion of the particle's motion x.
   In (b) through (e), assume that γ and α are small.
   (b) What is the energy stored in the spring?
   (c) What is the average power (over a period) dissipated by the spring?
   (d) Assuming that the spring reradiates as a simple dipole antenna (average radiated power, $P_{av} = (\eta_0/12\pi)\, k^2 p^2$), find the average power radiated into the fundamental ($e^{-i\omega t}$) mode. Ignore beating between different frequency components. How does the reradiated portion compare with the absorbed portion as a function of ω?
   (e) What percentage of the radiation is in the second harmonic ($e^{-2i\omega t}$), assuming small α? How does this percentage vary with frequency?
8. Recall the system of equations relating the amplitudes of atomic levels a and b:

$$\frac{\partial C_a}{\partial t} = \frac{1}{ih} V_{ab} e^{-i\omega t} C_b$$

$$\frac{\partial C_b}{\partial t} = \frac{1}{ih} V_{ab} e^{i\omega t} C_a$$

(a) For $V_{ab} = \mathcal{P} E_0 \cos vt$, show that the above system approximately (slowly vary-ing) reduces to

$$\frac{\partial C_a}{\partial t} = \frac{1}{2ih} \mathcal{P} E_0 e^{-i(\omega - v)t} C_b$$

$$\frac{\partial C_b}{\partial t} = \frac{1}{2ih} \mathcal{P} E_0 e^{i(\omega - v)t} C_a$$

(b) Assuming that $C_a(t) \approx e^{i\mu t}$, one can show that

$$C_a(t) = A e^{i\mu_1 t} + B e^{i\mu_2 t}$$

$$C_a(t) = 2 \left(\frac{\mathcal{P} E_0}{h} \right)^{-1} e^{i(\omega - v)t} [A\mu_1 e^{i\mu_1 t} + B\mu_2 e^{i\mu_2 t}]$$

Find the μ_1, μ_2, A, and B in the above expression.
(c) Use the above $C_a(t)$ and $C_b(t)$ to evaluate the time evolution of the wave function $\psi(t) = C_a(t)u_a(\mathbf{r})e^{i\omega_a t} + C_b(t)u_b(\mathbf{r})e^{-i\omega_a t}$. Sketch your answer. In the text, we say that it varied like e^{ivt}. Does it vary like that now? Why or why not?
(d) The z-directed microscopic dipole moment should vary something like $(1/\epsilon_0 \alpha)$ $p_z \approx E_z$ where E_z is the local field. Does it do that here? If so, what is α? What happens in the limit $\mathcal{P} E_0/h \to 0$?

9. This problem refers to the quantum dipole that was considered in the chapter.
(a) What is $C_b(t)$ for a D.C. field that is adiabatically applied (turned on very slowly)? Sketch the time evolution of $|\psi(r, t)|^2$.
(b) What is $C_b(t)$ for a D.C. field turned on suddenly at $t = 0$? Sketch the time evolution of $|\psi(r, t)|^2$.
(c) What is $C_b(t)$ for an applied field described by

$$f(t) = \frac{1}{\Delta_t} \text{rect} \left(\frac{t - t_0}{\Delta_t} \right)$$

Find the time evolution of $|\psi(r, t)|^2$, giving special attention to the limits of $\Delta_t \to \infty$ and $\Delta_t \to 0$. Interpret your answer. Recall that

$$\text{rect}(t) = \begin{cases} 0 & |t| > 1/2 \\ 1 & |t| < 1/2 \end{cases}$$

as was used also in problems 2-5 and 2-25.
(d) Let us say that $V(r, t) = V_0$, independent of both t and r. What is the evolution of $\psi(r, t)$ in this case?

10. Consider a two-level system, but one in which the excitation is such that it is the first and third levels that are important. For the assumed centro-symmetric potential,

this would mean that both coupled states are centro-symmetric; however, state 3 would have a null in the radial direction, but state 1 would not.

(a) Find an expression for a general wave function $\psi(\mathbf{r}, t)$ in terms of c_1, c_3, u_1, u_3, ω_1, and ω_3 where a stationary state is given by

$$\psi_i(\mathbf{r}, t) = u_i(\mathbf{r})e^{-i\omega_i t}$$

As was done in the text, sketch the evolution of the probability cloud

$$P(\mathbf{r}, t) = |\psi(\mathbf{r}, t)|^2$$

for several values of ωt, $\omega = \omega_3 - \omega_1$.

(b) Would a state such as that described in (a) radiate electromagnetic energy? That is, if one took a current density $j_\mu(\mathbf{r}, t)$ with the symmetry of the oscillating charge cloud in (a) and used it in microscopic Maxwell's equations, would it lead to time-varying solutions? Explain your answer.

11. Starting from the Schroedinger equation, we have

$$ih\frac{d\psi}{dt} = [H_0 + V(\mathbf{r}, t)]\psi$$

and let us say that $V(\mathbf{r}, t) = V_0$, independent of both t and $\mathbf{r}$. What is the evolution of $\psi(\mathbf{r}, t)$ in this case?

12. In cases where an atom is nearly unexcited ($C_a \approx 1$), it behaves classically like the Lorentz oscillator considered in the last section of this chapter. For strong excitations the electronic system behaves nonclassically, and one has to use the Schroedinger equation for a correct description of the system. Consider the nonclassical response of an atom due to a resonant electrical field pulse of the form $E_t = E_0(t)$ cos ωt in the rotating wave approximation. (This approximation was used to derive equation 3-48.)

(a) Solve the Schroedinger equation for an arbitrary envelope $E_0(t)$.

(b) By assuming that the atom is initially the goundstate, $C_a = 1$, $C_b = 0$, find the transition probability $|C_b(t)|^2$ for a step impulse

$$E_0(t) = \begin{cases} 0 & t \leq 0 \\ E_0 & t > 0 \end{cases}$$

Give a physical interpretation of what is happening.

(c) One can define the pulse envelope area as

$$\theta = \int_{-\infty}^{\infty} dt\, \frac{p}{2h} E_0(t), \quad p = er_{ab}$$

Consider a pulse of area $n\pi$. What is the transition probability in this case? (Hint: First solve for $C_a - C_b$ and $C_a + C_b$.)

13. In quantum mechanics, the mean value of an observable A is defined as

$$\langle A \rangle = \langle \psi | \hat{A} | \psi \rangle$$

where $|\psi\rangle$ is the state that the system is in. Calculate the mean value of the electric dipole moment $\mathbf{p} = -e\mathbf{r}$:
   (a) In the absence of an electric field.
   (b) In a uniform electric field $\mathbf{E}$.

14. Starting with equation (3-50), $\ddot{\mathbf{r}} + \gamma\dot{\mathbf{r}} + \omega^2\mathbf{r} = -(e/m)\,\mathbf{E}$ and letting $\mathbf{E} = \mathbf{E}_0 \sin \omega_0 t$, find $\mathbf{r}(t)$.

15. (a) Show that

$$x(t) = x_0 \cos (\omega t + \varphi)$$

$$+ \frac{eE_0}{m} \left[\frac{\cos (\nu t + \varphi_0)}{w^2 - \nu^2} - \frac{\cos (\omega t + \varphi_0)}{2\omega(\omega - \nu)} - \frac{\cos (\omega t + \varphi_0)}{2\omega(\omega + \nu)} \right]$$

is a solution to the oscillator equation of motion

$$\ddot{x}(t) + \omega^2 x(t) = \frac{e}{m} E_0 \cos (\nu t + f_0)$$

with $x(0) = x_0 \cos \phi$ and $\dot{x}(0) = \omega x_0 \sin \phi$.
   (b) Find the power averaged over one cycle, using

$$P(t) = \frac{\nu}{2\pi} \int_0^{2\pi/\nu} \dot{x}(t + \tau)eE_0 \cos [\nu(t + \tau)]d\tau$$

   (c) Using the expression you found for $P(t)$ and assuming that both E_0 and t are small, what conditions must be met so that $P(t) < 0$ (i.e., stimulated emission is occurring)?

16. Consider the often-given model for an atom in a field, as illustrated in Figure 3.26

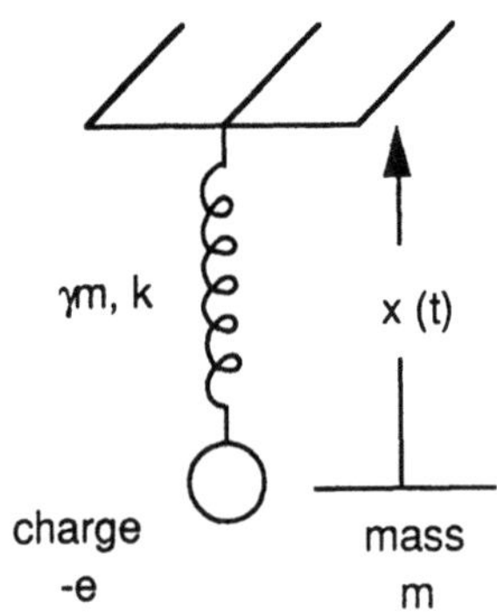

FIGURE 3.26.   A charged mass on a spring as in problems 7, 14, 15, 16, 17, 21, and 22, and 23.

and described by $\ddot{x} + \gamma\dot{x} + \omega_0^2 x = eE_x$. Note that this model really should not be valid for a plane wave although it is generally used. The Lorentz force law reads $\mathbf{F} = e\mathbf{E} + e\mathbf{v} \times \mathbf{B}$. For an incident x-polarized plane wave, E_x causes a v_x, which in turn causes a $\mathbf{v} \times \mathbf{B}$ force. Here we wish to study the consequences of such effects. In this model we must consider a three-dimensional spring that always exerts an inward force and damps along the direction of velocity.

(a) Write the vector equation describing the motion of the particle.

In (a) through (d), solve for and describe the off-resonance steady-state motion of the particle for the following incident waves:

(b) x-Polarized.

(c) Right-hand circularly polarized.

(d) Elliptically polarized.

17. Consider a (classical) spring attached to a (classical) particle of mass m and charge $-e$, as illustrated in Figure 3.26. Assume that the spring's motion is driven by a polychromatic field of amplitude spectrum $E(\omega)$ given by

$$E(\omega) = \frac{1}{N} \begin{cases} (\omega_0^2 - \omega^2)^2 + \gamma^2\omega^2 & \omega_0 - \dfrac{\Delta\omega}{2} < \omega < \omega_0 + \dfrac{\Delta\omega}{2} \\ 0 & \text{otherwise} \end{cases}$$

where N is a normalization factor.

(a) Determine N for normalization to unity field amplitude over the spectrum.

(b) Using the $x(t)$ derived in the text, find an expression for the particle's motion in the given incident field. Sketch the motion for $\Delta\omega \to 0$ and $\Delta\omega \to$ large.

(c) Assuming that the (classical) particles of a medium move according to our model, radiate like isotropic radiators (assume $E_{\text{local}} \sim E$), and lie in a one-dimensional straight line with element spacing γ_0, find the far field radiation pattern for the incident field described above, incident at an angle θ with respect to the normal to the array.

(d) Sketch what the radiation amplitude spectrum in the far field will look like.

18. Relations between α, n, and ϵ are derived in the chapter.

(a) Suppose that we know $\alpha = \alpha_r + i\alpha_i$. Find n and ϵ in terms of α_r and α_i.

(b) At the center of a resonance $\alpha_r \to 0$, what form does n take?

(c) Enough above a resonance, $\alpha_r \to 0$ and $\alpha_i \sim 0$. What are n and ϵ?

(d) Find α for $n = n_r + in_i$.

(e) What is α for $n_r < 0$, $n_i \sim 0$?

(f) What is α for $n_r \sim 0$, $n_i > 0$?

19. Consider the expansion for the k vector

$$k(\omega) = k(\omega_0) + (\omega - \omega_0) \left.\frac{\partial k}{\partial \omega}\right|_{\omega_0} + \frac{(\omega - \omega_0)^2}{2} D_2$$

$$+ \frac{(\omega - \omega_0)^3}{3!} D_3 + \frac{(\omega - \omega_0)^4}{4!} D_4$$

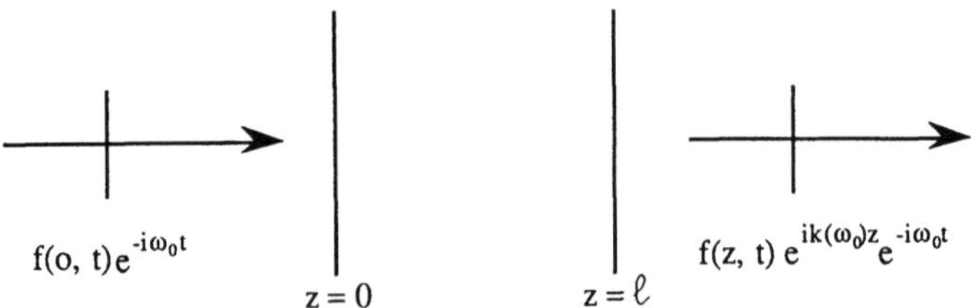

FIGURE 3.27.   A typical dispersion geometry, as used in problems 19, 20, and 21.

in a medium excited by an incident wave packet $f(0, t)$, as illustrated in Figure 3.27.

(a) For $D_2 = D_3 = D_4 = 0$, find $f(z, t)$ in the form of a convolution of $f(0, t)$ with a medium transfer function.

(b) Find $f(z, t)$ in the general case, in the form of a multiple convolution with the inverse Fourier transforms of the various parts of the medium transfer function.

In (c) through (e) one can take $f(0, t) = \delta(t)$. In (d) and (e) it will not be possible to find exact results for the inverse transforms. One must satisfy oneself with approximate (for example, stationary phase) results, and a rough sketch of the result. For (c) through (e), find and sketch $f(z, t)$ for:

(c) $D_3 = D_4 = 0$, $D_2 \neq 0$.

(d) $D_2 = D_4 = 0$, $D_3 \neq 3$.

(e) $D_2 = D_3 = 0$, $D_4 \neq 0$.

20. A pulse with an initial envelope function given by $f(0, t) = e^{-t^2/2\delta^2}$, and carrier $e^{-i\omega_0 t}$ is put into a medium in which second-order dispersion $(\partial^2 k / \partial \omega^2)$ is important.

(a) Find the Fourier transform of the input pulse.

(b) Write an expression for $f(z, t)$, the position-dependent pulse envelope function. Using

$$v_g = \frac{\partial k}{\partial \omega} \text{ and } D = \frac{\partial^2 k}{\partial \omega^2}$$

evaluate the necessary integrals to obtain an expression for $f(z, t)$.

(c) Take the limit $D \to 0$ in the expression for $f(z, t)$.

21. Suppose that there is a (plane phase front) pulse of finite duration propagating in a medium as is illustrated in Figure 3.27, with index $n(\omega)$, as illustrated in Figure 3.16.

(a) What is the phase velocity of the pulse?

(b) What is the group velocity of the pulse in terms of $k_0 = \omega_0/c$ and of $n(\omega)$? Make a sketch illustrating v_p and v_g.

(c) Calculate the dispersion of the pulse, that is, the amount of spreading in the pulse due to a small bandwidth $d\omega$ (dimensions should be length/time/bandwidth).

(d) The group index is often defined as $n_g(\lambda) = n(\lambda) - \lambda \, dn/d\lambda$, and therefore the average propagation delay is $t = l/c \, n_g$. Dispersion thus can also be expressed as $dt/d\lambda$. What happens around λ values where $d^2n/d\lambda^2 = 0$? Explain your answer.

22. Recall that the microscopic dipole moment $\mathbf{p}$ should satisfy

$$\frac{\mathbf{p}_0}{\epsilon_0} = \alpha \mathbf{E}_\ell + \beta \mathbf{E}_\ell \mathbf{E}_\ell + \gamma \mathbf{E}_\ell \mathbf{E}_\ell \mathbf{E}_\ell \cdots$$

where the α, β, and γ are tensors and, for a spring model, $\mathbf{p}_0/\epsilon_0 = -e\mathbf{r}$. Say that we have an x-directed spring, as illustrated in Figure 3.26, where $k/m = \omega^2 + d\omega^2 x^2$.

(a) Find the equation satisfied by x, the mass coordinate, under the influence of a field $E(t)$.

(b) Given that $E(t) = E_0 \cos \omega_0 t$, use the ansatz

$$x(t) = \sum_n a_n^c \cos n\omega_0 t + \sum_n a_n^s \sin n\omega_0 t$$

to find the response. The solution should have the form of recursion relations for the a_n's.

(c) Find the scalar γ and the microscopic $p(3\omega)$.

(d) What is the macroscopic $P(3\omega)$ and therefore the index of refraction for first and third order?

(e) Suppose that one has a chunk of the above-described material of thickness ℓ, as is illustrated in Figure 3.28. Qualitatively explain what the properties of the reflected and the transmitted fields would be.

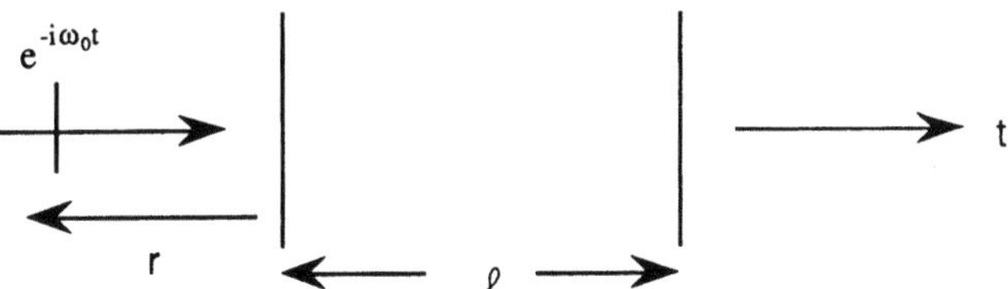

FIGURE 3.28.   Illustration of the transmission reflection problem as discussed in problem 22.

23. In this problem, we will consider both micro- and macroscopic models of an optically active medium, that is, a medium that has two indices of refraction, one for right-hand circularly polarized plane waves and one for left-hand circularly polarized plane waves.

(a) Discuss in some detail what a spring model (see Figure 3.26) for this medium would look like. Include the two harmonic oscillator equations that one finds for determining α_{right} and α_{left}.

(b) From the microscopic string model, derive the indices α_{right} and α_{left}, assuming spherically symmetric unit cells.

(c) Find a Mueller matrix for a slab of this material of thickness ℓ.

(d) Describe the evolution of a linear polarization in traveling through such a medium.

4

Wave Propagation in Anisotropic Media

4.1 INTRODUCTION

Chapter 3 concluded with a discussion of the dispersive effects that arise in isotropic materials due to the microscopic nature of the "springs" that make up the material. Another kind of dispersion arises when the wave functions that exhibit the transitions that give rise to the dielectric constant are not strictly symmetric. In this case, as will be seen in the coming paragraphs, the relation between the propagation constant and the optical frequency can be not only nonlinear but also multivalued. This multivaluedness can lead to very many interesting polarization effects, as we will presently see. Analysis of these effects will allow us to use some of the tools we developed in Chapter 2, as well as to come to a better understanding of how the polarization converting devices we discussed in that chapter work.

The plan for the present chapter is to first give some motivation, from microscopic considerations, for why the dielectric constant takes the form of a second-rank tensor in the general case. Discussion of the symmetries of the tensors in the absence of chiral symmetry will lead us to the construction of the Fresnel and the index ellipsoids. Consideration of plane wave propagation in such materials, with real dielectric tensors, will then lead to the normal and the ray surface constructions. With these tools in hand, it will be possible to consider a number of interesting effects, including double refraction and conical refraction. The last topic we will take up is an exposition of the salient features of polarization devices, including Nicol prisms, quarter-wave plates, and rotators.

136

4.2 MICROSCOPIC BASIS FOR THE EXISTENCE OF AN INDEX TENSOR

The microscopic polarization, $p(t)$, satisfied a differential equation of the form (3-58)

$$\frac{\partial^2 \mathbf{p}(t)}{\partial t^2} + \gamma \frac{\partial \mathbf{p}(t)}{\partial t} + \omega^2 \mathbf{p}(t) = K\, d(t)\mathbf{E}(t) \cdot \wp\wp. \qquad (4\text{-}1)$$

where $d(t)$ is the inversion parameter, given by the difference of the squared magnitude of the level occupances, $\mathbf{E}(t)$ is the time-varying component of the applied electric field, and $\wp$ is a microscopic strength factor given by the integral

$$\wp = -e \int u_a \mathbf{r} u_b d^3 r \qquad (4\text{-}2)$$

where u_a and u_b are the wave functions of the lower and upper states taking part in the index producing transition.

In the classical regime, $d(t)$ does not differ significantly from -1, as the ensemble averaged fraction of excited states is generally below 10^{-10} or so for usual strength electric fields. As $\wp$ does not vary with time, the derivative in (4-1) can be transferred to only the $\mathbf{E}(t)$, and, hence, one can write that

$$\frac{\partial^2 \mathbf{p}(t)}{\partial t^2} + \gamma \frac{\partial \mathbf{p}(t)}{\partial t} + \omega^2 \mathbf{p}(t) = -K\wp\wp \cdot \mathbf{E}(t) \qquad (4\text{-}3)$$

Now, somewhat paradoxically, from (4-2) it seems that the $\wp$ is dependent in direction and magnitude only on the nature of the atom. This seems to say that the direction of the polarizability $\mathbf{p}(t)$ from the form of (4-1), should be independent of the field direction. This must be nonsense, as in an isotropic medium $\mathbf{p}(t)$ must, from symmetry, be along the direction of $\mathbf{E}(t)$. The solution to this riddle lies in the definitions of the wavefunctions u_a and u_b. If the atom had a symmetric potential, then the ground state would be symmetric and the next highest state antisymmetric. However, antisymmetry requires the definition of an axis. The dependence $\cos \theta$ in a spherical coordinate system denotes an antisymmetric dependence, as $\cos (\theta + \pi) = -\cos \theta$. But, quite generally, and as is illustrated in Figure 4.1, the definition of an angle θ requires one to define a special axis z. If there were no external field $\mathbf{E}(t)$ applied, then there would be no unambiguous way to define this mystical z-axis. As soon as there is one, however, the field direction must become the direction to make the solution to the problem unique. In the symmetric case considered above, there-

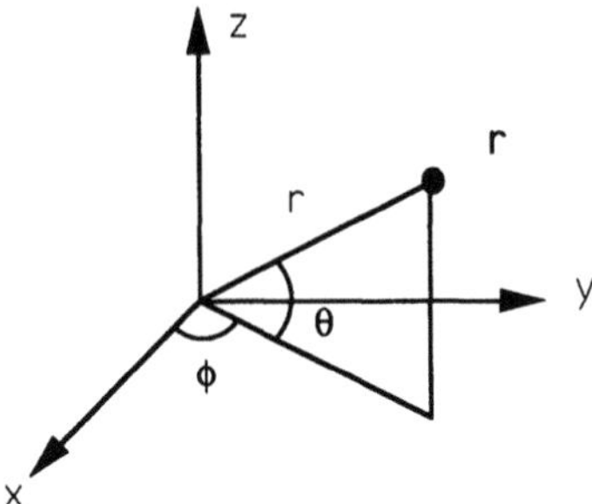

FIGURE 4.1.   Illustration of how the definition of an angle θ requires the existence of a special axis z.

fore, the existence of the directed electric field will cause the $\mathcal{P}$ also to become z-directed, that is, to take the direction of the field and, therefore, cause the microscopic polarization to take the direction of the field. However, if the potential were not symmetric, the electric field direction would still define a z direction, but the integral of (4-2) would not, in general, follow that direction. In this case, one can still simplify equation (4-1) somewhat in order to attempt to derive a form for the dielectric tensor ε, which reduces to the scalar dielectric constant ϵ in the fully symmetric case.

Now, as was stated in Chapter 2, equation (2-5), and has been reiterated by the derivations of Chapter 3, in general, what one means by the dielectric constant is the constant multiplying $\mathbf{E}_\omega$ in the expression for $\mathbf{D}_\omega$, that is,

$$\mathbf{D}_\omega = \epsilon(\omega)\mathbf{E}_\omega \qquad (4\text{-}4)$$

where $\mathbf{D}_\omega$ and $\mathbf{E}_\omega$ are defined by

$$\mathbf{E}(t) = \mathrm{Re}\ [\mathbf{E}_\omega e^{-i\omega t}] \qquad \text{(a)}$$
$$\mathbf{D}(t) = \mathrm{Re}\ [\mathbf{D}_\omega e^{-i\omega t}] \qquad \text{(b)}$$

$$(4\text{-}5)$$

It should be noted here that the time domain version of (4-4) would require a convolution. Using (4-5) (a) and a similar form for $\mathbf{p}_\omega$ in (4-3), one obtains

$$\mathbf{p}_\omega = f(\omega)\mathbf{E}_\omega \cdot \mathcal{P}\mathcal{P} \qquad (4\text{-}6)$$

where $f(\omega)$ is a scalar function of ω. Recalling that the direction $\mathcal{P}$ is linearly related to the direction of $\mathbf{E}_\omega$, one can express (4-6) in the form

$$\mathbf{p}_\omega = \alpha(\omega)\mathbf{E}_\omega \qquad (4\text{-}7)$$

where $\alpha(\omega)$ is a polarizability tensor (corresponding to the polarizability scalar $\alpha(\omega)$ of the Chapter 3) that is independent of $\mathbf{E}_\omega$. To obtain the polarization vector $\mathbf{P}_\omega$, one needs to perform an average over the microscopic coordinates of $\mathbf{p}_\omega$. To obtain $\mathbf{D}_\omega$, one then needs to use the relation

$$\mathbf{D}_\omega = \epsilon_0 \mathbf{E}_\omega + \mathbf{P}_\omega \tag{4-8}$$

Regardless of the details of the averaging process, the linearity assumption is sufficient to reduce (4-8) to the form

$$\mathbf{D}_\omega = \varepsilon(\omega)\mathbf{E}_\omega \tag{4-9}$$

where the important point is that $\varepsilon(\omega)$ is now a second-rank tensor, as was $\alpha(\omega)$. The operation of multiplication of $\mathbf{E}_\omega$ by $\varepsilon(\omega)$ will, in general, cause the direction of $\mathbf{D}_\omega$ to differ from that of $\mathbf{E}_\omega$. Actually there are many other ramifications of the existence of this dielectric tensor, as we will begin to see in the next section.

4.3 FRESNEL'S AND THE INDEX ELLIPSOIDS

One could brutally try to plug the $\varepsilon(\omega)$ of equation (4-9) into Maxwell's equations and try to see how far one could go. The use of a little finesse, as was employed as far back as the time of Fresnel (which was pre-Maxwell), can instill in one a lot of insight into the physics of wave propagation in such a tensor-defined medium. In this section, we will first consider the symmetries of $\varepsilon(\omega)$ required by energy considerations and then proceed to two geometrical constructions that can be used to great advantage in picturing propagation in anisotropic materials.

The form of (4-9) seems to indicate that $\varepsilon(\omega)$ must contain nine complex-valued components. The number is actually not nearly so high if we restrict consideration to frequency regimes far from the resonant frequencies of the materials under consideration. This is not really so unrealistic an approximation, as one would, in general, not want to use a nontransparent element in an optical system except as an attenuator. If one is operating in such a nonresonant regime, then it is possible from energy considerations to determine some symmetries of the $\varepsilon(\omega)$ tensor (see, for example, Sections 61 and 76 of Landau and Lifshitz 1960). As was discussed in Chapter 2 (equation 2-18), the divergence of the time-varying Poynting vector may be expressed in the form

$$\nabla \cdot \mathbf{S}(t) = -\mathbf{H}(t) \cdot \frac{\partial \mathbf{B}(t)}{\partial t} - \mathbf{E}(t) \cdot \frac{\partial \mathbf{D}(t)}{\partial t} - \mathbf{E}(t) \cdot \mathbf{J}(t) \tag{4-10}$$

A time-averaged version of (4-10) can be obtained by writing

$$\mathbf{V}(t) = \tfrac{1}{2}(\mathbf{V}_\omega e^{-i\omega t} + \mathbf{V}_\omega^* e^{i\omega t}) \tag{4-11}$$

for each of the field vectors in (4-9), and then explicitly performing a time average to obtain

$$-4 \langle \nabla \cdot \mathbf{S}(t) \rangle = i\omega(\mathbf{H}_\omega \cdot \mathbf{B}_\omega^* - \mathbf{H}_\omega^* \cdot \mathbf{B}_\omega + \mathbf{E}_\omega \cdot \mathbf{D}_\omega^* - \mathbf{E}_\omega^* \cdot \mathbf{D}_\omega)$$
$$+ \mathbf{E}_\omega \cdot \mathbf{J}_\omega^* + \mathbf{E}_\omega^* \cdot \mathbf{J}_\omega \tag{4-12}$$

Using the relations

$$\mathbf{B}_\omega = \mu_0 \mathbf{H}_\omega \tag{a}$$
$$\mathbf{J}_\omega = \sigma(\omega)\mathbf{E}_\omega \tag{b}$$

$$(4\text{-}13)$$

where $\sigma(\omega)$ is the tensor that corresponds to the scalar $\sigma(\omega)$ discussed in Chapter 2, yields the result that

$$4 \langle \nabla \cdot \mathbf{S}(t) \rangle = i\omega \mathbf{E}_\omega^* \cdot [\varepsilon^\dagger(\omega) - \varepsilon(\omega)]$$
$$\times \mathbf{E}_\omega + \mathbf{E}_\omega^* \cdot [\sigma^\dagger(\omega) + \sigma(\omega)] \cdot \mathbf{E}_\omega \tag{4-14}$$

where the † denotes Hermitian transpose. Now, as we have already seen, the divergence of the Poynting vector corresponds to an energy flow. The time average of this quantity is, therefore, an actual loss of electromagnetic energy. As one generally associates $\sigma(\omega)$ with loss and $\varepsilon(\omega)$ with propagation, one sees that from (14) the conditions on $\varepsilon(\omega)$ and $\sigma(\omega)$ to satisfy our intuitive definitions should be

$$\epsilon_{ij}(\omega) = \epsilon_{ji}^*(\omega) \tag{a}$$
$$\sigma_{ij}(\omega) = \sigma_{ji}(\omega) \tag{b}$$

$$(4\text{-}15)$$

that is, that $\varepsilon(\omega)$ should be Hermitian and $\sigma(\omega)$ real and symmetric. Indeed, a material with a zero $\sigma(\omega)$ and Hermitian $\varepsilon(\omega)$ is a lossless propagation medium. It is such materials that will be considered in this chapter, but with the remembrance that real materials inevitably exhibit resonances and therefore have losses. Further, because of anisotropy, different polarizations may ''see'' different resonances, and, therefore, the relative magnitudes of the tensor components of $\varepsilon(\omega)$ and $\sigma(\omega)$ will vary with frequency.

Now, for $\varepsilon(\omega)$ Hermitian, one could just as well write that

$$\varepsilon(\omega) = \varepsilon_r + i\gamma \tag{4-16}$$

where ε_r is real symmetric, and γ is real and antisymmetric. Now, the ellipsoid constructions arise from consideration of expressions for the stored electrical energy. A standard expression for the electric energy density, as is given in most (nonrelativistic) texts is (see, for example, the discussions leading to equations 6.16 and 4.88 of the book by Jackson, 1975)

$$W_e = \mathbf{E}(t) \cdot \mathbf{D}(t) \tag{4-17}$$

To be able to use expressions such as (4-15) (a) for materials that are in general dispersive,[1] one wants to transform (4-17) to the complex domain. A way to do this is to use the forms of (4-5) in the expression for the time-averaged electric energy density $\langle W_e \rangle$, to obtain

$$\langle W_e \rangle = \tfrac{1}{4}(\mathbf{E}_\omega \cdot \mathbf{D}_\omega^* + \mathbf{E}_\omega^* \cdot \mathbf{D}_\omega) \tag{4-18}$$

which, through the use of (4-9), can be simplified to yield

$$\langle W_e \rangle = \tfrac{1}{2}\mathbf{E}_\omega \cdot \varepsilon(\omega) \cdot \mathbf{E}_\omega^* \tag{4-19}$$

It should be pointed out here that assuming monochromatic fields in the derivation of the time-averaged energy density is a slightly strange thing to do. If the excitation is monochromatic, the energy density does not vary with time, and, therefore, the time average is, at best, redundant. Such, however, is the nature of the approximations involved in obtaining phase velocities, which we shall presently obtain. As we saw in the last chapter, phase velocity represents neither the transport of energy nor information, but relates to structure of the field, which is very important with respect to polarization characteristics.

To obtain a form for the time-averaged electric energy density that is of the same level of approximation as the concept of energy velocity, one needs to

---

[1]Although great care is being used in this text to differentiate the use of $\varepsilon(t)$ from $\varepsilon(\omega)$, the difference does become specious if one assumes a completely nondispersive medium. That is, the Fourier transform of (4-9) yields a relation $\mathbf{D}(t) = \int \varepsilon(t - t') \, \mathbf{E}(t')dt'$. Now, if ε were truly frequency independent from $D.C.$ through X rays (a physical impossibility but often a useful approximation), then one could write that $\mathcal{F}_t[\varepsilon(\omega)] = \varepsilon\delta(t)$, and the above expression for $\mathbf{D}(t)$ would reduce to $\mathbf{D}(t) = \varepsilon\mathbf{E}(t)$. If one is operating in a frequency regime where one need not worry about dispersion, then the constant ε assumption may be acceptable. There does seem to be something strange about using such an assumption in the calculation of a so-called dispersion relation, however.

start from a form for the time derivative of the electric energy density, such as that contained in equation (4-10) for the divergence of the Poynting vector. Picking out the second term on the right-hand side, one finds

$$\frac{\partial W_e}{\partial t} = \mathbf{E}(t) \cdot \frac{\partial \mathbf{D}(t)}{\partial t} \tag{4-20}$$

By starting from such a form, one can retain time history in the derivation, and thus a memory of the formation and transport of the energy packet (see, for example, Papas 1965, Section 6.2). What we wish to do here is to substitute the nonmonochromatic forms

$$\mathbf{V}(t) = \frac{1}{2}\left(\int \mathbf{V}(\omega + \alpha)e^{-i(\omega_0 + \alpha)t}\, d\alpha + \int \mathbf{V}^*(\omega_0 + \alpha)e^{i(\omega_0 + \alpha)t}\, d\alpha \right) \tag{4-21}$$

for $\mathbf{D}(t)$ and $\mathbf{E}(t)$. Assuming, however, that the excitation is quasi-monochromatic, one can then also assume that maximal extent of α is small and that

$$\mathbf{E}(\omega_0 + \alpha) \approx \mathbf{E}_t(\omega_0) \tag{4-22}$$

where $\mathbf{E}_t(\omega_0)$ is actually a slowly varying function of time (see, for example, Landau and Lifshitz 1960, Sections 61 and 62). Such an assumption is not tenable for $\varepsilon(\omega)$, however. Looking in some detail at $\partial \mathbf{D}/\partial t$, one finds that

$$\frac{\partial}{\partial t} \mathbf{D}(t) = -i\frac{1}{2} \int (\omega_0 + \alpha)\varepsilon(\omega_0 + \alpha)\mathbf{E}(\omega_0 + \alpha)e^{-i(\omega_0 + \alpha)t}\, d\alpha$$
$$+ i\frac{1}{2} \int (\omega_0 + \alpha)\varepsilon^*(\omega_0 + \alpha)\mathbf{E}^*(\omega_0 + \alpha)e^{i(\omega_0 + \alpha)t}\, d\alpha \tag{4-23}$$

Making the substitution

$$\mathbf{f}(\omega_0 + \alpha) = (\omega_0 + \alpha)\varepsilon(\omega_0 + \alpha) \tag{4-24}$$

and Taylor-expanding $\mathbf{f}(\omega_0 + \alpha)$ about ω_0, one can obtain (4-23) in the form

$$\frac{\partial \mathbf{D}(t)}{\partial t} = \frac{1}{2}\left[-i\omega_0\varepsilon(\omega_0)\mathbf{E}_t(\omega_0)e^{-i\omega_0 t} - i\frac{\partial \omega_0 \varepsilon(\omega_0)}{\partial \omega_0} \int \alpha\mathbf{E}(\omega_0 + \alpha)e^{-i(\omega_0 + \alpha)t}\, d\alpha \right.$$
$$+ i\omega_0\varepsilon^*(\omega_0)\mathbf{E}_t^*(\omega_0)e^{i\omega_0 t} + i\frac{\partial(\omega_0\varepsilon^*(\omega_0))}{\partial \omega_0}$$
$$\left. \cdot \int \alpha\mathbf{E}^*(\omega_0 + \alpha)e^{i(\omega_0 + \alpha)t}\, d\alpha \right] \tag{4-25}$$

The integrals are easily evaluated from a Fourier transform theorem, to yield

$$\frac{\partial \mathbf{D}(t)}{\partial t} \approx \frac{1}{2}\left[-i\omega\varepsilon(\omega)\mathbf{E}_t(\omega)e^{-i\omega t} + \frac{\partial(\omega\varepsilon(\omega))}{\partial\omega}\frac{\partial\mathbf{E}_t(\omega)}{\partial t}e^{-i\omega t}\right]$$
$$+ \frac{1}{2}\left[i\omega\varepsilon^*(\omega)\mathbf{E}_t^*(\omega)e^{i\omega t} - \frac{\partial}{\partial\omega}(\omega\varepsilon^*(\omega))\frac{\partial\mathbf{E}_t^*(\omega)}{\partial t}e^{i\omega t}\right] \tag{4-26}$$

where the subscript 0 has been omitted from the ω. Performing the operations indicated in (4-20), one finds that

$$\frac{\partial W_e}{\partial t} = i\omega\frac{1}{2}(\mathbf{E}\varepsilon^*\mathbf{E}^* - \mathbf{E}^*\varepsilon\mathbf{E})$$
$$+ \frac{1}{2}\left[\mathbf{E}^*\frac{\partial(\omega\varepsilon)}{\partial\omega}\frac{\partial\mathbf{E}}{\partial t} + \mathbf{E}\frac{\partial(\omega\varepsilon^*)}{\partial\omega}\frac{\partial\mathbf{E}^*}{\partial t}\right] \tag{4-27}$$

The first term indicates loss, and for a lossless medium, therefore, leads to the Hermitian requirement that we obtained in (4-15) (a). Using this Hermiticity in (4-27) yields

$$\frac{\partial W_e}{\partial t} = \frac{1}{2}\frac{\partial}{\partial t}\left[\mathbf{E}^*\frac{\partial(\omega\varepsilon)}{\partial\omega}\mathbf{E}\right] \tag{4-28}$$

which is equivalent to the form

$$\langle W_e\rangle = \frac{1}{2}\mathbf{E}^*\frac{\partial(\omega\varepsilon)}{\partial\omega}\mathbf{E} \tag{4-29}$$

It is interesting to note that this energy density reduces to the monochromatic one in the limit that $\varepsilon(\omega)$ is independent of ω. Later in the chapter, however, we will find that there is a more direct way to find group (energy) velocities than having to use Poynting vectors and energy densities. But, for the present, we will return to consideration of phase characteristics.

Now it can be of interest to see what geometrical structures may be associated with the expression (4-19). It is almost a quadratic form, apart from a conjugate sign, and one that should be positive definite because of our innate concept of energy. Using (4-16) and the following expressions for the components of the electric field vector

$$E_j = \sqrt{I_j}\,e^{i\varphi_j}\quad j = x, y, z \tag{4-30}$$

where from now on, the ω subscripts on the fields will be suppressed, one can brutally write out the separate terms in the implicit sums of (4-29) as

$$\langle W_e \rangle = \tfrac{1}{2}(\varepsilon_{xx}I_x + \varepsilon_{yy}I_y + \varepsilon_{zz}I_z)$$
$$+ \sqrt{I_xI_y}\,(\varepsilon_{xy}\cos\varphi_{xy} + \gamma_{xy}\sin\varphi_{xy})$$
$$+ \sqrt{I_xI_z}\,(\varepsilon_{xy}\cos\varphi_{xz} + \gamma_{xz}\sin\varphi_{xz}) \qquad (4\text{-}31)$$
$$+ \sqrt{I_yI_z}\,(\varepsilon_{yz}\cos\varphi_{yz} + \gamma_{yz}\sin\varphi_{yz})$$

where the ε_{ij} are the components of ε_r, and the γ_{ij} the components of γ.

Perhaps the most important thing to be gleaned from (4-31) is that (4-19) can be reduced to a positive definite quadratic form in a very practically important subcase. Clearly, if $\gamma = 0$, then $\langle W_e \rangle$ reduces to a quadratic form when the φ's are zero. But a linearly polarized plane clearly has all the φ's equal to zero, as there is only one polarization state, which cannot very well have a phase angle with respect to itself. And, indeed, we will find that when γ is zero, all the eigenstates of the real symmetric ε tensors are linear states. This is the case we will consider for the Fresnel and index ellipsoid.

The standard textbook treatment of the geometric approach to anisotropic wave propagation [see, for example, Sommerfeld (1972, Volume 4, Chapter 4); Born and Wolf (1975, Chapter 14); Yariv and Yeh (1984, Chapter 4) or Yeh (1988, Chapter 9)] begin with replacing the components of the electric field (or electric displacement) vector, by coordinates $x_1 = x$, $x_2 = y$, $x_3 = z$, such that

$$\langle W_e \rangle = \tfrac{1}{2}\sum \epsilon_{ij}x_ix_j \qquad (4\text{-}32)$$

As the right-hand side of equation (4-32) is a positive definite form, in general, it is the equation of an ellipse (Yeh 1988, Chapters 9–10). An ellipse can be defined by its three, orthogonal, principal axes. This is equivalent to saying that the six independent parameters of equation (4-32) (ϵ_{xx}, ϵ_{yy}, ϵ_{zz}, ϵ_{xy}, ϵ_{xz}, ϵ_{yz}) can be reduced to three free parameters through a general rotation, which is actually a product of three rotations, one about each of the three orthogonal axes of the original coordinate system. As each of the rotations is through an arbitrary angle, these three angles of the general rotation can be used to eliminate three of the six parameters of equation (4-32). The coordinate system that results is the principal axis system, in which the equation of the ellipse is

$$\langle W_e \rangle = \tfrac{1}{2}\sum_i \epsilon_i x_i^2 \qquad (4\text{-}33)$$

and in this rotated (1, 2, 3) coordinate system, the relationship between the components of $\mathbf{D}$ and $\mathbf{E}$ can be written as

$$D_i = \epsilon_i E_i \quad i = 1, 2, 3 \tag{4-34}$$

As equation (4-33) could as well have been written in terms of the principal indices of refraction n_i, one sees that the ellipse is expressible in the form

$$\langle W_e \rangle = \tfrac{1}{2}\Sigma\, n_i^2 x_i^2 \quad i = 1, 2, 3 \tag{4-35}$$

where one can now identify principal phase velocities v_i by

$$v_i = c/n_i \tag{4-35a}$$

where the principal phase velocities represent the phase velocities that linearly polarized plane waves polarized along each of the principal axes would exhibit.

The index ellipsoid construction can be easily obtained as a dual construction of the above-carried-out Fresnel's ellipsoid construction. In an arbitrary coordinate system, one can express $\mathbf{E}$ as a function of $\mathbf{D}$ by

$$\mathbf{E} = \mathbf{\eta}\mathbf{D} \tag{4-36}$$

where η must be given by the inverse of ε:

$$\mathbf{\eta} = \varepsilon^{-1} \tag{4-37}$$

where the -1 denotes matrix inverse. However, if one limits one's attention to linearly polarized states, one sees that the energy expression

$$\langle W_e \rangle = \tfrac{1}{2}\Sigma_{ij}\, \eta_{ij} D_i D_j \quad i, j = x, y, z \tag{4-38}$$

must also reduce to a form such as (4-33) in a principal axis system, such that

$$\langle W_e \rangle = \tfrac{1}{2}\Sigma\, \eta_i x_i^2 \quad i = 1, 2, 3 \tag{4-39}$$

where, in this principal axis system, one must have the relation that

$$\eta_i = 1/\epsilon_i \quad i = 1, 2, 3 \tag{4-40}$$

In this system, therefore, one can write that

$$\langle W_e \rangle = \tfrac{1}{2}\Sigma \, x_i^2 / n_i^2 \qquad (4\text{-}41)$$

to see that the principal axes in the index ellipsoid system are just proportional to the principal indices, unlike the case with Fresnel's ellipsoid where the axes were proportional to the principal velocities.

The ellipsoids tell quite a lot about propagation of phases for fields polarized along the principal axes. However, the more comprehensive study of the next section is necessary to determine relative polarization directions allowed by the dielectric tensor for a given propagation direction.

4.4 THE NORMAL SURFACE AND THE RAY SURFACE

The purpose of this section is primarily to find and interpret the so-called dispersion relation, that is, the relation between the frequency of operation ω and the allowed $\mathbf{k}$ vectors for an ω in an anisotropic crystal. This will require a study of Maxwell's equations in an anisotropic medium for plane wave solutions. The wave normal equation serves as the desired dispersion relation. Among other things, it tells us that for a given frequency and $\mathbf{k}$ vector direction two distinct polarization solutions with different phase velocities exist. These solutions can be most easily visualized by constructing a ray normal surface, a two-sheeted surface that allows one, for a given $\mathbf{k}$ vector direction, to picture the two different phase velocities. The phase velocities, however, turn out not to be the whole story. We found in the last chapter that for lossless isotropic media, although the directions of the phase and group (energy) velocity were identical, their magnitudes were not. Here we will find that the directions do not even coincide. Therefore, to determine the characteristics of the energy transfer we will require the derivation of an equation for the ray normals and the use of a double-sheeted ray normal surface to help visualize the two solutions of the equation for each of the given two directions. We will see that there are symmetry axes, known as optic axes, for both the wave and the ray surfaces, but that these directions, in general, are not the same for rays and $\mathbf{k}$ vectors.

We begin the discussion of the present section by considering Maxwell's equations specialized to the case of monochromatic plane waves. Assuming that all of the field vectors can be expressed as

$$\mathbf{V}(x, y, z, t) = \mathrm{Re}\,[\mathbf{V}(x, y)e^{i\mathbf{k}\cdot\mathbf{r}}e^{-i\omega t}] \qquad (4\text{-}42)$$

Maxwell's equations for a lossless, charge-free medium (Chapter 2, equation 2-1) take on the form

$$\mathbf{k} \times \mathbf{E} = -\omega\mathbf{B} \qquad (a)$$

$$\mathbf{k} \times \mathbf{H} = -\omega\mathbf{D} \qquad (b)$$

$$\mathbf{k} \cdot \mathbf{D} = 0 \qquad (c)$$

$$\mathbf{k} \cdot \mathbf{B} = 0 \qquad (d)$$

(4-43)

Perhaps the first information to be derived from (4-43) is information concerning the relative directions of the fields. Clearly, $\mathbf{B}$ and $\mathbf{D}$ are perpendicular to the propagation direction $\mathbf{k}$. Further (4-13) (a), together with (4-43) (b), indicates that $\mathbf{H}$ is parallel to $\mathbf{B}$, which is, in turn, perpendicular to $\mathbf{D}$. Equation (4-43) (a) then gives that $\mathbf{B}$ and $\mathbf{E}$ are also perpendicular. The situation can be depicted as in Figure 4.2. The point there is that one has no information about the relative directions of $\mathbf{E}$ and $\mathbf{D}$ except that they are both perpendicular to the $\mathbf{B}$–$\mathbf{H}$ direction. It remains to be calculated what the angle α between these fields is. But an important consequence of the existence of this angle α is that the unit vector $\hat{\mathbf{e}}_k$ in the $\mathbf{k}$ direction will not be in the same direction as the unit vector $\hat{\mathbf{e}}_s$ in the direction of the Poynting vector. If one takes the $\hat{\mathbf{e}}_s$ direction to be that of energy flow, then this is to say that the phase of the wave increases in a direction different from that of energy flow. Consequences of this statement will become evident later in this section and the next.

To get a dispersion relation, one needs to simplify (4-43) further. Taking $\mathbf{k} \times$ (4-43) (a) and substituting from (4-43) (b) yields the relation

$$\mathbf{k} \times \mathbf{k} \times \mathbf{E} = -\omega^2 \mu_0 \mathbf{D} \qquad (4\text{-}44)$$

Now, using the vector identity

$$\mathbf{A} \times \mathbf{B} \times \mathbf{C} = \mathbf{B}\mathbf{A} \cdot \mathbf{C} - \mathbf{C}\mathbf{A} \cdot \mathbf{B} \qquad (4\text{-}45)$$

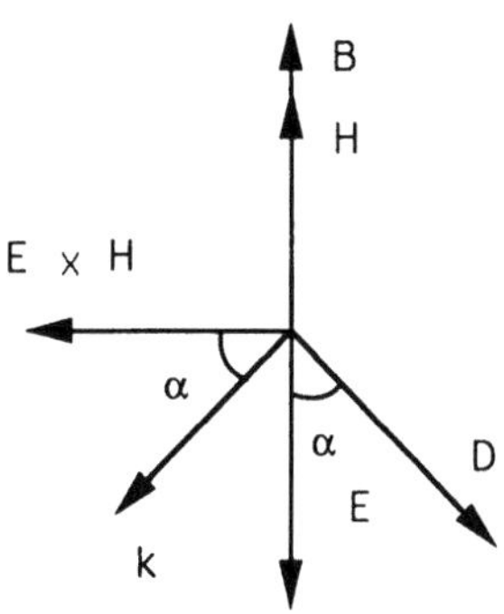

FIGURE 4.2.   Schematic depiction of the directions of the field vectors, $\mathbf{k}$ vector, and ray ($\mathbf{E} \times \mathbf{H}$) vector in an optically anisotropic medium.

and denoting

$$v_p = \frac{\omega}{|\mathbf{k}|} \tag{4-46}$$

one finds that

$$-\mu_0 v_p^2 \mathbf{D} = \frac{\mathbf{k}}{k^2}(\mathbf{k} \cdot \mathbf{E}) - \mathbf{E} \tag{4-47}$$

Now, in the principal axis system of coordinates, one can write that

$$E_i = D_i/\epsilon_i \quad i = 1, 2, 3 \tag{4-48}$$

where the principal light velocities are given by equation (4-35a), and one can write (4-47) in the form

$$(v_p^2 - v_i^2)D_i = k_i K \tag{4-49}$$

where K is defined by

$$K = -\frac{1}{k^2} \sum_i v_i^2 k_i D_i \tag{4-50}$$

Now, multiplying (4-49) by $k_i/(v_p^2 - v_i^2)$, summing over i, and using (4-43) (c) yields the desired dispersion relation

$$\sum_i \frac{k_i^2}{v_p^2 - v_i^2} = 0 \tag{4-51}$$

which could as well be written in the alternative form

$$(v_p^2 - v_2^2)(v_p^2 - v_3^2)k_1^2 + (v_p^2 - v_1^2)(v_p^2 - v_3)^2 k_2^2$$
$$+ (v_p^2 - v_1^2)(v_p^2 - v_2^2)k_2^3 = 0 \tag{4-52}$$

which shows explicitly that the dispersion relation is a second-order equation in v_p^2.

Equation (4-52), being a second-order equation in v_p^2, will have two solutions for v_p^2 for each direction of $\mathbf{k}$; that is, for a value of ω and a direction in space, $\mathbf{k}$ can take on two distinct magnitudes. This means that the dispersion relation is multivalued. Each of these two solutions, in turn, yields a forward- and a backward-propagating phase velocity. Now, as (4-49) is a homogeneous linear system of equations for the vector $\mathbf{D}$, and (4-52) is essentially the condition that the determinant of the coefficient matrix be zero, then it follows that there will be a vector corresponding to each of the solutions of the determinant. Perusal of (4-49) indicates that it could be recast in the form

$$\mathbf{MD} = v_p^2 \mathbf{D} \tag{4-53}$$

which manifestly shows that the values of v_p^2 are eigenvalues, and the corresponding directions of $\mathbf{D}$ are eigenvectors of the propagation problem. As the matrix $\mathbf{M}$ is symmetric, in the absence of mathematical pathologies, the eigenvectors are orthogonal, and, therefore, for a given propagation direction, there will be two transverse polarization states with different phase velocities. As the matrix $\mathbf{M}$ is real and symmetric, these eigenvectors are real, and, therefore, the polarization states are linear. Further calculation [see any elementary geometry or linear algebra text, for example, Noble (1969, Chapter 12)] would further show that these polarization directions correspond to the projections of the principal axes onto the plane perpendicular to the propagation direction, and that the phase velocities are just the principal phase velocities in these directions.

Although it is easy to make the above arguments about the structure of the plane wave solutions for $\mathbf{D}$, and straightforward to carry out the calculations, it is very useful to develop a tool with which to visualize the solution. To do this, we wish to plot the solutions for the two phase velocities versus the $\mathbf{k}$ vector direction. What will be obtained, therefore, will be a two-sheeted, closed surface in $\mathbf{k}$ space. To do this, we define coordinates ξ_i by

$$\xi_i = \frac{k_i}{|\mathbf{k}|} v_p \tag{4-54}$$

such that

$$\sum_i \xi_i^2 = v_p^2 \tag{4-55}$$

Substituting (4-54) into (4-51), leads to the relation

$$\sum_i \frac{\xi_i^2}{(v_p^2 - v_i^2)} = 0 \tag{4-56}$$

which leads to the relation

$$v_p^6 - v_p^2(\xi_1^2(v_1^2 + v_3^2) + \xi_2^2(v_1^2 + v_3^2) + \xi_3^2(v_1^2 + v_2^2))$$
$$+ \xi_1^2 v_2^2 v_3^2 + \xi_2^2 v_1^2 v_3^2 + \xi_3^2 v_1^2 v_2^2 = 0 \tag{4-57}$$

which, in general, represents a sixth-order surface. What are really wish to do is look at the intersection of this sixth-order, two-sheeted surface with the planes $\xi_1 = 0$, $\xi_2 = 0$, $\xi_3 = 0$. From (4-56), one can see that in the plane $\xi_1 = 0$ one of two conditions must be satisfied, that is, either

$$v_p^2 = v_1^2 \tag{4-58}$$

or

$$\frac{\xi_2^2}{v_p^2 - v_2^2} + \frac{\xi_3^2}{v_p^2 - v_3^2} = 0 \tag{4-59}$$

where the first condition arises from the fact that a limiting process could allow v_p to approach v_1 as fast as ξ_1 approaches zero, thereby allowing $0/0$ to be whatever it needs to be to satisfy (4-56). Now, for $\xi_1 = 0$, (4-55) gives that $\xi_2^2 + \xi_3^2 = v_p^2$, giving that (4-58) and (4-59) can be written as

$$\xi_2^2 + \xi_3^2 = v_1^2 \tag{4-60}$$

or

$$(\xi_2^2 + \xi_3^2)^2 = v_3^2\xi_2^2 + v_2^2\xi_3^2 \tag{4-61}$$

Equation (4-60) is the equation of a circle, whereas that of (4-61) is that of an oval, a surface that happens to be a pedal surface to the ellipsoid we soon will see to represent the ray surface. More detailed discussion of this surface is given in Sommerfeld (1972, Chapter 4) and Landau and Lifshitz (1960, Sections 61 and 76). Now, one could assume that $\xi_1 < \xi_2 < \xi_3$ and, therefore, that $v_1 > v_2 > v_3$. The surfaces corresponding to (4-60) and (4-61) for the planes $\xi_2 = 0$ and $\xi_3 = 0$ will also be a circle and an oval, but their scales will be different. The situation is summed up in Figure 4.3. The point in the diagram is that the two surfaces do not intersect in the $\xi_1 = 0$ or $\xi_3 = 0$, but only in the $\xi_2 = 0$ plane, because of the assumed condition that $v_1 > v_2 > v_3$. In the $\xi_2 = 0$ plane, there are, in general, four intersection points that can pairwise be joined by lines that pass through the origin of the ξ_1, ξ_3 coordinate plane. These lines are often referred to as the optic axes. They are, indeed, symmetry axes, as, for propagation along these directions, the phase velocity becomes independent of polarization. Now, earlier, we discussed how equation (4-49) was actually of the form of equation (4-53), and, therefore, for each distinct polarization solution v_p^2 there existed a distinct polarization direction. For the optic axes directions, however, the eigenvalues of (4-53) become degenerate so that the eigenvectors are no longer unique, and, in fact, the polarization directions can

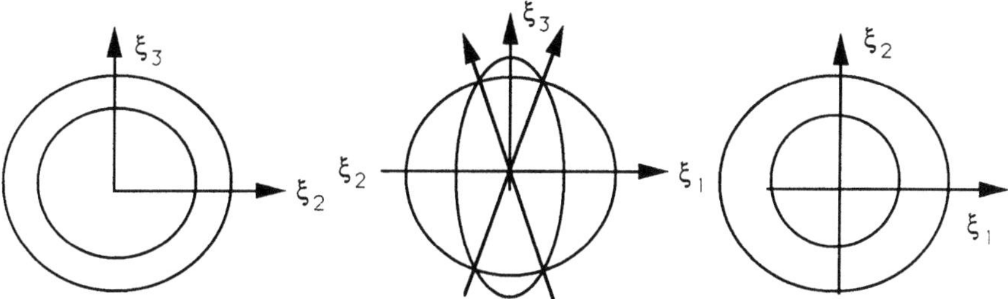

FIGURE 4.3.   The intersections of the normal surface with each of the planes (a) $\xi_1 = 0$, (b) $\xi_2 = 0$, (c) $\xi_3 = 0$.

cover the whole cross-sectional plane. After the discussion of the ray surface, we will see that this degeneracy has some quite interesting consequences.

Now, as we know from the presentation of the last section of Chapter 3 as well as the results of (4-19) and (4-29), the phase velocity is by no means the whole story. Any form of dispersion will cause the phase and group velocities to differ, as any loss will cause the group (information) and energy velocities to differ. As in the present anisotropic case we are assuming lossless propagation, it is probably safe to assume that the group and energy velocities are the same. However, as the **k** vector differs from the Poynting vector, it is clear that the group and phase velocities still differ in direction, if not magnitude, and what will be presented next is a construction, analogous to the wave normal surface, that allows one to visualize the relation between the ray direction (that is, the Poynting vector direction) and the group velocity.

As was seen in Figure 4.2, the **E** and **D** vectors both lie in a plane perpendicular to the directions of **B** and **H**. As the unit normal in the direction of the **k** vector is given by

$$\hat{\mathbf{e}}_k = \frac{\mathbf{D} \times \mathbf{B}}{|\mathbf{D} \times \mathbf{B}|} \tag{4-62}$$

and the unit vector in the direction of the Poynting vector $\hat{\mathbf{e}}_s$, is given by

$$\hat{\mathbf{e}}_s = \frac{\mathbf{E} \times \mathbf{H}}{|\mathbf{E} \times \mathbf{H}|} \tag{4-63}$$

then both of these vectors must also lie in the plane defined by **E** and **D**. If one were, therefore, to define $\mathbf{D}_1$ as the part of **D** perpendicular to the ray direction, $\hat{\mathbf{e}}_s$, one could write

$$\mathbf{D}_1 = \mathbf{D} - \hat{\mathbf{e}}_s(\hat{\mathbf{e}}_s \cdot \mathbf{D}) \tag{4-64}$$

Relating $\mathbf{D}_1$ to $\mathbf{E}$ is the problem at hand. In an isotropic medium where $\mathbf{D}$ and $\mathbf{E}$ are parallel, the proportionality is just the dielectric constant. The dielectric constant, however, is also given by

$$\varepsilon = \frac{(\sqrt{\varepsilon_0 \mu_0})^2}{\mu_0} = \frac{1}{\mu_0 v_p^2} \tag{4-65}$$

In the case of an anisotropic medium, one can deduce from (4-44) that the relation between $\mathbf{E}_1$, as defined by

$$\mathbf{E}_1 = \mathbf{E} - \hat{\mathbf{e}}_k(\hat{\mathbf{e}}_k \cdot \mathbf{E}) \tag{4-66}$$

and $\mathbf{D}$ is given by

$$\mathbf{E}_1 = \mu_0 v_p^2 \mathbf{D} \tag{4-67}$$

To relate $\mathbf{D}_1$ to $\mathbf{E}$, however, v_p cannot be the correct velocity to use. One may recall that in an isotropic medium

$$v_g = \frac{1}{(\partial k / \partial \omega)} \tag{4-68}$$

and that

$$k = \frac{\omega n(\omega)}{c} \tag{4-69}$$

which give that

$$v_g = \frac{c}{n_g} \tag{4-70}$$

where n_g is the group index, given by

$$n_g(\omega) = n(\omega) + \omega \frac{\partial n}{\partial \omega} \tag{4-71}$$

In the isotropic medium, n_g reduces to $n_p = n$ in the limit of dispersionless propagation. In the anisotropic medium, this cannot be the case because of the different directions of the two velocities. However, one can use the relation between $\mathbf{D}_1$ and $\mathbf{E}$ as a defining equation for v_g, by writing

$$\mathbf{E} = \mu_0 v_g^2 \mathbf{D}_1 \tag{4-72}$$

as v_g is just the velocity at which the triad involving $\mathbf{E}$, $\mathbf{H}$, and $\mathbf{S}$ is moving, in direct analog to v_p being the velocity of the triad $\mathbf{D}$, $\mathbf{B}$, and $\mathbf{k}$.

With the substitution of (4-72) and the use of a principal axis coordinate system, one can rewrite (4-64) in the form

$$\frac{E_i}{\mu_0 v_g^2} = \varepsilon_i E_i - e_i (\hat{\mathbf{e}}_s \cdot \mathbf{D}) \tag{4-73}$$

Equation (4-73) is evidently just the so-called dual equation to equation (4-47). Steps analogous to those that led from (4-47) to (4-51) lead to the ray normal equation

$$\sum \frac{s_i^2}{\dfrac{1}{v_g^2} - \dfrac{1}{v_i^2}} = 0 \tag{4-74}$$

where the v_i are the principal velocities, given as the lengths of the major axes of Fresnel's ellipsoid.

As was done with the wave normal equation, one can define a ray normal surface to attempt to visualize what the solutions to the ray normal equation may look like. To do this, one defines coordinates

$$\xi_i = s_i v_g \tag{4-75}$$

to convert (4-75) to the form

$$\sum \frac{\xi_i^2 v_i^2}{v_i^2 - v_g^2} = 0 \tag{4-76}$$

which can be written out in the form

$$v_g^4(v_1^2 \xi_1^2 + v_2^2 \xi_2^2 + v_3^2 \xi_3^2) - v_g^2\{\xi_1^2 v_1^2(v_2^2 + v_3^2) + \xi_2^2 v_2^2(v_1^2 + v_3^2)$$

$$+ \xi_3^2 v_3^2(v_1^2 + v_2^2)\} + v_1^2 v_2^2 v_3^2(\xi_1^2 + \xi_2^2 + \xi_3^2) = 0 \tag{4-77}$$

where the identity

$$\sum_i \xi_i^2 = v_g^2 \tag{4-78}$$

was used to reduce (4-77) to a fourth-order equation. Equation (4-78) is the defining equation for the so-called ray surface.

As we did with the normal surface, we will try to understand the characteristics of the ray surface by investigating the shape of its intersection with each of the planes $\xi_1 = 0$, $\xi_2 = 0$, and $\xi_3 = 0$. The intersection with the plane $\xi_1 = 0$ is best analyzed with reference to equation (4-76) and with the use of relation $\xi_2^2 + \xi_3^2 = v_p^2$ for $\xi_2 = 0$. In analogous fashion to the results of (4-60) and (4-61), the resulting two sheets of the fourth-order ray normal surface are found to be described by

$$\xi_2^2 + \xi_3^2 = v_1^2 \tag{4-79}$$

and

$$\frac{\xi_2^2}{v_2^2} + \frac{\xi_3^2}{v_3^2} = 1 \tag{4-80}$$

If, as in the case of wave normal surface, we further assume that $v_1 \geq v_2 \geq v_3$, we find a situation analogous to the one depicted in Figure 4.3, except that ovals depicted in that figure are replaced by the ellipses described by equation (4-80) and its permutations. As before, it is found that only for the intersection with $\xi_2 = 0$ do the two sheets of the fourth-order surface defined by (4-77) cross. These intersection points define ray optic axes, which, as with the optic axes discussed in relation to the wave normal surface, represent propagation directions in which all the possible linear polarization states share the same velocity; and in the ray optic axis case, that velocity is a group velocity. The ray optic axis and the optic axis do not have to coincide, and in a low-symmetry crystal, do not. The fact that there is no preferred polarization direction, thus meaning that there are an infinite number of directions, means that there are a corresponding infinite number of directions for **D**, as defined by equation (4-47). The converse situation holds for the optic axis, with the corresponding infinite number of allowed directions for **D** and the **E** associated with the given **D** through equation (4-72). There will be more about this in the next section, where the effects of internal and external conical refraction will be discussed.

4.5 SOME PROPAGATION EFFECTS
IN CRYSTALS

In the present section, we wish to discuss some of the salient features of wave propagation in media that contain anisotropic slabs. First, the different types of materials will be discussed, at least to the extent of mentioning the differences between isotropic, uniaxial, and biaxial media, with reference made to the literature, which discusses crystal classification in much greater detail. Then attention will be given to the simplest of boundary value problems, that of a plane wave incident from a homogeneous isotropic medium onto a finite slab of the anisotropic material. What happens in the case of an isotropic slab will be reviewed briefly to develop a frame of reference. Transmission of the slab then will be discussed when the incident direction is the optic axis, as well as the practical case of when the incidence is normal to both the surface and the optic axis. The normal-incidence case then leads us to a consideration of the effect of double refraction. We go on to demonstrate that propagation in biaxial crystals is never so simple, as even in the case of normal incidence in an optic axis direction either external or internal conical refraction is observed, depending on whether the propagation direction is along an optic axis or a ray optic axis.

We will here take a very pedestrian view of the world of crystal symmetry, if only because optical propagation is really a very poor probe of internal crystal structure. As can be read in any number of books on crystals, (Azároff and Brophy 1963, Chapter 2; Kittell 1971, Chapter 1; Juretschke 1974, Chapter 1) crystal symmetry is a very rich field. The seven definable crystal systems (cubic, tetragonal, hexagonal, triagonal, orthorhombic, monoclinic, and triclinic) are made up of 32 different crystal classes, which are identified with 32 different point groups that in turn could be classified into 230 different space groups. However, as we have seen in this chapter, if we ignore chiral symmetries (which are not present in the 230 space groups), a succinct description of optical propagation can be given by Fresnel's ellipsoid, in which the described dielectric tensor contains only three elements. This description, however, is not at odds with Neumann's principle (Nye 1957, Chapter 1 and Appendix 3), which states that the symmetry elements of any physical property of a crystal must include the symmetry elements of the point group of the crystal. In the case of optical propagation properties, the crystals exhibit a higher form of symmetry that thereby includes all of the elements of the more elemental symmetries. It is with these higher symmetries that the seven crystal systems are reduced to just three cases: the case in which there is but one independent constant (ellipsoid degenerates to a sphere), the case of two independent constants (ellipsoid of revolution), or the case where all three constants are different (general ellipsoid). Cubic crystals form the isotropic, single-constant case; tetragonal, hexagonal, and trigonal crystals form the two-constant, so-called uniaxial case; and orthorhombic, monoclinic, and triclinic crystals form the third optical class of biaxial crystals.

Although we have already analyzed in some detail what happens in the case of a linearly polarized plane wave incident on a slab of isotropic material, let us review these results in order to build a framework with which to analyze the more exotic uniaxial and biaxial cases. The situation is as depicted in Figure 4.4. Without any great loss of generality, one can assume that incident and

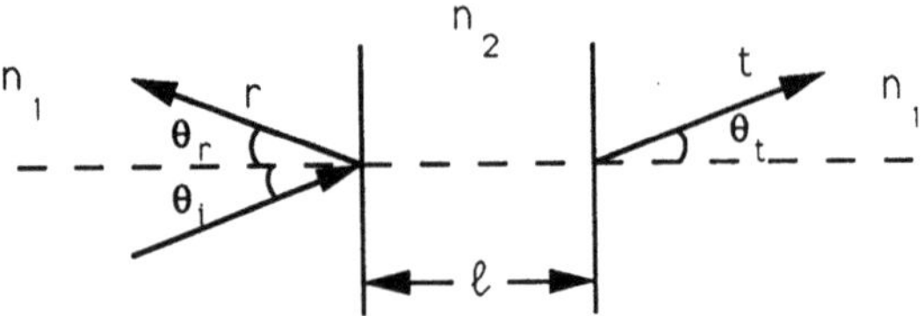

FIGURE 4.4.   A schematic depiction of the boundary value problem, which we will analyze for isotropic, uniaxial, and biaxial media in this section.

transmitted media have the same index. Essentially the problem we are analyzing here is that of a beam splitter, and in the following paragraph successively more exotic beam splitters. Now, from Section 2.6, we know how to solve such a boundary value problem. Indeed, for a TE polarization incident, the solution for the reflection and transmission can be given by equation (2-164):

$$r = \frac{(m_{11} + n_1 \cos \theta_i m_{12})n_1 \cos \theta_i - (m_{21} + m_{22}n_1 \cos \theta_i)}{(m_{11} + n_1 \cos \theta_i m_{11})n_1 \cos \theta_i + (m_{21} + n_1 \cos \theta_i m_{22})} \quad \text{(a)}$$

$$t = \frac{2n_1 \cos \theta_i}{(m_{11} + n_1 \cos \theta_i m_{12})n_1 \cos \theta_i + (m_{21} + n_1 \cos \theta_i m_{22})} \quad \text{(b)}$$

(4-81)

where the m_{ij}'s are the elements of the characteristic matrix M of the sandwiched medium, which in our present case of an isotropic slab, are given by

$$M = \begin{bmatrix} \cos (k_s l \cos \theta_2) & \dfrac{i}{n_2 \cos \theta_2} \sin (k_2 l \cos \theta_2) \\ -i\eta_2 \cos \theta_2 \sin (k_2 l \cos \theta_2) & \cos (k_2 l \cos \theta_2) \end{bmatrix} \quad (4\text{-}82)$$

where θ_2 is related to θ_i by Snell's law.

Perhaps the most important salient feature of the propagation problem depicted in Figure 4.4 is that the transmitted wave exits the slab at the same angle at which it entered, but with a slightly (or maybe greatly) reduced amplitude because of the wave that is reflected back from the front interface. The details differ with respect to reflected amplitude for the TE and TM cases, so the reflected and transmitted polarization states will differ from an arbitrary incident one, but the phase and energy propagation directions are not affected by the plate.

Now, let us consider that the plate of Figure 4.4 is made up of a uniaxial medium, that is, one in which there are two different dielectric constants in the principal axis system. Thus there exists an optic axis, that is, an axis that is transverse to the two equal index principal axes. This means that a plane wave propagating along this axis has no preferential polarization direction, as the indices of both of the transverse states are the same. Now, further, as the crystal is uniaxial, symmetry is also going to require that the optical ray axis be identical to the optic axis. This symmetry can hold (although it does not necessarily) even in highly dispersive media, as the symmetry between the directions transverse to the optic axis can remain symmetric with wavelength, and thus wavelength-dependent derivatives such as in (4-71) can also retain symmetry.

The existence of the optic axis greatly simplifies a large number of propagation problems, but by no means all of them. Now, as it did in the isotropic case, and as it will in the biaxial case, Snell's law

$$n_1 \sin \theta_i = n_2 \sin \theta_2 \tag{4-83}$$

will still hold. The problem in the anisotropic case is that we do not know n_2 without knowing both θ_2 and the details of the incident polarization. However, in the case where θ_i is picked such that θ_2 is the optic axis and n_2 the index transverse to the optic axis, then the problem reduces to the isotropic case. That is, both the TE and the TM incident waves will be refracted to propagate, in the crystal, down the optic axis. The reflection and transmission will be given in equations (4-81) and their TM counterparts.

For angles that do not satisfy the conditions discussed in the last paragraph, the result will exhibit an effect known as double refraction. To be definite, let us consider the situation illustrated in Figure 4.5, where a TE-polarized wave is incident at angle θ_i on a medium where the optic axis is parallel to the surface, but oriented at an angle θ to the incident polarization direction.

The incident wave is assumed to be TE-polarized and incident at an angle θ_i, whereas the optic axis is parallel to the surface but oriented at an angle θ to the y (TE polarization) axis. Now, it can be of interest to plot the intersection of the normal surface with the plane $z = 0$, which is done in Figure 4.6 for a somewhat special θ, where the so-called extraordinary index maxima lie along

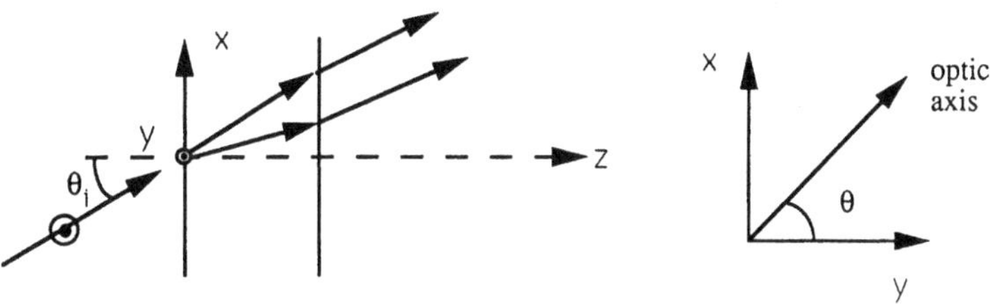

FIGURE 4.5.   Schematic depiction of the situation where a TE-polarized wave is incident at an angle θ_i on a medium where the optic axis is parallel to the surface, but oriented at an angle θ to the incident polarization direction.

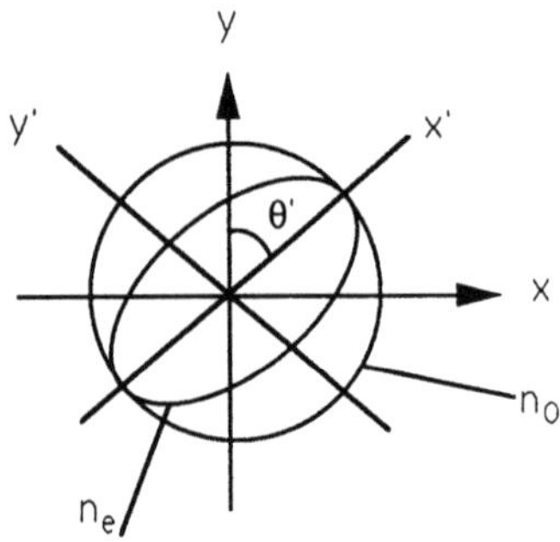

FIGURE 4.6.   Sketch of the normal surface for a uniaxial crystal, with the optic axis oriented at an angle θ to the incident polarization direction.

the X' axis and its minima along the Y' axis. It is assumed that the optic axis lies at an angle θ from the X axis such that the X'–Y' coordinate system is rotated from the X–Y system by the angle θ. We note that, in general, the incident TE wave will see two different indices n_e and n_0, depending on the relative orientations θ_i and θ. For this reason, there will be two Snell's law relations

$$n_1 \sin \theta_i = n_e \sin \theta_{2e} \qquad \text{(a)}$$
$$\text{(4-84)}$$
$$n_1 \sin \theta_i = n_0 \sin \theta_{20} \qquad \text{(b)}$$

for the two possible propagation angles in the material, which correspond to two polarization directions. At the second interface, the plane waves will return to their incident propagation angles. The details of the boundary value problem can be solved by mapping the incident polarization state into the two inside the crystal and solving the two problems separately, each problem having been previously solved in Chapter 2. The most important feature of the solution, however, will be the splitting that occurred at the surface. Although we have assumed plane wave incidence, if the incident wave were actually a pencil-like beam (needless to say of many wavelengths in cross section), then the beam would actually be split in two. Conversely, if one were to place a slab of calcite over a sample of writing, one would generally see the writing double when viewed through the calcite, as one's brain would extrapolate the separated images as coming from different locations.

Now, if we pick a different plane for the optic axis on a mixed TE–TM input polarization state, the problem solution will be qualitatively the same although the details of affecting the solution can become more complex. The solution will consist of mapping the incident polarization state into two polarization states in the crystal, each corresponding to a different propagation direction in the crystal, with that propagation direction being determined by the condition that Snell's law must be satisfied for the incident direction. The complexity arises from the fact that the polarization state is determined by the direction of propagation, which in turn is determined by the index of refraction, which is determined by the polarization state and the direction of propagation. Iterative or graphical solutions are necessary. But perhaps the main point is that the problem can be reduced to a boundary value problem, whose solution we studied in Chapter 2.

Now, in the case of a biaxial medium, that is, one in which all the principal indices are different, for incidence at angles different from the optic axis or optical ray axis directions, the qualitative picture is not too different from that described above for uniaxial media. The incident direction and the polarization state will be mapped onto two different directions and polarization states inside the crystal. But an interesting difference between the uniaxial and biaxial crystal is that in the biaxial crystal the optic axes and optical ray axes need not be in the same directions. This can have a dramatic effect on unpolarized incident light, leading to the effects known as internal and external conical refraction.

When unpolarized light is incident on a biaxial crystalline slab at an angle (see Figure 4.4) such that the wave direction will be refracted into the direction of an optic axis, then internal conical refraction will occur. The situation is as depicted in Figure 4.7. The point is that although the phase velocity direction is unique inside the crystal, the group velocity is not. For each polarization state, it is necessary to solve equation (4-47) to determine the corresponding direction of $\mathbf{E}$. It can be shown that (Born and Wolf 1975, Section 14.3.4; Yariv and Yeh 1984, Section 4.8) the corresponding $\mathbf{E}$ vector directions form a cone. The energy, therefore, while still in the crystal, propagates outward from the phase velocity direction on the surface of this cone. Now, as the phase velocity

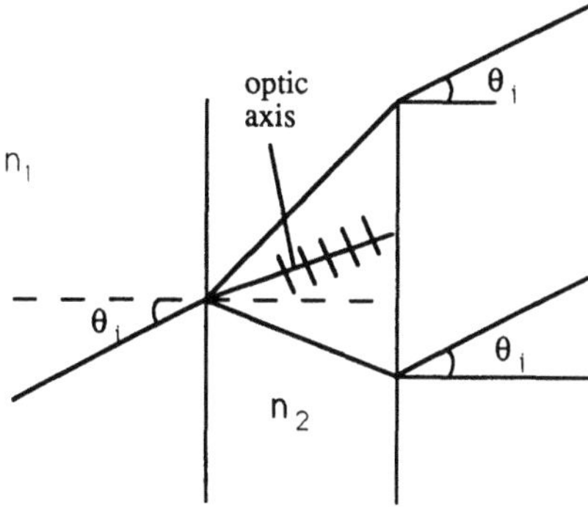

FIGURE 4.7.   A schematic depiction of what occurs in internal conical refraction, that is, when an incident unpolarized wave is incident at such an angle that its propagation direction is along the optic axis inside the crystal.

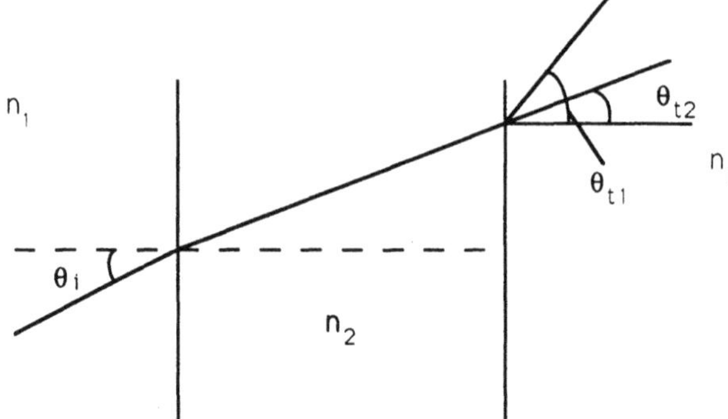

FIGURE 4.8.   Schematic depiction of the phenomenon that occurs when an unpolarized wave is incident on a slab at an angle such that its group velocity direction inside the crystal is identical with that of an optical ray axis.

direction is unchanged because of the internal conical refraction, the wave returns to its initial propagation angle upon exiting the crystal. However, if the incident wave were actually a pencil-type ray and the crystal were long enough, then the exiting beam could form the surface of a cone—hence the designation conical refraction.

When unpolarized light is incident on a biaxial crystalline slab at an angle (see Figure 4.4) such that the ray (group velocity) direction will be refracted into the direction of an optical ray axis, then external conical refraction will occur. The situation is the dual of that discussed in the previous paragraph, as schematically depicted in Figure 4.8. Although the direction of the group velocity is well determined inside the crystal, one must solve equation (4-72) to find the direction of **D** and therefore the phase velocity. As it turns out, the phase velocity directions form the surface of a cone. Therefore, if one were to draw the intersection of a plane with the propagation picture, as has been done in Figure 4.8, one would see that there are two different phase velocity directions (in three dimensions an infinite number) associated with the group velocity direction. The wave, however, does not conically refract, per se, until it reaches the output facet. In this external conical refraction case, unlike the case of the previous internal conical one, the phase velocity directions have been perturbed, and, therefore, the initial propagation direction will not be restored after the second interface. The rays thus will propagate out into the second medium along a set of directions defined by the surface of a cone.

4.6 SOME POLARIZATION DEVICES

In this section, we will investigate some devices based on slabs of lossless, nonchiral, anisotropic media that affect the incident polarization. Here it will be assumed that the slab is placed in a configuration such as that depicted in Figure 4.4, with the incident angle set at $\theta_i = 0$. The devices involved here are the phase retarder and the polarization rotator, as based on the adiabatic rotation of the polarization axis.

The phase retarder was discussed in Chapter 2, to the extent that the Mueller and Jones matrices were found given the effect that the phase retarder had on an incident complex polarization vector. Here we wish to concentrate more on the actual propagation effect. Now, the basic idea is that, whether the crystal be uniaxial or biaxial, it is to be cut so that the optic axis lies perpendicular to the crystal surface and, therefore, in the direction that one of the two possible incident polarization states can take. For maximal effect, one would want the normally incident wave to be polarized at 45° to the optic axis, as is illustrated in Figure 4.9. The point is that the incident state will be split into equal X and Y components, will then propagate a distance l in the crystal, and will emerge in a new state, for which the **E** field can be expressed in the form

$$\mathbf{E}_{\text{out}}(l) = [a_x e^{ikn_x l}\hat{\mathbf{e}}_x + a_y e^{ikn_y l}\hat{\mathbf{e}}_y] \tag{4-85}$$

where the a_x and a_y are the (complex) incident amplitudes. The output state is in general elliptical, as is the incident state. It should, of course, be mentioned that the problem of propagation from the first to the third medium will in general require one to solve two boundary value problems to solve for the resulting amplitudes a_x and a_y transmitted. In most practical cases, however, the relation

$$\frac{|n_x - n_y|}{|n_x + n_y|} \ll 1 \tag{4-86}$$

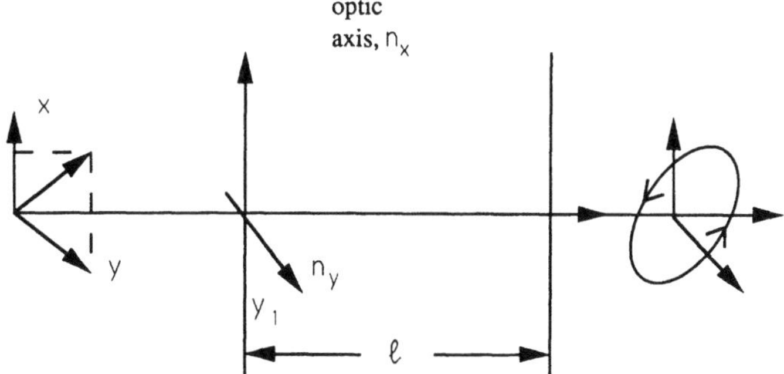

FIGURE 4.9.   An illustration of the direction of incident polarization relative to the optic axis, with a schematic depiction of the generally elliptical output polarization state of a phase retarder.

holds, and, therefore, the amplitudes will both be affected in approximately the same manner by the reflection from the slab. Now, in the case that

$$\frac{2\pi}{\lambda}(n_x - n_y)l = \frac{\pi}{2} \tag{4-87}$$

where it is tacitly assumed that $n_x > n_y$ (positive crystal), then the phase retarder becomes a quarter-wave plate in which the transformation of (4-85) becomes

$$\mathbf{E}_{\text{out}}(l) = e^{ikn_x l}[a_x\hat{\mathbf{e}}_x - ia_y\hat{\mathbf{e}}_y] \tag{4-88}$$

which is the transformation that takes linear, 45° states to circular states and vice versa, as was discussed in Chapter 2.

Now, an archetypical polarization rotator is a nematic liquid crystal; so we will consider the propagation of polarized light through a twisted nematic liquid crystal. Using Jones calculus, this can be described as a continuum of wave-plates with rotation $d\varphi/dz = \alpha$ and retardence $d\gamma/dz = k_0(n_e - n_0)$ per unit length. The transmission through a unit length of the crystal at position z can be written

$$\mathbf{E}(z + \Delta z) = R(-\varphi)R(-\alpha\Delta z)W\left(\frac{d\gamma}{dz}\Delta z\right)R(\alpha\Delta z)R(\varphi)\mathbf{E}(z) \tag{4-89}$$

or

$$[R(\varphi)\mathbf{E}(z + \Delta z)] = R(-\alpha\Delta z)W\left(\frac{d\gamma}{dz}\Delta z\right)R(\alpha\Delta z)[R(\varphi)\mathbf{E}(z)] \tag{4-90}$$

where $R(\cdot)$ is the rotation matrix and $W(\cdot)$ is the matrix for the retardation. Using the well-known expressions for R and W, and the fact that the rotation $\alpha\Delta z$ and the retardance $(d\gamma/dz)\Delta z$ are small, we find

$$R(\varphi)\mathbf{E}(z + \Delta z) = (I + T\Delta z)R(\varphi)\mathbf{E}(z) \tag{4-91}$$

where

$$T = \begin{bmatrix} -i\dfrac{d\gamma/dz}{2} & -\alpha \\[2ex] \alpha & i\dfrac{d\gamma/dz}{2} \end{bmatrix} \tag{4-92}$$

We can write this as (in the limit $\Delta z \to 0$)

$$\frac{d\,\boldsymbol{R}(\varphi)\mathbf{E}(z)}{dz} = \boldsymbol{TR}(\varphi)\mathbf{E}(z) \tag{4-93}$$

Solving the differential equation results in

$$\boldsymbol{R}(\varphi(z))\mathbf{E}(z) = e^{Tz}\boldsymbol{R}(\varphi(0))\mathbf{E}(0) \tag{4-94}$$

and the Jones matrix follows as

$$\boldsymbol{M} = \boldsymbol{R}(-\varphi(z))e^{Tz}\boldsymbol{R}(\varphi(0)) \tag{4-95}$$

The matrix e^{Tz} can be evaluated by expanding the exponential

$$e^{Tz} = 1 + \boldsymbol{T}z + \tfrac{1}{2}(\boldsymbol{T}z)^2 + \cdots \tag{4-96}$$

Using the properties that

$$Z^2T^2 = \left[\left(\frac{\gamma(z)}{2}\right)^2 + \varphi(z)\right]I = X^2I \tag{4-97}$$

where $\gamma(z) = (1/z)(\partial\gamma/\partial z)$ and $\varphi(z) = (1/z)(d\varphi/dz)$, and we have defined the new quantity X, we find

$$e^{Tz} = \left(1 + \frac{1}{2}X^2 + \frac{1}{4!}X^4 + \cdots\right)I + \frac{\boldsymbol{T}z}{X}\left(X + \frac{1}{3!}X^3 + \cdots\right) \tag{4-98}$$

or

$$e^{Tz} = \cos X + \frac{\boldsymbol{T}z}{X}\sin X \tag{4-99}$$

The Jones matrix then follows as

$$\boldsymbol{M} = \boldsymbol{R}(-\varphi)\begin{bmatrix} \cos X - i\,\dfrac{\gamma}{2}\,\dfrac{\sin X}{X} & -\varphi\,\dfrac{\sin X}{X} \\[2mm] \varphi\,\dfrac{\sin X}{X} & \cos X + i\,\dfrac{\gamma}{2}\,\dfrac{\sin X}{X} \end{bmatrix} \tag{4-100}$$

where we have set $R(0) = I$. If $\gamma \gg \varphi$, we find

$$M \approx R(-\varphi) \begin{bmatrix} e^{-i(\gamma/2)} & 0 \\ 0 & e^{i(\gamma/2)} \end{bmatrix} \tag{4-101}$$

That is, if the input polarization is along one of the crystal axes in the input plane, the polarization will rotate by an angle φ.

References

Azároff, L. V. and J. J. Brophy, *Electronic Processes in Materials*, McGraw-Hill, New York (1963).

Born, M. and E. Wolf, *Principles of Optics*, Fifth edition, Pergamon Press, New York (1975).

Jackson, J. D., *Classical Electrodynamics*, Second edition, John Wiley and Sons, New York (1975).

Juretschke, H. J., *Crystal Physics; Macroscopic Physics of Anisotropic Solids*, Benjamin, Reading, MA (1974).

Kittell, C., *Introduction to Solid State Physics*, John Wiley and Sons, New York (1971).

Landau, L. D. and E. M. Lifshitz, *Electromagnetics of Continuous Media*, Pergamon Press, Oxford (1960).

Noble, B., *Applied Linear Algebra*, Prentice-Hall, Englewood Cliffs, NJ (1969).

Nye, J. F., *Physical Properties of Crystals*, Clarendon Press, Oxford (1985).

Papas, C. H., *Theory of Electromagnetic Wave Propagation*, Dover Publications Inc., New York (1988).

Sommerfeld, A., *Optics*, Volume 4 of the *Lectures on Theoretical Physics*, Academic Press, New York (1972).

Yariv, A. and P. Yeh, *Optical Waves in Crystals*, John Wiley and Sons, New York (1984).

Yeh, P., *Optical Waves in Layered Media*, John Wiley and Sons, New York (1988).

Problems

1. Consider an anisotropic medium that is a uniaxial medium whose permitivity E takes the form

$$\mathbf{E} = \begin{bmatrix} \varepsilon_x & 0 & 0 \\ 0 & \varepsilon & 0 \\ 0 & 0 & \varepsilon \end{bmatrix}$$

Find the associated wave equations for such media and their solutions.

2. Here we wish to consider the polarization eigenstates of z-directed plane waves in homogeneous anisotropic media. As for homogeneous isotropic media, we a priori know that z-directed plane wave solutions exist; we need only describe the media

by their transverse refractive index tensors

$$\varepsilon_t = \begin{bmatrix} \varepsilon_{xx} & \varepsilon_{xy} \\ \varepsilon_{yx} & \varepsilon_{yy} \end{bmatrix}$$

Find the form of the wave equation and write down the polarization eigenstates for the media described by the following transverse tensors:

$$\text{(a)} \quad \varepsilon_t = \begin{bmatrix} \varepsilon_{xx} & 0 \\ 0 & \varepsilon_{yy} \end{bmatrix} \qquad \text{(b)} \quad \varepsilon_t = \varepsilon_s \begin{bmatrix} 1 & 1 \\ 1 & 1 \end{bmatrix}$$

$$\text{(c)} \quad \varepsilon_t = \varepsilon_s \begin{bmatrix} 1 & -i \\ i & 1 \end{bmatrix} \qquad \text{(d)} \quad \varepsilon_t = \varepsilon_s \begin{bmatrix} 1 & i \\ i & 1 \end{bmatrix}$$

3. For the following anisotropic and inhomogeneous ε's, find a wave-equation-like form (i.e., something like $(\nabla^2 + k^2)\mathbf{E} + \text{junk} = 0$, where you are to explicitly find the junk) and discuss the nature of the solutions, giving attention to such matters as whether there are z-directed plane-wave-like solutions, and if there are, what the polarization states are, etc.

$$\text{(a)} \quad \varepsilon = \begin{bmatrix} \varepsilon_{xx} & 0 & 0 \\ 0 & \varepsilon_{yy} & 0 \\ 0 & 0 & \varepsilon_{zz} \end{bmatrix} \quad \text{independent of coordinate.}$$

$$\text{(b)} \quad \varepsilon = \begin{bmatrix} \varepsilon_{xx} & \varepsilon_{xy} & 0 \\ \varepsilon_{yx} & \varepsilon_{yy} & 0 \\ 0 & 0 & \varepsilon_{zz} \end{bmatrix} \quad \text{independent of coordinate.}$$

(c) $\varepsilon = \varepsilon(z)$.

(d) $\varepsilon = \varepsilon(x)$.

4. Consider the complex representation of a plane wave

$$\mathbf{E}_\omega(\mathbf{r}) = E_1 \hat{\mathbf{e}}_1 + E_2 \hat{\mathbf{e}}_2$$

which is propagating in an anisotropic medium.

(a) Find expressions for E_x and E_y in a birefringent medium, where $\hat{\mathbf{e}}_1 = \hat{\mathbf{e}}_x$ and $\hat{\mathbf{e}}_2 = \hat{\mathbf{e}}_y$, and the y component is progressively retarded with respect to the x component. Sketch the progression of the polarization state for several values of z, where $E_x(z=0) = a_x = a_0$ and $E_y(z=0) = a_y = a_0/2$.

(b) Repeat (a) but for a chiral medium in which $\hat{\mathbf{e}}_1 = \hat{\mathbf{e}}_x + i\hat{\mathbf{e}}_y$ and $\hat{\mathbf{e}}_2 = \hat{\mathbf{e}}_x - i\hat{\mathbf{e}}_y$.

5. Consider a medium in which the amplitudes of the two electric field states propagate according to the law

$$a_y(z,\, t) = a_y(0,\, t)e^{-\gamma z}$$

$$a_y(z,\, t) = a_y(0,\, t)$$

and the phases according to the law

$$\varphi(t) = \varphi(0) + \delta z$$

where γ and δ are given (positive) constants.

(a) Find a Mueller matrix that describes propagation in this medium.

(b) If the incident light is unpolarized, how long would the medium have to be to have linear polarized light, with less than -40 dB cross polarization, exit the material?

(c) Sketch the evolution of the polarization ellipse of a linearly polarized wave at $45°$, while traveling through such a medium, where $\delta \sim \gamma$.

6. Consider propagation along the z-axis in a crystal which extends from $z = 0$ to $z \to \infty$. The crystal is uniaxial and has $\varepsilon_x \neq \varepsilon_y$. Find expressions for the evolution of the fields and Stokes parameters for the following normally incident fields, assuming that measurement of fields and Stokes parameters are made outside the crystal:

(a) x-Polarized monochromatic light.

(b) $45°$ polarized monochromatic light.

(c) Unpolarized light.

(d) Unpolarized light passing through a $45°$ polarizer at the input.

7. In much of the work in Chapter 3, our model for dispersion was based on a two-level system operating in the classical regime, that is, with no depletion at all of the lower state. A natural extension to this description would be to consider an N-level system, where transitions from the ground state to each of the excited states need to be considered. The composite spring motion $x(t)$ to be used in calculation of polarizability would therefore be the sum of individual x_i's of each of the transition springs, such that

$$x(t) = \sum_{i=1}^{N} x_i(t)$$

What we wish to consider here, however, is a three-level system that is kept at a high-enough temperature that we must take into account a finite population in the first excited level, but a system in which ground state to second excited state transition is forbidden. What we have then is a coupled oscillator problem in which the independent oscillators $x(t)$ and $x_2(t)$ of the two transition will have different masses, damping γ, and frequencies ω_i, and are coupled through terms we will call ω_{xy}^2 and ω_{yx}^2.

(a) Write down the equations of motion of the springs.

(b) Assuming monochromatic excitation, find the steady state form of $x(\omega)$.

(c) Sketch the dielectric constant of this material as a function of frequency.

(d) Try to make realistic estimates of the values of the parameters in the equations of (a) from quantum-mechanical arguments.

8. Consider a medium whose microscopic polarizability can be described by an anisotropic oscillator; that is, a wave with transverse polarization propagating in the z-direction will have different polarizabilities α_x and α_y. The system is operated at a particular optical frequency that is near the resonance of α_y but well below the resonance for α_x.

   (a) Find the index for both polarizations x and y.

   (b) How far will a 45° polarized wave have to propagate in order to have the x-polarized component exceed the y-polarized component by 3 dB in power, taking $\dfrac{N\alpha_x}{\varepsilon_0} \sim \dfrac{3}{2}$, $\dfrac{N\alpha_y}{\varepsilon_0} \sim -150i$?

9. Consider a z-propagating plane wave in a medium described by the refractive index tensor $\mathbf{n} = \begin{bmatrix} \hat{n}_x & 0 \\ 0 & \hat{n}_y \end{bmatrix}$ where $\hat{n}_x$ and $\hat{n}_y$ are in general complex quantities, and are related to the complex dielectric constants by

$$\hat{n}_x = \sqrt{\frac{\hat{\varepsilon}_x}{\varepsilon_0}} \quad \text{and} \quad \hat{n}_y = \sqrt{\frac{\hat{\varepsilon}_y}{\varepsilon_0}}$$

where the dielectric constants can be related to microscopic polarizability by

$$\hat{\varepsilon}_x = \varepsilon_0 \frac{1 + \frac{2}{3}N\alpha_x}{1 - \frac{1}{3}N\alpha_x} \quad \text{and} \quad \hat{\varepsilon}_y = \varepsilon_0 \frac{1 + \frac{2}{3}N\alpha_y}{1 - \frac{1}{3}N\alpha_y}$$

In general the field can be expressed in the form

$$\mathbf{E} = [a_x(z)\hat{\mathbf{e}}_x + a_y(z)e^{i\delta(z)}\hat{\mathbf{e}}_y]e^{i\varphi(z)}$$

   (a) Find the $a_x(z)$, $a_y(z)$, and $\delta(z)$ in terms of $a_x(0)$, $a_y(0)$, and $\delta(0)$, where $z = 0$ is the boundary of the medium.

   (b) Find $\hat{n}_x$ and $\hat{n}_y$ in the limits where

   (i) $\alpha_{ix}, \alpha_{iy} \to 0$, $N\alpha_{rx}, N\alpha_{ry} \ll 1$.

   (ii) $\alpha_{rx} \ll \alpha_{ix}$, $\alpha_{ry} \ll \alpha_{iy}$, $N\alpha_{ix}, N\alpha_{iy} \gg 1$.

   (c) Find the Stokes vector at the plane z in the limits stated in (b).

10. Consider a medium that can be modeled as an anisotropic harmonic oscillator, that is, such that a wave with transverse polarization propagating in the z-direction will "see" different spring constants, ω_x and ω_y, in the x and y directions. As usual, take the friction to be γ and the mass and the charge m and $-e$, respectively.

   (a) Say that we drive the system with an optical frequency ω that matches the ω_y but is far below the ω_x. If $\dfrac{N\alpha_x}{\varepsilon_0} \sim \dfrac{3}{2}$ and $\left|\dfrac{N\alpha_y}{\varepsilon_0}\right| \sim 100$, how far will a 45° polarized wave have to propagate before the magnitude of its x-polarized component exceeds that of the y-polarized by 40 dB (in power carried)?

(b) Find a Mueller-Stokes matrix that describes the medium described in (a). Leave the distance l propagated into the medium as a parameter in the matrix.

(c) Say that we now drive the medium with an optical frequency such that $\dfrac{N\alpha_x}{\varepsilon_0} \sim \dfrac{3}{2}$ and $\dfrac{N\alpha_y}{\varepsilon_0} \sim 1 + i$. Find the two principal refractive indices.

(d) Find a Mueller-Stokes matrix to describe propagation into a medium such as in (c). Leave the propagation distance l into the medium as a parameter in the matrix.

11. Consider a medium whose microscopic polarizability can be described by a harmonic oscillator (read as spring) model, where the mass is m, charge $-e$, and damping γ, but the natural spring frequency ω_0^2 can be modified by a D.C. field at right angles (z-direction) to the wave's polarization vector (in the x-y plane). This relation is expressible as $\omega_0^2 = \omega_0^2(E_z = 0) + \Delta\omega^2 E$. The electro-optic coefficient, r, of such a medium can be defined by the expression $n = n_0 + rE_z$.

(a) Find an expression for the index of refraction, $n(\omega)$ of an isotropic medium of density N of such oscillators.

(b) Find the electro-optic coefficient for this medium.

(c) For an applied field $E_z = a \cos \Omega t$, where $\Omega \ll \omega$ (the applied optical frequency), find an expression for the molecular polarizability α. This expression should be in the form of a sum with coefficients defined recursively.

(d) Find the index of refraction $n(\omega, \Omega, t)$, for $\omega \gg \omega_0$ to first-order Ω/ω.

(e) What does the spectrum of a monochromatic wave entering such a medium become on exiting it?

12. Here we will consider the microscopic model of an electro-optic medium where the spring frequency ω_x in the x spring is given by $\omega_x^2 = \omega_{x0}^2 - rE_y$, where ω_{x0} is a constant, r is an electro-optic coefficient, E_y is the y-directed field, and $\omega_y^2 \neq \omega_{x0}^2$ is independent of the applied field.

(a) What is the spring motion for an incident field of $\mathbf{E} = E_y \hat{\mathbf{e}}_y e^{i\omega_0 t} e^{ikz}$?

(b) What is the spring motion for an incident wave of $\mathbf{E} = E_0(\hat{\mathbf{e}}_x + \hat{\mathbf{e}}_y)e^{i\omega_0 t} e^{ikz}$? (*Hint:* The x motion can be expanded in the form $x(\omega) = \sigma x_n e^{-in\omega_0 t}$).

(c) In (b), what does the motion look like? Sketch it.

(d) Derive the n_x and n_y associated with case (b).

13. Consider the following optical arrangement: Say that a voltage V applied across the crystal (see Figure 4.10) causes a pure rotation of the a_x and a_y components of polarization, that is

$$\begin{pmatrix} a_x \\ a_y \end{pmatrix}_{\text{out}} = \begin{bmatrix} \cos\theta(V) & -\sin\theta(V) \\ \sin\theta(V) & \cos\theta(V) \end{bmatrix} = \begin{pmatrix} a_x \\ a_y \end{pmatrix}_{\text{in}}$$

where $\theta = aV$ and a is a known material constant. In the following we will consider $V = V_0 \cos\omega_m t$ where

$$t_m = \frac{2\pi}{\omega_m} \gg \tau_d \text{ and } aV_0 \ll 1.$$

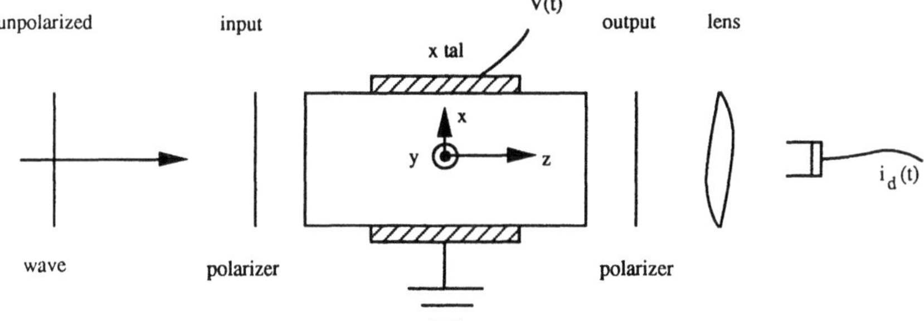

FIGURE 4.10.   Figure for problem 13.

In what follows we will always be interested in determining the V_0 and ω_m from the detector current $i_d(t)$.

(a) If the input polarizer is at $45°$ and the output is x-oriented, what is the detector current in terms of V_0 and ω_m?

(b) If the input and output polarizers are both at $45°$, what is the output current $i_d(t)$? Is it easier or harder to determine V_0 and ω_m from this arrangement or that of (a)?

(c) Repeat (b) for this case where the output polarizer is at $-45°$.

(d) What would happen if the input state were circularly polarized, and the input polarizer were removed? Is there enough information given in the problem to determine the answer?

14. The linear electro-optic effect is the change in the indices of the ordinary and extraordinary rays that is caused by and is proportional to an applied electric field. In general, in the presence of an electric field, the equation of the index ellipsoid can be expressed as

$$\left(\frac{1}{n^2}\right)_1 x^2 + \left(\frac{1}{n^2}\right)_2 y^2 + \left(\frac{1}{n^2}\right)_3 z^2 + 2\left(\frac{1}{n^2}\right)_4 yz + 2\left(\frac{1}{n^2}\right)_5 xz + \left(\frac{1}{n^2}\right)_6 xy = 1$$

The x, y, z can be chosen as parallel to the principal dielectric axes of the crystal, that is:

$$\left(\frac{1}{n^2}\right)_1 \Bigg|_{E=0} = \frac{1}{n_x^2} \quad \left(\frac{1}{n^2}\right)_2 \Bigg|_{E=0} = \frac{1}{n_y^2} \quad \left(\frac{1}{n^2}\right)_3 \Bigg|_{E=0} = \frac{1}{n_z^2} \quad \left(\frac{1}{n^2}\right)_4 \Bigg|_{E=0}$$

$$= \left(\frac{1}{n^2}\right)_5 \Bigg|_{E=0} = \left(\frac{1}{n^2}\right)_6 \Bigg|_{E=0} = 0$$

The linear change in the coefficients due to the applied electric field $\vec{E}(E_x, E_y, E_z)$ is defined by

where r_{ij} are elements of the electro-optic tensor. The crystal of potassium dihydrogen phosphate KH_2PO_4, also known as KDP, has non-zero elements of its electro-optic tensor: $r_{41} = r_{52}$, r_{63}, and $\left(\dfrac{1}{n^2}\right)_1 = \left(\dfrac{1}{n^2}\right)_2 = \dfrac{1}{n_0^2}$, $\left(\dfrac{1}{n^2}\right)_3 = \dfrac{1}{n_e^2}$. What is the effect of the applied electric field $\vec{E}$ parallel to z on a plane wave propagating in the z direction and originally polarized along the x direction?

5

Geometrical Optics

5.1 INTRODUCTION

In this chapter we will touch on some of the salient features of the geometrical optical theory. Geometrical optics is a mature and broadly applied discipline, and a thorough coverage of its applications would be voluminous. Lens design, for example, is practically exclusively within the realm of geometrical optics. As less than a volume on any aspect of lens design would be doing that subdiscipline a disservice, the purpose here is not to give an overview of the applications of geometrical optics. The idea is rather to illuminate some of the theoretical structure underlying the geometrical optics approximation and thereby gain some insight into the kinds of problems that are amenable to geometrical analysis, as well as to, by analogy, show how the set of problems amenable to geometrical analysis relates to other problems in mechanics and other disciplines.

The chapter is organized as follows: The second section gives a brief review of the WKB approximation as well as Keller's multidimensional generalization of it, before discussing the basic approximation of geometrical optics. The goal is to allow the reader to be able to use the crutch of understanding the simple scalar one-dimensional WKB theory to help in grasping the fundamental limitations of the much richer three-dimensional vector geometrical optics theory. (The WKB approach will appear again in discussions of coupled mode equations in Chapter 7.) The third section includes details of the derivation of the eikonal equation along with many of the important discussions unearthed by the derivation, including the structure of the local plane wave solutions. Some examples of the few cases in which the eikonal equation is solvable are discussed. The fourth section follows directly from the third, in that the structure of the plane wave solutions gives the impetus for a thorough discussion of the electromagnetic energy flow within the geometrical optics approximation. The story is not quite complete, however, without some discussion of the evolution of the polarization vector along the gradient of the eikonal. This is the second topic of the section. The polarization evolution equation, however, is not especially useful without knowledge of the details of the ray, and as the eikonal equation itself is so complex, it is easier to proceed through deriving an equation for the rays themselves. The second-order nature of the ray equation leads naturally to the concept of phase space, as well as the concept of the point characteristic, which is quite useful in the analysis of optical systems. Further analysis of optical systems, however, is deferred to the sixth section of the chapter, where it can be taken up under the further simplification of the paraxial approximation. The fifth section of the chapter then particularizes the discussion to paraxial ray optics. This further approximation greatly simplifies the mathematics, making examples more easily presentable. Further, as is shown early in the section, in the paraxial limit the ray equation reduces to Newton's Law for a particle and allows a plethora of analogies to be drawn. Well-known solutions to mechanical problems can be straightforwardly applied to an understanding of much more advanced problems in guided wave propagation. In the sixth section of the chapter, the linear system approach to paraxial ray optics is presented, along with the attendant *ABCD* matrices. The section closes with numerous examples of actual optical systems. The seventh section of the chapter closes the chapter with a discussion of how Liouville's theorem of mechanics defines the Lagrange invariant of optics and leads to the attendant brightness theorem of optics.

5.2 THE WKB APPROXIMATION AS IT RELATES TO GEOMETRICAL OPTICS

Almost immediately after the publications of Schrödinger (1926) and Eckart (1926) which presented the wave-mechanical equations of quantum mechanics, it became patently clear that very few quantum-mechanical problems could be solved analytically. Therefore, within the same year that the Schrödinger equation appeared in the literature, Wentzel(1926), Kramers (1926), and Brillouin (1926) independently developed perturbation techniques that could be used to derive approximate solutions to the Schrödinger equation in cases where the problem separated into one-dimensional problems. The approach of these authors is generally referred to as the WKB or semiclassical method. The word ''semiclassical'' appears in its name because the technique acts somewhat as a bridge between classical particle mechanics and quantum mechanics. The lowest-order WKB term, indeed, gives classical particle trajectories, while each successive term "bootstraps" one toward the quantum-mechanical solution. Now it should be borne in mind that Maxwell's equations in many circumstances reduce mathematically to a Schrödinger equation. The fact is that Schrödinger's equation is the equation describing the wave behavior of a spinless particle, whereas Maxwell's equations are the equations describing the wave behavior of a particle with spin 1 (a photon). In the limit where the wave nature of a spin 0 (spinless) particle is unimportant, one can show that the results of a calculation based on Schrödinger's equation will reduce to a classical calculation of particle trajectories based on classical mechanics. Likewise, as we will soon see in some detail, in cases where the wave nature of light becomes unimportant, the solutions of Maxwell's equations will reduce to the calculation of ray trajectories, as could be predicted by classical mechanics. As the wave nature of light was known long before the wave nature of particles, it is not surprising that WKB techniques date back into the 1800s, in the work of Liouville (1837), as well as having reappeared in work of Lord Rayleigh (1912) and Jeffries (1923).

Here we wish to bring out some of the salient features of geometrical optics by studying some simple examples of the application of the WKB method to propagation equations in one independent variable. Einstein (1917), Brillouin (1926), and Keller (1958) succeeded in generalizing the WKB to multidimensional, nonseparable problems, the kind that are the general case encountered in attempting to solve Maxwell's equations in three dimensions. A brief description of the resulting theory is presented. However, by first giving a simple presentation of WKB and then applying the same principles to the most interesting features of Maxwell's equations, we will find that we can get an understanding of geometrical optics without having to resort to the more complicated formulations of the so-called EBK method.

We will now give a rapid overview of the subject. There are various good treatments of the WKB method in the literature. See, for example, Bender and Orszag (1978, Chapter 6). Consider an equation of the form

$$\epsilon^2 \frac{\partial^2 \psi(\alpha)}{\partial \alpha^2} + f(\alpha)\psi(\alpha) = 0 \tag{5-1}$$

where ϵ is a dimensionless parameter, α a dimensionless coordinate and $\psi(\alpha)$ the function we wish to "approximately" determine. Now, if $f(\alpha)$ were actually a constant, the solution to (5-1) could be written as complex exponentials. If $f(\alpha)$ doe not vary too much (or too rapidly), then one would expect that

$$\psi(\alpha) = Ae^{iS(\alpha)/\epsilon} \tag{5-2}$$

might be a good choice for a form of the solution. Plugging (5-2) into (5-1), one finds

$$i\epsilon \frac{\partial^2 S}{\partial \alpha^2} - \left(\frac{\partial S}{\partial \alpha}\right)^2 + f(\alpha) = 0 \tag{5-3}$$

Now, if one could assume ϵ to be small, one could conceivably look for a perturbation solution whereby

$$S(\alpha) = \sum_i \epsilon^i S_i \tag{5-4}$$

where, by virtue of the smallness of α, successive terms in the sum should get progressively smaller. When the sum is truncated at the first term, the result is usually referred to as the geometrical optics limit. When the first two terms are retained, one often refers to the physical optics limit. Replacing $S(\alpha)$ by the first two terms of (5-4) (the physical optics limit), one finds (to the first order in ϵ) that

$$i\epsilon \frac{\partial^2 S_0}{\partial \alpha^2} - 2\epsilon \frac{\partial S_1}{\partial \alpha} \frac{\partial S_0}{\partial \alpha} - \left(\frac{\partial S_0}{\partial \alpha}\right)^2 + f(\alpha) = 0 \tag{5-5}$$

Now if ϵ is really small, then one can solve (5-5) in two pieces, that is, by first satisfying (5-5) to the zeroth order in ϵ and then to the first order. This procedure yields the pair of equations

$$\left(\frac{\partial S_0}{\partial \alpha}\right)^2 = f(\alpha) \tag{a}$$

$$\frac{\partial S_1}{\partial \alpha} = \frac{i}{2} \frac{\partial^2 S_0}{\partial \alpha^2} \frac{1}{\dfrac{\partial S_0}{\partial \alpha}} \tag{b}$$

(5-6)

which can be solved to yield

$$S_0 = \pm \int f^{1/2}(\alpha)\, d\alpha \tag{a}$$

$$S_1 = \frac{i}{4} \ln\left(f(\alpha)\right) \tag{b}$$

(5-7)

The resulting $\psi(\alpha)$ can, therefore, be written in the form

$$\psi(\alpha) = A_+ \frac{e^{i/\epsilon} \int f^{1/2}(\alpha)\,d\alpha}{f^{1/4}(\alpha)} + A_- \frac{e^{-i/\epsilon} \int f^{1/2}(\alpha)\,d\alpha}{f^{1/4}(\alpha)} \tag{5-8}$$

where the A_+ and A_- are to be determined from the two necessary boundary conditions.

Now that we have obtained an approximate solution to (5-1) in the form of (5-8), a valid question to ask is, what is the regime of validity of (5-8)? Clearly, the answer must lie in the expression of (5-4), that this series must converge and converge rapidly, meaning that successive orders must decrease or

$$|\epsilon S_{i+1}| \ll |S_i| \tag{5-9}$$

One can analytically perform the operations required by (5-9) in the case $i = 0$, to find that

$$\left| \frac{\epsilon}{4} \ln f(\alpha) \right| \ll \left| \int_0^\alpha f(\alpha')^{1/2}\,d\alpha' \right| \tag{5-10}$$

Clearly, independent of ϵ, (5-10) will not hold when $f(\alpha) \to 0$. A second condition can be derived by taking the derivative of (5-10), to obtain

$$\left| \frac{\epsilon}{4} \frac{\partial f(\alpha)}{\partial \alpha} \right| \ll \left| f(\alpha)^{3/2} \right| \tag{5-11}$$

which is a condition on the rate of change of $f(\alpha)$ with α. It is quite interesting to note that none of the conditions of (5-10) or (5-11) requires that ϵ be small, even though we thought we used that in the derivation. Actually it was not used, and ϵ can take on any value as long as (5-10) and (5-11) are satisfied.

A serious problem with solving equations such as (5-1) that have all dimensionless parameters and variables is that these are the equations one finds in mathematics texts but not in practical application. In practical applications, dimensions are valuable tools and should be treated as such. For this reason, we will now turn attention to the equations

$$\frac{\partial^2 \psi}{\partial z^2} + k^2(z)\psi \tag{5-12}$$

where z is a coordinate and $k(z)$ a wave number. There is no nice expansion parameter in (5-12) as there was in (5-1). However, we can make one. Recalling that one can always write

$$k^2(z) = k_0^2 n^2(z) \tag{5-13}$$

one could express (5-12) in the form

$$\frac{1}{k_0^2} \frac{\partial^2 \psi}{\partial z^2} + n^2(z)\psi = 0 \tag{5-14}$$

and identify ϵ with $1/k_0$. Now k_0 is not dimensionless, and we thus cannot say it is large or small because the next person could always change the dimensions of k, thereby transforming it from big to small or vice versa. But the last time we solved the problem, we found out that ϵ did not have to be small. Therefore, we can copy the previous derivation and proceed directly to results analogous to it, that is,

$$\psi(z) = A_+ \frac{e^{i \int k(z')dz'}}{k^{1/2}(z)} + A_- \frac{e^{-i \int k(z')dz'}}{k^{1/2}(z)} \tag{5-15}$$

with the corresponding validity conditions

$$\left| \frac{1}{2} \ln n(z) \right| << \left| \int k(z')dz' \right| \tag{a}$$

$$\left| \frac{1}{2n(z)} \frac{\partial n}{\partial z} \right| < |k(z)| \tag{b}$$

$$\tag{5-16}$$

where the first condition states that the solution becomes invalid near zeros of the "effective" index profile, and the second condition states that the index profile cannot vary appreciably over lengths corresponding to the wavelength of the light.

Generally, an equation of the form of (5-12) is the result of some separation, and the $n(z)$ of equation (5-13) is not really a physical index but an effective one. For example, in waveguiding problems, one generally encounters forms

$$\frac{\partial^2}{\partial x^2} \psi + (k^2(z) - \beta^2)\psi = 0 \tag{5-17}$$

where now the effective wave number $\kappa^2(z) = k^2(z) - \beta^2$ can be either real or imaginary. Regions of real κ will be separated from regions of imaginary κ by zeros of the $\kappa^2(z)$ function where the WKB solution becomes invalid. Some possible $\kappa^2(z)$'s are plotted in Figure 5.1. In Figure 5.1(a), we see that in the regions to the right and left of the zero of $\kappa^2(z)$, the solutions will be propagation

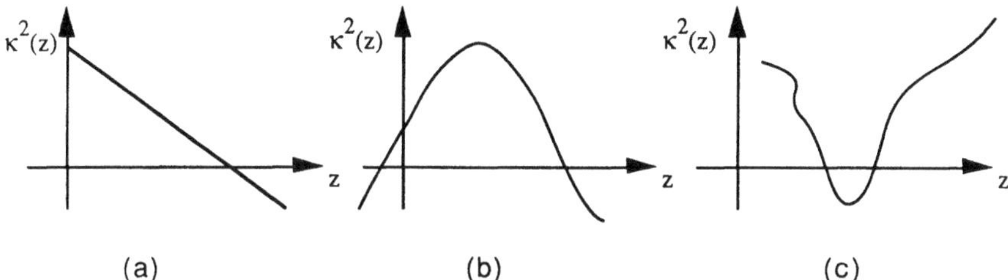

FIGURE 5.1.   Some possible shapes for effective $k^2(z)$ functions: (a) a totally reflecting barrier, (b) a waveguiding region, and (c) a tunneling barrier.

and dying exponentials, respectively. A wave packet incident on the so-called caustic point, that is, the zero of $\kappa^2(z)$, will therefore be totally reflected back, as it cannot propagate into the region to the left of the caustic. To find the exact form of the fields throughout space one would have to solve the problem exactly in the neighborhood of the caustic and tie this exact solution to the asymptotic forms that are valid to the right and left. Indeed, by taking the $\kappa^2(z)$ to have a simple form in the neighborhood of the caustic, this can be done, and it is done in texts such as Bender and Orszag (1978, Chapter 6) that treat the WKB method. Simple isolated zeros such as that in Figure 5.1(a) or those in Figure 5.1(b) are treated by using linear approximations to $\kappa^2(z)$ and yield Airy functions, whereas closely spaced zeros as in Figure 5.1(c) can be treated by using parabolic cylinder functions. In Figure 5.1(b), we see that the solutions will be dying exponentials to both the left and the right of the central guiding region, and therefore this shape of $\kappa^2(z)$ is that of a waveguide. Figure 5.1(c) has propagating regions surrounding a nonpropagating region. This is the form of a scattering potential. A wave packet incident from the left, for example, would be split up by the negative $\kappa^2(z)$ region into a reflected and a transmitted component. As is evident from Figure 5.1, most of the interesting things that can happen to a propagating wave happen in the neighborhoods of caustics. Without caustics, an initially propagating disturbance will just continue to propagate. Indeed, it is the treatment of the caustics that forms the difference between the physical optics and the geometrical optics limits.

The WKB method is not limited to the wave equation; any ordinary differential equation can be treated. The general case can be treated by considering the following system of ordinary equations:

$$\frac{\partial \psi_i}{\partial z} = \sum_{j=1}^{N} M_{ij}(z)\psi_j \quad i = 1, \cdots, N \tag{5-18}$$

where the ψ_i's are the (coupled) wave functions, and the M_{ij}'s are the elements of an $N \times N$, z-dependent, coupling matrix. Without proof, we will state here that the WKB solution of these coupled equations can be written in the form

$$\psi_i = \frac{a_{ij}}{\sqrt{\lambda_j(z)}} \, e^{\pm i \int \sqrt{\lambda_j(z')}\,dz'} \quad j = 1, \cdots, N \tag{5-19}$$

where the λ's are the eigenvalues of $M_{ij}(z)$, defined by

$$\det \lceil M - \lambda I \rceil = 0 \tag{5-20}$$

which has N solutions for $\lambda_i(z)$, and the a_{ij} are constants to be determined from the boundary conditions. Now, evidently, the caustics will be determined by the zeros of the λ_i's, which are the points z where the matrix M becomes singular. In regions where M is nonsingular, therefore, there is mode-coupling-free propagation. These regions, however, must be patched together through matrix singularity regions in which mode coupling takes place. We will find occasion to use some of these matrix results later in the book when diffraction and modulation are treated.

Now the next simplest case to the scalar, one-dimensional case is that of the scalar multidimensional case. This case was treated by Keller (Keller and Rubinow 1960; Keller 1985) under the assumption that the scalar function $\psi(x, y, z)$ satisfied a wave-type equation

$$\nabla^2 \psi + k^2(x, y, z)\psi = 0 \tag{5-21}$$

Making the assumption that $\psi(x, y, z)$ can be expressed in the form

$$\psi(x, y, z) = A(x, y, z)\,e^{ikS(x, y, z)} \tag{5-22}$$

it is straightforward to show that $A(x, y, z)$ and $S(x, y, z)$ must satisfy the equations

$$|\nabla S(x, y, z)|^2 = 1 \quad \text{(a)} \tag{5-23}$$
$$2\nabla S \cdot \nabla A + A\nabla^2 S = 0 \quad \text{(b)}$$

where (5-23) is a form of the eikonal equation, an equation we will soon take up in some detail. Assuming that one can find that vector $\nabla S(x, y, z)$, which can be considered a ray direction, one can rewrite equation (5-23) (b) in the form

$$2\frac{\partial A}{\partial s} + A\nabla^2 S = 0 \tag{5-24}$$

where s is an arc length measured along the vector $\nabla S(x, y, z)$, and $\partial A/\partial s$ is therefore a directional derivative of $A(x, y, z)$. Clearly (5-24), being a first-order equation, can be integrated to yield a form such as

$$A(s) = A_0 \left[\frac{\rho_1 \rho_2}{(\rho_1 + s)(\rho_2 + s)} \right]^{1/2} \tag{5-25}$$

therefore leading to a form for $\psi(x, y, z)$ as expressed by (5-22) that is strictly analogous to the form of (5-15) [assuming that there are both positive and negative roots to (5-23) (a)]. Indeed, the solution of (5-25) will have the same problems with singularities as (5-15), and many of the same propagation properties. However, rather than continue with cataloging mathematical properties of solutions that are not quite what we wish to deal with, we will turn our primary attention to the vector problem at hand, and begin to consider the physical properties of the corresponding equations. In geometrical optics, one assumes the field to be of the form

$$\mathbf{E}(x, y, z) = \mathbf{e}(x, y, z)e^{ik_0 S(x,y,z)} \tag{5-26}$$

where $\mathbf{e}$ is the geometrical optics field, and S is referred to as the eikonal. One then expands Maxwell's equations in a dimensionless expansion parameter that looks like $\nabla k/2k^2$ and assumes

$$\frac{1}{2k^2} |\nabla k| \ll 1 \tag{5-27}$$

to allow one to pick off the first term. The higher-order terms carry more information but require so much effort to obtain that they are usually not worth the effort. In cases where the approximation does not hold, it is probably best to go back to Maxwell's equations and start over. Note that this expansion parameter and validity condition correspond exactly to those discussed surrounding equation (5-16) in the WKB discussion.

One can picture the situation described by equation (5-26) as shown in Figure 5.2. Here, lines of equiphase (lines of $S(x, y, z) = $ constant) are drawn, and

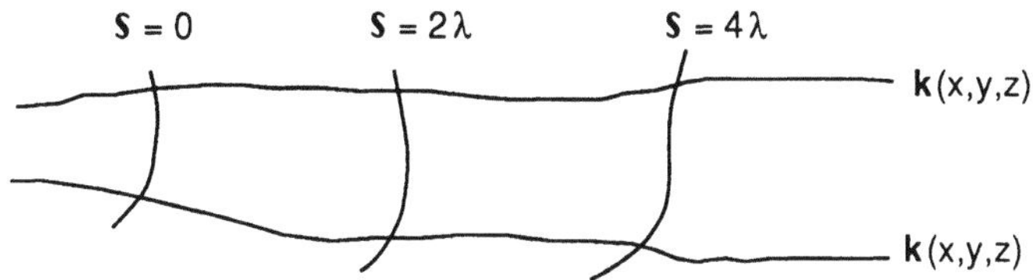

FIGURE 5.2.   A depiction of the physical situation described by equation (5-1).

the vectors perpendicular to these surfaces are drawn and labeled by $\mathbf{k}(x, y, z)$. As we know that for a plane wave the propagation vector is always perpendicular to the phase fronts, we would think that perpendiculars to the planes of constant phase should have the significance of some kind of local energy flow, as if we could break the wavefront up into local plane waves. Later, we will see that this indeed is the case.

The important question here is, what really is the meaning of (5-27)? Clearly, it says that the index must be slowly varying with respect to the wavelength of the medium, as could be clearly indicated by rewriting (5-27) in the form

$$\frac{1}{2k}\,|\nabla n| \ll 1 \tag{5-28}$$

Were the medium to vary appreciably over a wavelength, distributed reflections would be set up, and it would become impossible to represent the field as single rays. One might worry that this variation could preclude interfaces from being considered by geometrical optics, but this is not really true, and the restriction turns out to be more subtle. To illustrate this, consider Figure 5.3, where an interface whose maximal radius of curvature r_c, which is very much greater than λ, is depicted. If we believed the above discussion about local plane waves, it should be easy to visualize how one could apply the Fresnel relations at each (greater than wavelength-sized) point along the interface. So one can, indeed, treat interfaces geometrically. The problem that arises, however, is quite analogous to an earlier problem that was associated with Poynting vectors. We found earlier (Chapter 2) that the Poynting vector worked very well as an energy flow indicator as long as one always had forward- and backward-traveling waves and therefore did not need to attempt to use the Poynting vector with standing wave distributions. As can be seen from our present cursory investigation of

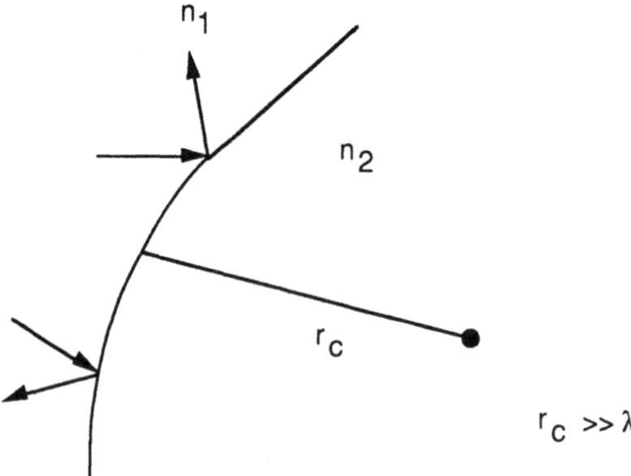

FIGURE 5.3.   Depiction of a gently curving interface between two media of differing indices.

the geometrical optic **k** vector, it acts much like a local Poynting vector and so will probably suffer some of the same diseases. The point is that geometrical optics will not (and cannot be made to) do an acceptable job of predicting interference phenomena.

Although interference phenomena can be quite important, they do not have to be. Everyday experience takes place in white light, which can hardly be made to interfere (except by Father Grimaldi[1]). Therefore, geometrical optics (together with our knowledge of plane waves) can be used to explain all kinds of everyday optical phenomena. These phenomena certainly include all the well-known cases of incoherent imaging; so geometrical optics can be used to describe microscopes, telescopes, and most non-diffraction- and non-interference-based optical instruments. Although waveguides usually are excited by coherent sources owing to considerations of coupling loss, these structures are designed only to transmit energy in one direction, and thus the majority of salient features of common waveguiding structures can be elucidated by geometrical optics analysis. Another interesting application area that one normally would not consider for geometrical optics is the description of transient wave phenomena. If a transient is emitted from a point-like source, the resultant electromagnetic phase fronts will travel out from the source along equiphase surfaces whose time evolution is described by geometrical optics, at least as long as the medium satisfies (5-28), and thus the wavefronts will not fold in upon each other. Although we will not consider electromagnetic transients in this chapter, we will consider both imaging systems and waveguides. First we give some consideration to deriving the equations of geometrical optics and then relating the behavior of these solutions to the behavior of local plane waves and local Poynting vectors.

5.3 THE EIKONAL EQUATION

We will assume that the field is sufficiently monochromatic that we can use time harmonic Maxwell's equations.[2] The medium is assumed to be charge and current–free and nonconducting but inhomogeneous such that

$$\epsilon = n^2(\mathbf{r})\epsilon_0 \tag{5-29}$$

---

[1]Father Francesco Maria Grimaldi (1618–63) sometimes is referred to as the father of diffraction because of his extensive experimentation with diffraction phenomena. See, for example, Grimaldi's treatise (Grimaldi 1665). Although Galileo and Snell had done work with refraction using only natural or candle light sources, Grimaldi was the first to consider diffraction phenomena. For a historical perspective on optics, it can be worthwhile to read the historical introduction in the book by Born and Wolf (1975).

[2]The following derivation closely resembles that in Born and Wolf (1975). Other general undergraduate-level texts that also treat geometrical optics include Hecht (1987) and Klein and Furtak (1986).

and therefore Maxwell's equations can be written in the form

$$\nabla \times \mathbf{E}(\mathbf{r}) = i\omega\mu_0 \mathbf{H}(\mathbf{r}) \qquad \text{(a)}$$

$$\nabla \times \mathbf{H}(\mathbf{r}) = -i\omega\epsilon \mathbf{E}(\mathbf{r}) \qquad \text{(b)}$$

$$\nabla \cdot (\epsilon \mathbf{E}(\mathbf{r})) = 0 \qquad \text{(c)}$$

$$\nabla \cdot \mathbf{H}(\mathbf{r}) = 0 \qquad \text{(d)}$$

$$(5\text{-}30)$$

where

$$\mathbf{E}(\mathbf{r}, t) = \text{Re}\,[\mathbf{E}(\mathbf{r})e^{-i\omega t}] \qquad \text{(a)}$$

$$\mathbf{H}(\mathbf{r}, t) = \text{Re}\,[\mathbf{H}(\mathbf{r})e^{-i\omega t}] \qquad \text{(b)}$$

$$(5\text{-}31)$$

Now, one wants to pass to the geometrical optics limit by first substituting the forms

$$\mathbf{E}(\mathbf{r}) = \mathbf{e}(\mathbf{r})e^{ik_0 S(\mathbf{r})} \qquad \text{(a)}$$

$$\mathbf{H}(\mathbf{r}) = \mathbf{h}(\mathbf{r})e^{ik_0 S(\mathbf{r})} \qquad \text{(b)}$$

$$(5\text{-}32)$$

in Maxwell's equations (5-30) to find

$$\nabla S \times \mathbf{e} - \eta_0 \mathbf{h} = -\frac{1}{ik_0}\,\nabla \times \mathbf{e} \qquad \text{(a)}$$

$$\nabla S \times \mathbf{h} + \frac{n^2}{\eta_0}\,\mathbf{e} = -\frac{1}{ik_0}\,\nabla \times \mathbf{h} \qquad \text{(b)}$$

$$\mathbf{h} \cdot \nabla S = -\frac{1}{ik_0}\,\nabla \cdot \mathbf{h} \qquad \text{(c)}$$

$$\mathbf{e} \cdot \nabla S = -\frac{1}{ik_0\epsilon}\,[\epsilon\nabla \cdot \mathbf{e} + \mathbf{e} \cdot \nabla\epsilon] \qquad \text{(d)}$$

$$(5\text{-}33)$$

What we wish to do now is to use the approximation of (5-28) to simplify (5-33) to a solvable form. Unfortunately, the parameter of (5-28) does not seem to show up in (5-33). Actually it does, but not explicitly. For example, consider the last term on the right-hand side of (5-33) (d). Clearly, if one were to consider only the magnitude of this term, one could write that

$$\left|\frac{1}{ik_0\epsilon}\,\mathbf{e} \cdot \nabla\epsilon\right| \sim \frac{2}{k_0}\,|\nabla n||\mathbf{e}| \qquad (5\text{-}34)$$

and now the parameter of (5-28) shows up manifestly (apart from a factor of 4), informing us that the last term on the right-hand side of (5-33) (d) is small compared to the left-hand terms. The other terms on the right-hand side, however, must be equally small. The argument for this is as follows. If the index of refraction were exactly constant, then **e** and **h** would become independent of the coordinate, as the exact solutions would be plane waves. In this case, the whole right-hand side of (5-33) would be zero. This is not quite the case, as n can vary slowly. This slow variation is the cause of the nonconstancy of **e** and **h**. Clearly, however, **e** and **h** can vary no faster than n, as the variation in n is causing the variation in **e** and **h**. One could therefore write that

$$\left| -\frac{1}{ik_0 \epsilon} \, \nabla \times \mathbf{e} \right| \sim \frac{1}{2k} \, |\nabla n| |\mathbf{e}| \qquad (5\text{-}35)$$

and as in (5-34), the parameter of (5-28) has shown up explicitly in (5-35). Similar arguments could hold for all the terms on the right-hand side of (5-33), leading one to the zeroth-order expansion in the parameter of (5-28) and therefore the lowest order of approximation system

$$\nabla S \times \mathbf{e} = \eta_0 \mathbf{h} \qquad \text{(a)}$$

$$\nabla S \times \mathbf{h} = \frac{-n^2}{\eta_0} \, \mathbf{e} \qquad \text{(b)}$$

$$\mathbf{h} \cdot \nabla S = 0 \qquad \text{(c)}$$

$$\mathbf{e} \cdot \nabla S = 0 \qquad \text{(d)}$$

(5-36)

The system of (5-34) contains a wealth of information. Equations (5-36) (c) and (d) say clearly that **h** and **e** are transverse to the direction ∇S. Coupling this transversality with (5-34) (a) and (b) tells us also that **e** and **h** are mutually orthogonal. The situation is illustrated in Figure 5.4. As ∇S is the gradient of the eikonal, it will be a vector that pierces the surfaces of $S = $ constant at right

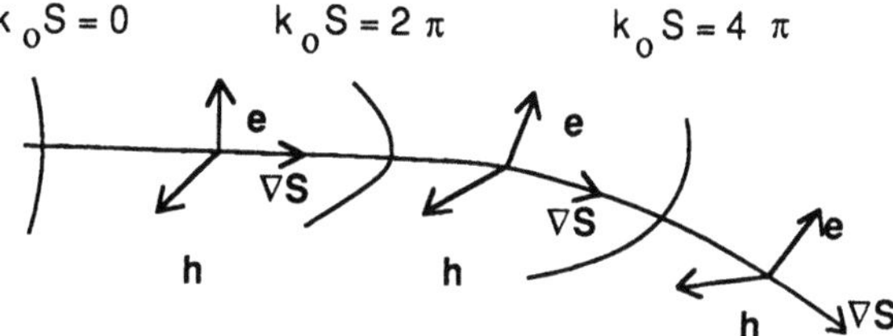

FIGURE 5.4.   A depiction of the evolution of the polarization vectors **e** and **h** along an eikonal.

angles along the directions of maximal change. As **e** and **h** are mutually orthogonal and transverse to ∇S, they are so-called parallel-transported along the back of the ∇S vector. The situation looks suggestively like the situation in a plane wave, with the clear exception that the propagation vector of this plane wave does not remain constant in direction or, as we will soon see, in magnitude either.

The geometrical relationships between the quantities in (5-36) thus are easy to visualize, but we would like to have a more formal solution to the problem. To do this, we note that it actually comprises six independent equations [(5-36) (c) and (d) are obtainable from dotting ∇S into (5-36) (a) and (b)] in seven unknowns, **e**, **h**, and ∇S. This situation is not hopeless, as can be seen from taking $\nabla S \times$ (5-36) (a) and using (5-36) (b) to obtain

$$\nabla S \times \nabla S \times \mathbf{e} + n^2 \mathbf{e} = 0 \tag{5-37}$$

Using a vector identity for the double cross product and invoking (5-36) (d) in the result, one obtains the eikonal equation

$$|\nabla S|^2 = n^2 \tag{5-38}$$

If one writes out this equation in more detail, one finds that it is expressible as

$$\left(\frac{\partial S}{\partial x}\right)^2 + \left(\frac{\partial S}{\partial y}\right)^2 + \left(\frac{\partial S}{\partial z}\right)^2 = n^2 \tag{5-39}$$

From (5-39), it is clear that the eikonal is a rather difficult, nonlinear, nonconstant-coefficient partial differential equation. But we will not be deterred, and, instead, will proceed to some examples where it can be solved.

Example. n^2 = constant

We do not know how to solve an equation of the form of (5-39), in general, but here we can at least assume a solution and see if we can satisfy both the equation and any of the (at least three) necessary boundary conditions. Trying

$$S = n(\alpha_1 x + \alpha_2 y + \alpha_3 z) \tag{5-40}$$

in (5-39), we find that

$$1 = \alpha_1^2 + \alpha_2^2 + \alpha_3^2 \tag{5-41}$$

and, as (5-41) can be satisfied, that

$$\nabla S = n(\alpha_1 \hat{\mathbf{e}}_x + \alpha_2 \hat{\mathbf{e}}_y + \alpha_3 \hat{\mathbf{e}}_z) \tag{5-42}$$

The situation is that of a plane wave propagating through a homogeneous medium in a direction determined by the directions cosines α_1, α_2, and α_3; that is, the normalized propagation vector of the wave is expressible in the form

$$\hat{s} = \frac{\nabla S}{n} = [\alpha_1 \hat{e}_x + \alpha_2 \hat{e}_y + \sqrt{1 - \alpha_1^2 - \alpha_1^2}\ \hat{e}_z] \tag{5-43}$$

Example. $n^2 = n^2(z)$

The form obtained in the above example suggests that we try a form here such as

$$S = n_0(\alpha_1 x + \alpha_2 y) + \int_0^z \sqrt{n^2(z) - (\alpha_1^2 + \alpha_2^2)n_0^2}\ dz' \tag{5-44}$$

and indeed one finds that the equation can be satisfied and that

$$\hat{s} = \frac{n_0(\alpha_1 \hat{e}_x + \alpha_2 \hat{e}_y) + \sqrt{n^2(z) - (\alpha_1^2 + \alpha_2^2)n_0^2}\ \hat{e}_z}{n(z)} \tag{5-45}$$

which can be depicted as in Figure 5.5.

For a wave entering the medium with non-zero values of α_1 and/or α_2, the ray path will bend in and out from the z-axis as the index increases and decreases, respectively.

Example. Parabolic index

Here we wish to consider a two-dimensional medium in which one can assume the refractive index to take the form

$$n^2(x) = n_0^2 \left[1 - 2\Delta \left(\frac{x}{a} \right)^2 \right], \quad |x| \le a \tag{5-46}$$

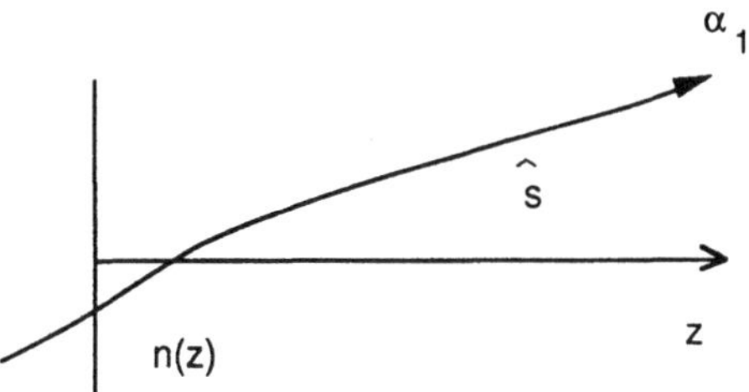

FIGURE 5.5.   Illustration of a possible ray path in a medium with a longitudinally varying index of refraction.

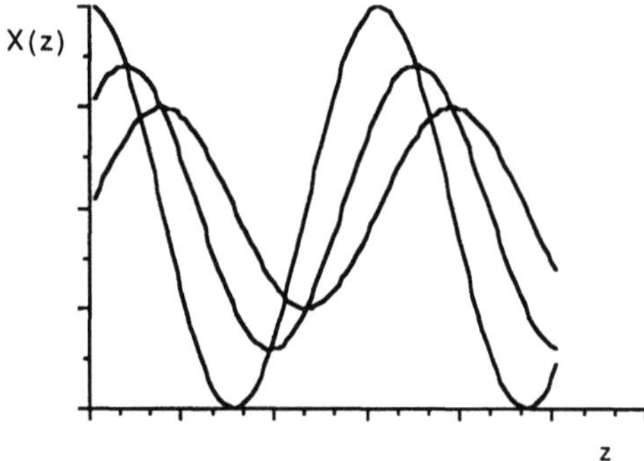

FIGURE 5.6.   Examples of a pair of possible ray paths for different initial conditions of a guiding medium such as that described by the index of (5-46).

at least for the x coordinate not too far from the z-axis (that is, so that the index remains positive). One could always try a solution of the form

$$S(x, z) = f(x) + \alpha_3 z \tag{5-47}$$

where $f(x)$ is to be determined from the eikonal equation

$$\left(\frac{\partial f}{\partial x}\right)^2 + \alpha_3^2 = n_0^2 \left[1 - 2\Delta \left(\frac{x}{a}\right)^2\right], \quad |x| \le a \tag{5-48}$$

which is a differential equation that can be solved when subjected to initial conditions. Here, we will just mention the form of the solution and leave the details of the solution as an exercise for the reader. It is clear that the ray directions will be defined by

$$\nabla S = \frac{\partial f}{\partial x} \hat{\mathbf{e}}_x + \alpha_3 \hat{\mathbf{e}}_z \tag{5-49}$$

Possible ray paths are illustrated in Figure 5.6. The basic idea here is that the index of refraction is larger in the middle, and, therefore, for a class of initial conditions, rays will be guided transversely.

5.4 ENERGY FLOW AND RADIOMETRY

Having found the eikonal, we have essentially cleared the way for finding the variation of the **e** and **h** vectors. The eikonal (and its gradient) we determined by setting the determinant of the coefficients of the **e** and **h** fields, expanded to first order, to zero. It remains, then, for us to plug the solution for the eikonal back into the system of (5-36). Rather than carry out the rather complex details of this calculation, here we will give an outline of some of the steps of the derivation and refer one to the literature for more of the details (Babič and Buldyrev, 1991, Section 2.5).

Now, given that one has found the eikonal, one should also be able to find the normalized gradient of the eikonal, which we have denoted by $\hat{s}$. This amounts to finding all of the ray paths in space. Later in this section, we will see that there are easier ways to find the ray paths than to solve the eikonal equation, but for the present polarization argument, we will just assume the principle is correct. Once one has the tangent $\hat{s}$ to the ray path, one can find the principal normal $\hat{n}$ and binormal $\hat{b}$ to this path. These three quantities are, then, propagated through space according to the Frenet-Serret relations (Spiegel, 1959):

$$\frac{\partial \hat{s}}{\partial s} = \kappa \hat{n} \tag{a}$$

$$\frac{\partial \hat{n}}{\partial s} = \tau \hat{b} - \kappa \hat{s} \tag{b} \quad (5\text{-}50)$$

$$\frac{\partial \hat{b}}{\partial s} = -\tau \hat{n} \tag{c}$$

where s is the arc length measured along a ray path, κ is the radius of curvature, and τ is the torsion. In the absence of torsion, note that $\hat{b}$ is conserved, and $\hat{s}$ and $\hat{n}$ satisfy a coupled equation. The addition of torsion greatly complicates the propagation of these vectors, and, as we will soon see, also has an effect on the transport of the field vectors.

Now in light of equations (5-36) (c) and (d), it is clear that one can express the $\mathbf{e}$ and $\mathbf{h}$ vectors in terms of $\hat{b}$ and $\hat{n}$ by

$$\mathbf{e} = A(s)\hat{n} + B(s)\hat{b} \tag{a}$$
$$\mathbf{h} = C(s)\hat{n} + D(s)\hat{b} \tag{b}$$

$$(5\text{-}51)$$

Now, as one can again use (5-36) to interrelate $\mathbf{e}$ and $\mathbf{h}$, let us presently concentrate on the form of $\mathbf{e}$. In particular, let us assume that A and B can be expressed in the form

$$A(s) = |e_0| \cos \theta(s) \tag{a}$$
$$B(s) = |e_0| \sin \theta(s) \tag{b}$$

$$(5\text{-}52)$$

Again, referring the interested reader to the literature for details, we will state that it can be shown that $\theta(s)$ varies according to the rule that

$$\frac{\partial \theta}{\partial s} = \frac{1}{\tau} \tag{5-53}$$

which is to say that the presence of torsion caused power to couple between the orthogonal polarization states. This excess rotation is sometimes referred to as Berry's phase (Berry 1984, 1985; Chiao and Wu 1986), and it turns out is a manifestation of a symmetry of gauge theories, but we will go no further with this except to mention that the rotation has been verified experimentally (Tomita and Chiao 1986).

Now we wish to find an equation for the ray paths directly, such that it is not necessary to find the eikonal first. In the paraxial limit, at least, we will later find that these ray equations really are considerably simpler than the eikonal equation. Consider a ray path such as the one depicted in Figure 5.7. Let us say that the distance measured along a ray path is called s, and the unit vector for the ray path at any coordinate $\mathbf{r}$ is called $\hat{\mathbf{s}}(\mathbf{r})$ as we have been doing. It is clear that

$$\frac{d\mathbf{r}}{ds} = \hat{\mathbf{s}}(\mathbf{r}) \tag{5-54}$$

That is, the rate of change of the radius vector at the point $\mathbf{r}$ is just the normal. Using the definition of the eikonal and the eikonal equation (5-38), one can further write that

$$\hat{\mathbf{s}}(\mathbf{r}) = \frac{\nabla S}{n} \tag{5-55}$$

which we used previously in equation (5-43) and thus we find that

$$n\frac{d\mathbf{r}}{ds} = \nabla S \tag{5-56}$$

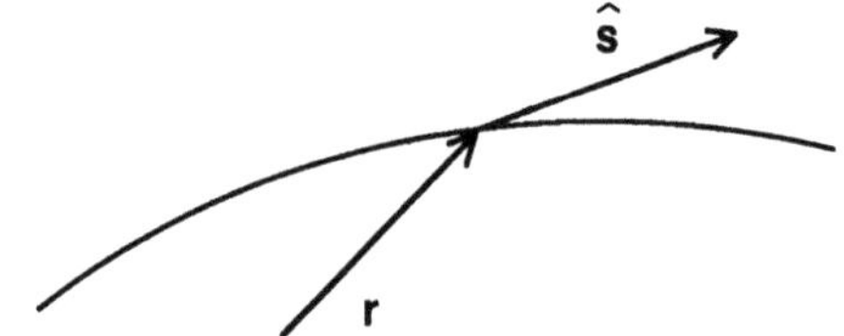

FIGURE 5.7.   A ray path defined by a unit vector $\hat{\mathbf{s}}(\mathbf{r})$.

As d/ds is the comoving derivative along the eikonal, one can write for any arbitrary function f that

$$\frac{df(\mathbf{r})}{ds} = \frac{d\mathbf{r}}{ds} \cdot \nabla f \qquad (5\text{-}57)$$

and therefore that

$$\frac{d}{ds}\left(n\,\frac{d\mathbf{r}}{ds}\right) = \frac{d\mathbf{r}}{ds} \cdot \nabla\,(\nabla S) \qquad (5\text{-}58)$$

Using 5-56 and the vector identity that

$$\nabla(\mathbf{f} \cdot \mathbf{f}) = 2(\mathbf{f} \cdot \nabla)\mathbf{f} \qquad (5\text{-}59)$$

for arbitrary $\mathbf{f}$ and (5-38) in (5-58), one finds the ray equation

$$\frac{d}{ds}\left(n\,\frac{d\mathbf{r}}{ds}\right) = \nabla n \qquad (5\text{-}60)$$

The meaning of this equation is best illuminated by some examples.

Example. $n = $ constant

Here we find that the ray equation reduces to

$$\frac{d^2\mathbf{r}}{ds^2} = 0 \qquad (5\text{-}61)$$

which has solutions

$$\mathbf{r} = \mathbf{a}s + \mathbf{b} \qquad (5\text{-}62)$$

where $\mathbf{a}$ and $\mathbf{b}$ are determined from the initial conditions. The situation is as depicted in Figure 5.8

Example. $n = n\,(r)$

We could try to brutally solve the ray equation here, but it is sufficiently complicated that instead we will steal from the physicists and find a "constant of the motion" that will allow us to say something about the possible ray paths. Actually, we will not even find the constant of motion, as we already know what one is, but we will show that it is invariant. Now, a constant of the motion

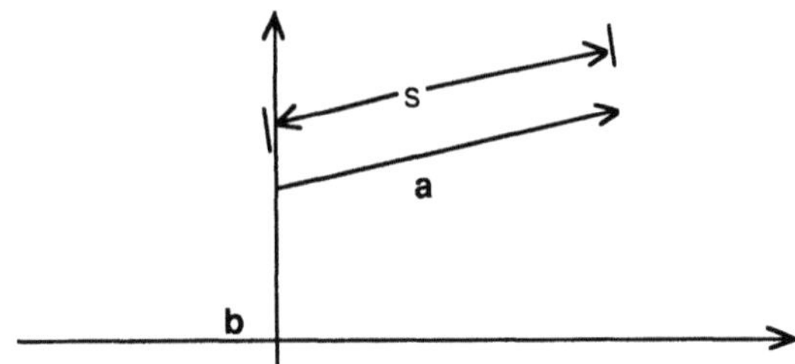

FIGURE 5.8.   A depiction of the solution (5-59) to the eikonal equation in a homogeneous medium.

is something that does not vary along the ray path, and therefore its comoving derivative is zero. Consider

$$\frac{d}{ds}(\mathbf{r} \times n\hat{\mathbf{s}}) = \frac{d\mathbf{r}}{ds} \times n\hat{\mathbf{s}} + \mathbf{r} \times \frac{d}{ds}(n\hat{\mathbf{s}}) \qquad (5\text{-}63)$$

As $d\mathbf{r}/ds$ is just $\hat{\mathbf{s}}$, the first term on the right-hand side of (5-63) is identically zero. The quantity with which $\mathbf{r}$ is crossed in the second term on the right-hand side of (5-63) is, by the ray equation, just ∇n. However, for a radially varying index, this vector must be parallel to $\mathbf{r}$, and therefore the second term must also be zero. We can therefore conclude that

$$\mathbf{r} \times n\hat{\mathbf{s}} = nr \sin \Phi = \text{constant} \qquad (5\text{-}64)$$

where Φ, defined by (5-64) and illustrated in Figure 5.9, must be a constant of the motion.

We could probably think of applications where we have spherically varying optics, but probably the best applications lie in astrophysics, where the sphe-

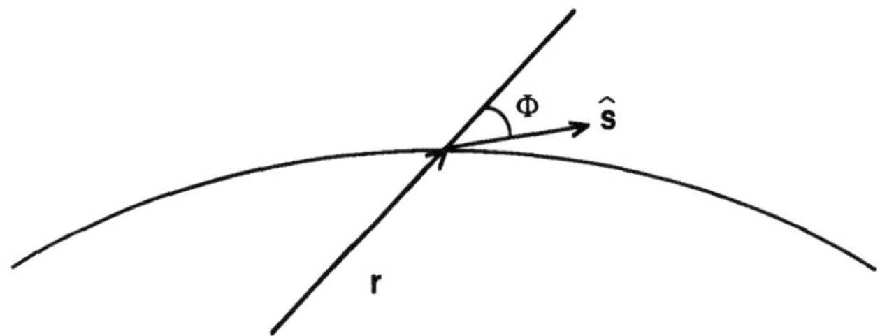

FIGURE 5.9.   A depiction of the relationships between the vectors $\mathbf{r}$ and $\mathbf{s}$ and the angle Φ for a radially varying medium.

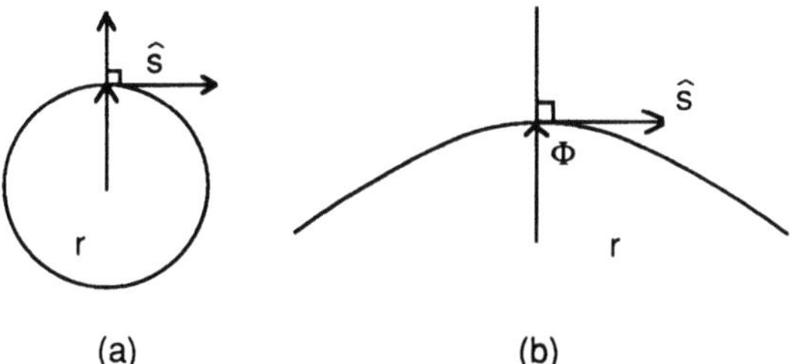

FIGURE 5.10.   Illustration of two possible ray paths for rays propagating past very large masses.

ricity of stellar structures tends to cause spherical symmetric index of refraction distributions. In fact, it was Einstein who noted that gravity causes an effective change in the index of refraction for electromagnetic waves, which causes ray paths to curve in the neighborhood of large masses, as is illustrated in Figure 5.10. In the gravitational field of a black hole, it has been predicted that there are bound rays where not only $nr \sin \Phi$ is constant, but n, r, and $\Phi = \pi/2$ are all constant; but no one has traveled to a black hole to verify this. One prediction of Einstein's general relativity was, however, that light would bend when passing the limb of the sun. Indeed, in 1919, Eddington showed that Einstein's result was correct to within experimental error. The experimental error was greatly reduced when Dicke tried to disprove this result in the 1970s.

Now, in general, to solve a second-order differential equation such as the ray equation of (5-60) we will require either initial conditions on both the ray coordinates and their derivatives or boundary conditions on the rays. In practice, both of these approaches are employed. These approaches can be generically referred to as the phase space approach and the point characteristic approach, respectively. In the phase space approach, one labels a ray by its position $\mathbf{x}$ along with a vector quantity $\mathbf{p}$ that has the direction of the vector but the magnitude of the index n at that point. Perhaps the most commonly used phase space technique is that of ray tracing, because this technique is easily amenable to numerical solution. The $\mathbf{p}$ is often referred to as generalized momentum. In the point characteristic approach, one defines two planes, x and x', and then writes out a function $V(\mathbf{x}, \mathbf{x}')$ of the two-dimensional vector coordinates in the two planes that gives the optical length for any ray joining the two planes. This point characteristic function has many interesting properties. For example, the field due to a point source at a source coordinate $\mathbf{x}$, is given by, to the WKB approximation (Arnaud 1971):

$$E(\mathbf{x}, \mathbf{x}') = \pm i/\lambda \left| \frac{\partial^2 V(\mathbf{x}, \mathbf{x}')}{\partial x_i \partial x'_j} \right|^{1/2} \exp\left[-ikV(\mathbf{x}, \mathbf{x}')\right] \qquad (5\text{-}65)$$

Further, the generalized momenta can be given by (Arnaud 1971):

$$\mathbf{p} = \nabla_x V(\mathbf{x}, \mathbf{x}') \qquad (a)$$
$$\mathbf{p}' = \mathbf{p} = \nabla_x V(\mathbf{x}, \mathbf{x}') \qquad (b)$$

$$(5\text{-}66)$$

In the next section, on paraxial ray optics, additional discussion will be given on phase space. However, our discussion next will turn to energy flow and propagation of specific intensity.

Returning to equation (5-38), we see that this equation has many implications for energy flow, which we will presently discuss. Equation (5-38) tells us that the unit vector $\hat{\mathbf{e}}_{\nabla S}$ in the ∇S direction can be expressed as

$$\hat{\mathbf{e}}_{\nabla S} \equiv \hat{\mathbf{s}} = \frac{\nabla S}{n} \qquad (5\text{-}67)$$

Using (5-38) in (5-36), one readily sees that

$$|\mathbf{h}(\mathbf{r})| = \frac{|\mathbf{e}(\mathbf{r})|}{\eta(\mathbf{r})} \qquad (5\text{-}68)$$

where $\eta(\mathbf{r})$ is given by the familiar-looking expression

$$\eta(\mathbf{r}) = \sqrt{\frac{\mu_0}{\epsilon_0 n^2(\mathbf{r})}} \qquad (5\text{-}69)$$

Writing down the expression for the Poynting vector of $\mathbf{E}(\mathbf{r}, t)$, $\mathbf{H}(\mathbf{r}, t)$,

$$\mathbf{S}_{av} = \langle \mathbf{E}(\mathbf{r}, t) \times \mathbf{H}(\mathbf{r}, t) \rangle \qquad (5\text{-}70)$$

one rapidly finds that

$$\mathbf{S}_{av}(\mathbf{r}, t) = \frac{|\mathbf{e}(\mathbf{r})|^2}{2\eta(\mathbf{r})} \hat{\mathbf{s}} = I(\mathbf{r})\hat{\mathbf{s}} \qquad (5\text{-}71)$$

which is the expression that would be obtained by a plane wave that was (locally) traveling in the $\hat{\mathbf{s}}$ direction in a medium of index $n(\mathbf{r})$. It should further be pointed out here that we could have considered either of the two possible polarization states, and, had we considered both, we would have come to the same conclusion as in Chapter 2 (2-25) that energy flow in the polarization states is independent. This further says that geometrical optics fully well han-

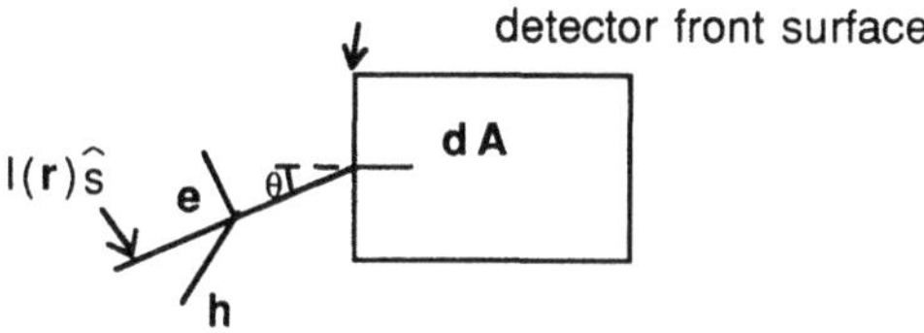

FIGURE 5.11.   Depiction of a single ray impinging on a detector surface.

dles problems with polarized light, contrary to some popular wisdom on the topic.

Rather than use the optical intensity $I(\mathbf{r})$ of equation (5-71), one often uses another quantity $I(\mathbf{r}, \Omega)$, which is sometimes referred to as the specific intensity. [See, for example, almost any text on radiative transfer, such as Mihalas (1978) or Motz (1970).] The idea behind the definition is as follows: The Poynting vector is not the important quantity; the amount of the Poynting vector crossing a closed surface is. Consider Figure 5.11. If we wished to calculate the energy entering a detector, due to an incident "plane wave," we could write

$$\text{energy in } dt = \mathbf{S}_{av} \cdot \mathbf{dA} \, dt = I(\mathbf{r})\hat{\mathbf{s}} \cdot \mathbf{dA} \, dt$$
$$= I(\mathbf{r}) \cos \theta \, dA \, dt \tag{5-72}$$

This is fine, but it assumes a nice, flat eikonal whose gradient is all precisely in one direction. In general, the eikonal is curved, and the rays come out in some solid angle, Ω, as is depicted in Figure 5.12. The idea here is that many ray paths, as will be defined more rigorously soon, can emanate from a point. This may be a little hard to conceive of until one realizes that our geometrical optics points must be bigger than a couple of wavelengths because of our slowly varying approximation. If we had used the specific intensity in (5-72) instead of the intensity, we would have found

$$\text{energy in } dt = \int_{\text{angle}} I(\mathbf{r}, \Omega) \cos \theta \, d\Omega \, dA \, dt \tag{5-73}$$

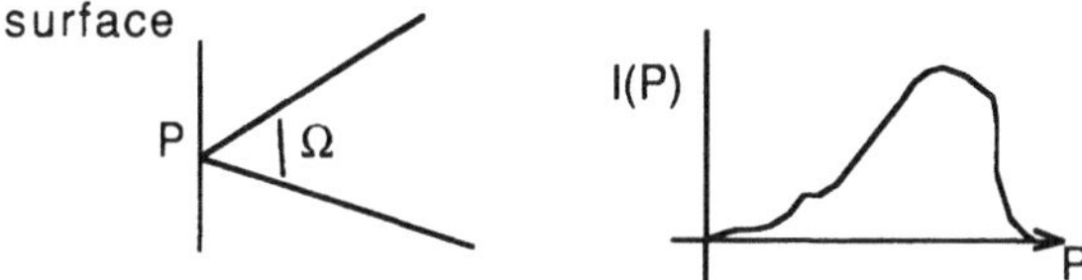

FIGURE 5.12.   Depiction of the fact that each point on a source surface will be defined not only by a pointwise intensity but also by a solid angle into which the radiation disperses.

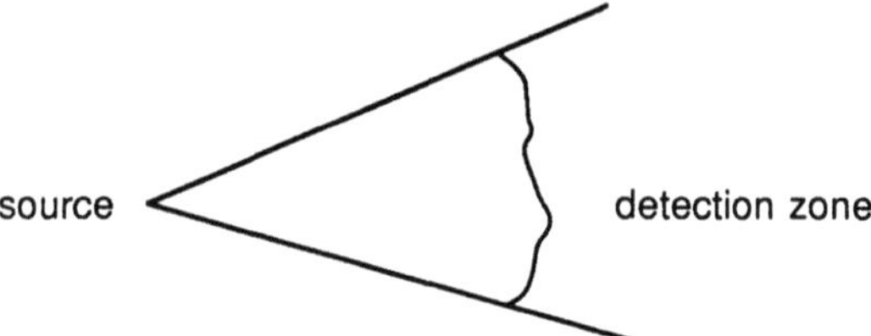

FIGURE 5.13.   A depiction of the far zone of a source.

where one sometimes defines the near-field intensity $I(\mathbf{r})$ as

$$I(\mathbf{r}) = \int I(\mathbf{r}, \Omega) \cos \theta \, d\Omega \tag{5-74}$$

The specific intensity can be used analogously in the far field, where one is so far from the source that the source dimensions appear tiny, a situation depicted in Figure 5.13. As one cannot discern details of the source surface from the far field, one often defines a far-field intensity $I(\Omega)$ by

$$I(\Omega) = \int I(\mathbf{r}, \Omega) \cos \theta \, d^2\mathbf{r} \tag{5-75}$$

where the integral is to be performed over the same coordinates.

Recall that we previously wrote Poynting's theorem in the form

$$\nabla \cdot \mathbf{S}(t) = -\frac{\partial}{\partial t} W_e - \frac{\partial}{\partial t} W_m - \mathbf{E} \cdot \mathbf{J} \tag{5-76}$$

where W_e is the stored electrical energy and W_m is the stored magnetic energy. If we consider a "steady-state" wave phenomenon, that is, a wave that has been as it was for a long-enough time (nontransient) in a nonconductive ($\sigma \sim 0$) medium, then (5-76) reduces to

$$\nabla \cdot \mathbf{S}(t) = 0 \tag{5-77}$$

Applying the divergence theorem over a region as sketched in Figure 5.14, where S_1 and S_2 are eikonals, and dS_2 is chosen such that all rays that passed through dS_1 will also pass through dS_2, we find that

$$\int I\hat{s} \cdot \mathbf{dA} = 0 \tag{5-78}$$

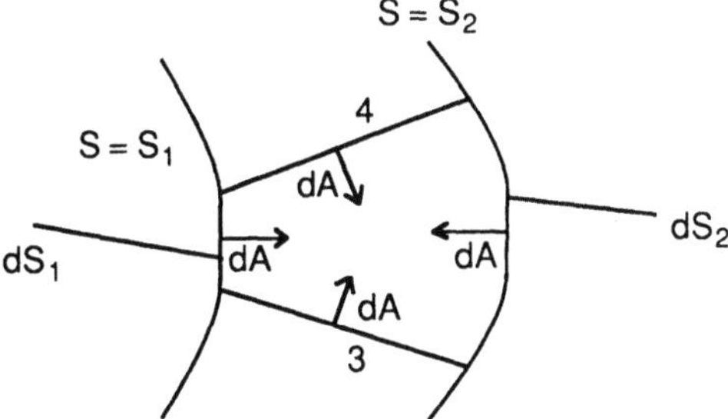

FIGURE 5.14.   A closed region A with a front face of area dS_1 and back face of area dS_2.

will give us the result (the area theorem) that

$$I_1 \, dS_1 = I_2 \, dS_2 \tag{5-79}$$

where I_1 is the average intensity over the region dS_1, and I_2 is that over dS_2. The idea here is that if the eikonal curvature is convex, the intensity will decrease proportionally with the area of the phase front. A special case of this, of course, could be the case of an isotropic spherical source radiating into free space. Here the intensity would follow the so-called inverse square law such that $I(\mathbf{r}) = I_0/r^2$.

Now it is clear from (5-79) that the optical intensity is not conserved along a ray path. As will soon be shown, however, the specific intensity $I(\mathbf{r}, \Omega)$ is. Let us aga consider Figure 5.14. Equation (5-79) shows that Ids is constant along a ray bundle, a result sometimes referred to as the intensity law of geometrical optics. The result is not too surprising, as the way that dS_2 is picked, such that all the energy crossing dS_1 must also cross dS_2, together with energy conservation, requires that IdS, which is the energy carried by the ray bundle, be conserved, at least in the absence of absorption or scattering. We could also state the energy contained in a ray bundle in terms of the specific intensity of the ray bundle. For simplicity, we will do this in free space first. Now the energy passing through a surface dS will be proportional to dS and the solid angle $d\Omega$ that the radiation subtends at dS with the specific intensity $I(\mathbf{r}, \Omega)$ as the constant of proportionality. In free space, the rays are straight lines, and the $d\Omega$ can be given by the area of the source divided by an effective radius of curvature r_c, such that

$$dE = I(\mathbf{r}, \Omega) \, dS_2 \, d\Omega = I(\mathbf{r}, \Omega) \frac{dS_1 \, dS_2}{r_c^2} \tag{5-80}$$

Now applying an argument similar to the situation depicted in Figure 5.14, and taking the distance from dS_1 to dS_2 to be given by r, one finds that

$$I_1(\mathbf{r}, \Omega) \frac{dS_1\, dS_2}{r^2} = I_2(\mathbf{r}, \Omega) \frac{dS_1\, dS_2}{r^2} \qquad (5\text{-}81)$$

which gives the result that

$$I_1(\mathbf{r}, \Omega) = I_2(\mathbf{r}, \Omega) \qquad (5\text{-}82)$$

or that the specific intensity, in the absence of absorption or scattering, remains constant along a ray bundle. Although the mathematics of deriving the transfer equation for a general medium is beyond the scope of this work, an understanding of the propagation of the specific intensity is explainable. As one recalls from Chapter 2, the velocity of propagation of radiation varies with the index of refraction of the medium. This causes the optical intensity to scale with index. However, refraction also causes the area of a ray bundle to scale such that the intensity law of (5-79) is satisfied. The specific intensity does not scale in the same manner as the intensity and therefore is not conserved. Moreover, quite in general, one makes their measurements in a medium whose characteristics are a lot like free space. If one were to take S_1 to lie in free space and S_2 also to lie in free space, (5-82) would still be satisfied, as long as the medium was between passive and bilateral. Any imaging system will satisfy such a criterion; so (5-82) constitutes the so-called brightness theorem of optics, which states that one cannot increase the brightness (another name for the specific intensity, which is also sometimes called the radiance) of a ray bundle with passive devices. This brightness theorem is further just another statement of the second law of thermodynamics, which tells one that one cannot decrease entropy without adding energy. An increase in the specific intensity would correspond to an increase in the degree of ordering of the field and thus would correspond to a decrease in entropy.

5.5 PARAXIAL RAY OPTICS

For a general index of refraction, the ray equation of (5-60) is a very complicated one to solve. Investigators have developed techniques, generically known as ray tracing, to effect computer solutions of the equation, and these techniques are used extensively in such areas as lens design. Here, we will not treat these techniques, as our purpose is to cover physical optics and not become overly entangled in engineering techniques; so we now will go to the paraxial approximation to elucidate some of the salient features of waveguides and imaging systems.

By the paraxial approximation, we mean an approximation in which we limit

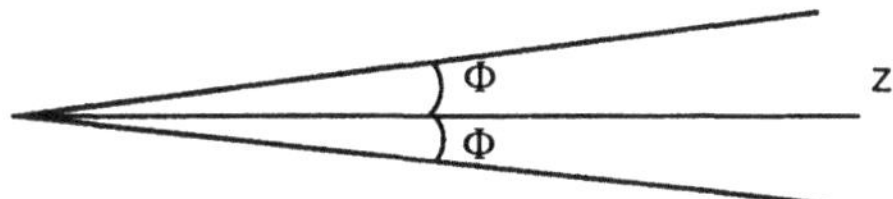

FIGURE 5.15.   The region of space to which we limit our attention when we work in the paraxial approximation.

our attention to rays that propagate "almost" parallel to the optical axis. The region in which the paraxial rays can propagate is sketched in Figure 5.15. In this approximation,

$$ds = \sqrt{dx^2 + dy^2 + dz^2} = dz \sqrt{1 + \left(\frac{dx}{dz}\right)^2 + \left(\frac{dy}{dz}\right)^2} \approx dz \quad (5\text{-}83)$$

as dx/dz and dy/dz are the angles at which the ray direction differs from the optic axis, and therefore are assumed to be small. With this approximation, we note that we can rewrite the ray equation in the form

$$\frac{d}{dz}\left(n \frac{d\mathbf{r}}{dz}\right) = \nabla n \quad (5\text{-}84)$$

If we further restrict our attention to a class of problems in which we need only consider a single transverse coordinate x, then the vector equation of (5-84) reduces to the single scalar equation

$$\frac{d}{dz}\left(n(x,\, z) \frac{dx(z)}{dz}\right) = \frac{dn}{dx} \quad (5\text{-}85)$$

Equation (5-85) still is not a simple equation, as in general it is highly nonlinear, and we do not generally know how to solve nonlinear equations. However, we will soon see that for a large class of practical problems we will be able at least to sketch the solutions to (5-85) rapidly and easily. Before we do this, we will consider the simple but important example in which (5-85) reduces to a linear equation.

Consider the case where

$$n^2(x,\, z) = \begin{cases} n_0^2 \left[1 - 2\Delta \left(\dfrac{x}{a}\right)^2\right] & |X| < a \\[2ex] n_0^2(1 - 2\Delta) & |X| > a \end{cases} \quad (5\text{-}86)$$

where $\Delta \ll 1$ (weakly guiding) (see, for example, Marcuse 1982).

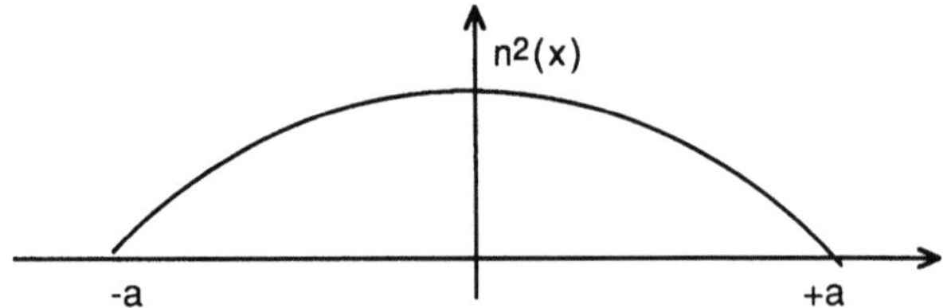

FIGURE 5.16.   A sketch of a parabolic index function.

The index described above appears as it is depicted in Figure 5.16. Using this index in the paraxial ray equation of equation (5-85) and expanding for small Δ, one finds

$$\frac{d^2x}{dz^2} = \begin{cases} -\dfrac{2\Delta}{a^2}\,x & |x| < a \\ \\ 0 & |x| > a \end{cases} \tag{5-87}$$

subject to initial conditions that $x(0) = x_0$ and $dx/dz(0) = x_0'$. For solutions within the guiding region ($|x| < a$), we find that

$$x(z) = x_0 \cos \frac{\sqrt{2\Delta}}{a}\,z + \frac{ax_0'}{\sqrt{2\Delta}} \sin \frac{\sqrt{2\Delta}}{a}\,z \tag{5-88}$$

We see immediately that, if $x_0 > a$ or $x_0' > \sqrt{2\Delta}$, the solution will not remain in the guide. If the ray leaves the guide, we see from 5-87 that its motion will be governed by the equation

$$x(z) = A + Bz \tag{5-89}$$

where A and B are constants that can be determined from (5-89) and the initial conditions. The situation is as depicted in Figure 5.17. There can be rays for which

$$\left(x_0^2 + \frac{a^2 x_0'^2}{2\Delta} \right)^{1/2} < a \tag{5-90}$$

which are bound rays, rays for which $x_0 < a$ but (5-90) is violated, which are initially (for less than one period) bound rays and then radiate, and rays for which $x_0 > a$, which are not bound from the outset.

One would like to think that, for profiles not differing very much from that in the preceding example, the solution should not differ greatly. But if we plug

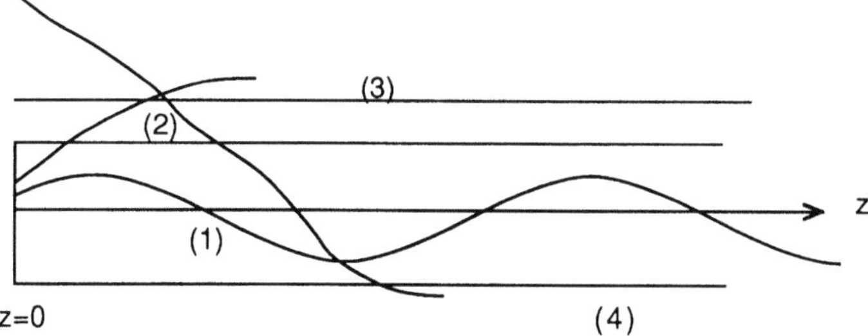

FIGURE 5.17.   A sketch of some possible ray solutions, where (1) is fully bound, (2) had too high an initial angle and became unbound, (3) had too high an initial x_0 and was not bound from the outset, and (4) represents rays that are not bound from the outset and pass through the guide without becoming bound.

in a slightly changed profile, for example, one with $x^{2.5}$ instead of x^2, we find that equation (5-85) becomes a horribly nonlinear equation. There are ways also to calculate approximate ray paths (see, for example, Evans and Rosenquist 1986). Let us consider an index profile of the form

$$n^2(x) = \begin{cases} n_0^2\left[1 - 2\Delta f\left(\dfrac{x}{a}\right)\right] & |x| < a \\ n_0^2(1 - 2\Delta) & |x| > a \end{cases} \tag{5-91}$$

in which we will again assume that $\Delta \ll 1$, $f(1) = f(-1) = 1$, and f is always less than or equal to 1. Using (5-91) in (5-87), we find

$$\frac{d^2x}{dz^2} = \begin{cases} -\Delta\dfrac{d}{dx} f\left(\dfrac{x}{a}\right) & |x| < a \\ 0 & |x| > a \end{cases} \tag{5-92}$$

subject to the initial conditions that $x(0) = x_0$ and $dx/dz(0) = x_0'$. By recalling some facts from elementary mechanics, we will soon see how such an equation can be solved.

Let us consider a mass on a nonlinear spring, as is pictured in Figure 5.18. We know that the kinetic energy, E_{kin}, in the particle is expressible as $\frac{1}{2}m\dot{x}^2$ where here the dot will be used to denote a time derivative (see, for example, problem 5 on page 77) but later will be used interchangeably with the dash that has been used to denote longitudinal derivative. For our spring, let us say that its potential energy, E_{pot}, is given by $\Delta mf(x/a)$, where, for a linear spring, $f(x/a)$ would be given by $(Kx^2)/(2\Delta m)$, where K is the spring constant. Our spring is more general. Energy conservation requires that the time derivative of

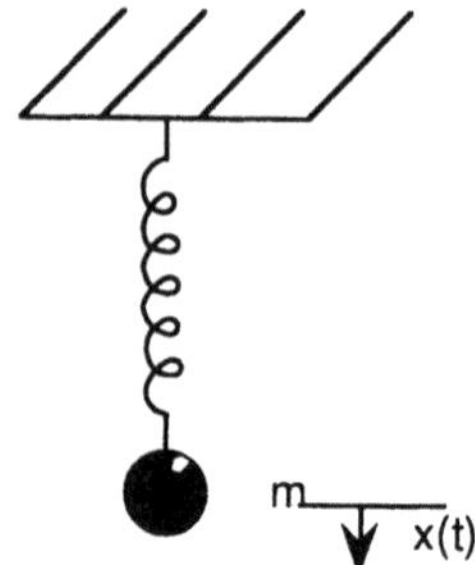

FIGURE 5.18.　A mass m attached to a nonlinear spring.

the total energy, $E_{\text{tot}} = E_{\text{pot}} + E_{\text{kin}}$, be zero. Writing this out,

$$\frac{d}{dt}\left[\frac{1}{2}m\dot{x}^2 + \Delta mf\left(\frac{x}{a}\right)\right] = \left[m\ddot{x} + \Delta m\frac{df\left(\frac{x}{a}\right)}{dx}\right]\dot{x} = 0 \qquad (5\text{-}93)$$

we see that the solution is either $\dot{x} = 0$ or

$$\ddot{x} + \Delta\frac{df\left(\frac{x}{a}\right)}{dx} = 0 \qquad (5\text{-}94)$$

which we see is strikingly similar to equation (5-92), with the z of (5-92) replaced by t in (5-94). What this tells us is that we can define for our waveguide parameters that are analogous to the potential and kinetic energies of a mass point, and use these quantities to show how a ray will be guided (again, see, for example, Evans and Rosenquist 1986). Indeed, we see that we can define a mode parameter R by

$$R^2 = f\left(\frac{x}{a}\right) + \frac{x'^2}{2\Delta} \qquad (5\text{-}95)$$

where the prime now denotes longitudinal derivative, such that R^2 represents the normalized total ''energy'' of a ray, which must be conserved along any ray. Recalling our earlier discussion of the parabolic guide, it becomes clear by analogy that rays with $R < 1$ are bound, and those with $R > 1$ will become unbound. Perhaps a good way to visualize the guiding process is by considering the motion of a marble in a gently sloping hole, as is depicted in Figure 5.19. Here we will implicitly replace the z dependence of (5-95) in the temporal de-

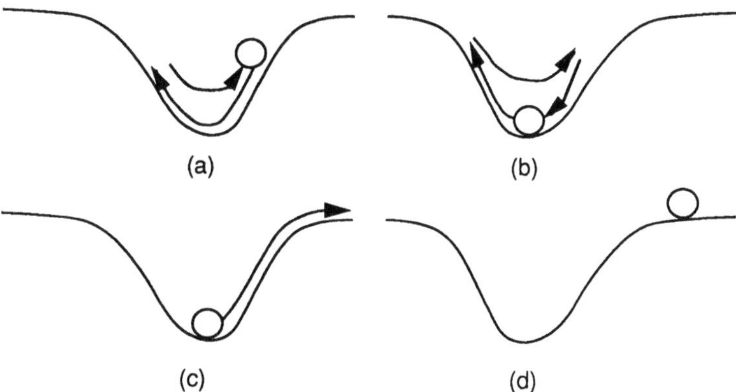

FIGURE 5.19.   Examples of the motion of a marble in a hole. (a) A marble that is dropped from a height less than the total hole height with no initial velocity; its motion remains in the hole. (b) A marble starting from the bottom of the hole with a small-enough velocity to remain bound. (c) A marble with a high-enough initial velocity that it flies out of the hole. (d) A marble that was initially placed outside the hole and does not move at all.

pendence assumed in Figure 5.19. The (a) part of the figure corresponds to a ray that has an initial angle of zero but is excited at some distance from the guide center, so that it oscillates back and forth as it propagates down the guide. The (b) part of the figure corresponds to a ray that is incident on the center of the guide but at an angle that does not exceed the guide's numerical aperture. The (c) part is like (b) but for a ray whose initial angle exceeds the guide's numerical aperture. Figure 5.19(d) corresponds to a ray such as (3) of Figure 5.17, that is, a ray that is excited in the cladding of the guide and remains there. What should be clear from this discussion is that one does not need actually to solve (5-85) to determine whether or not a ray will couple to a guiding structure, which is generally the problem one wants to solve in practice. Whether considering waveguides or imaging systems, one generally wants to maximize one's coupling efficiency. The mode parameter R is the one that determines this efficiency and, further, even gives one an idea of the shape of the ray paths in the guiding region.

5.6 ABOUT OPTICAL INSTRUMENTS

Our next topic is the use of $ABCD$ matrices to analyze (paraxial) imaging systems (see, for example, the discussion in Yariv 1985, Chapter 2). As we just saw with waveguides, the natural way to treat them is via the mode parameter. We will now see that the natural way to treat imaging systems is via the $ABCD$ matrix.

Let us consider what happens to the coordinates $(x, x' = dx/dz)$ of a ray

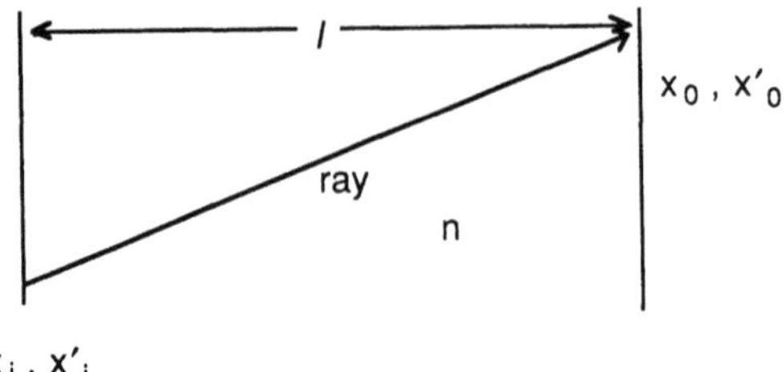

FIGURE 5.20.   Depiction of a ray path in a homogeneous medium.

on traversing a length l of a medium with index n, such as is depicted in Figure 5.20. Implicitly assuming the paraxial approximation to hold, one can write that

$$x_0 = x_i + lx_i' \qquad \text{(a)}$$
$$x_0' = x_i' \qquad \text{(b)}$$

$$(5\text{-}96)$$

which also can be written in the matrix form

$$\begin{bmatrix} x_0 \\ x_0' \end{bmatrix} = \begin{bmatrix} 1 & l \\ 0 & 1 \end{bmatrix} \begin{bmatrix} x_i \\ x_i' \end{bmatrix} \qquad (5\text{-}97)$$

where the matrix is the *ABCD* matrix for the constant index medium. Now we can start generating *ABCD* matrices for all kinds of media. For example, let us recall our parabolic index medium of the last section. If we put in a bound ray at x_i, x_i', equation (5-88) tells us that we will get a ray

$$\begin{bmatrix} x_0 \\ x_0' \end{bmatrix} = \begin{bmatrix} \cos \kappa l & \kappa^{-1} \sin \kappa l \\ -\kappa \sin \kappa l & \cos \kappa l \end{bmatrix} \begin{bmatrix} x_i \\ x_i' \end{bmatrix} \qquad (5\text{-}98)$$

where $\kappa = \sqrt{2\Delta}/a$, and l is the guide length.

We can observe here that if we define a medium, we then can calculate what happens to the coordinates of a ray on traversing the medium. If we consider only media in which the paraxial ray equations reduce to linear equations, we will find that the medium can always be represented as a matrix transformation of the incident ray coordinates. For gradient index media, this occurs essentially only for the parabolic gradient. However, in media that are piecewise homogeneous (i.e., like discrete optical components in a free space configuration), the representation holds for each of the media and therefore for the composite. This linear nature of the transformation makes the *ABCD* matrix approach very powerful, as very complicated systems of piecewise homogeneous media can be represented by products of the matrices representing the constituent components of the system. In this way, all kinds of optical systems may be easily analyzed. It should be remembered, though, that *ABCD* matrices only apply under restrictive approximations. To analyze systems that violate these approximations, one must use more general techniques such as ray tracing.

Figure 5.21 lists *ABCD* matrices for some simple optical media and interfaces. Just using these matrices, one can generate the matrices of some more interesting components. For example, let us first consider the effect of a glass plate of thickness l and index n_2 on an incident ray, as is depicted in Figure

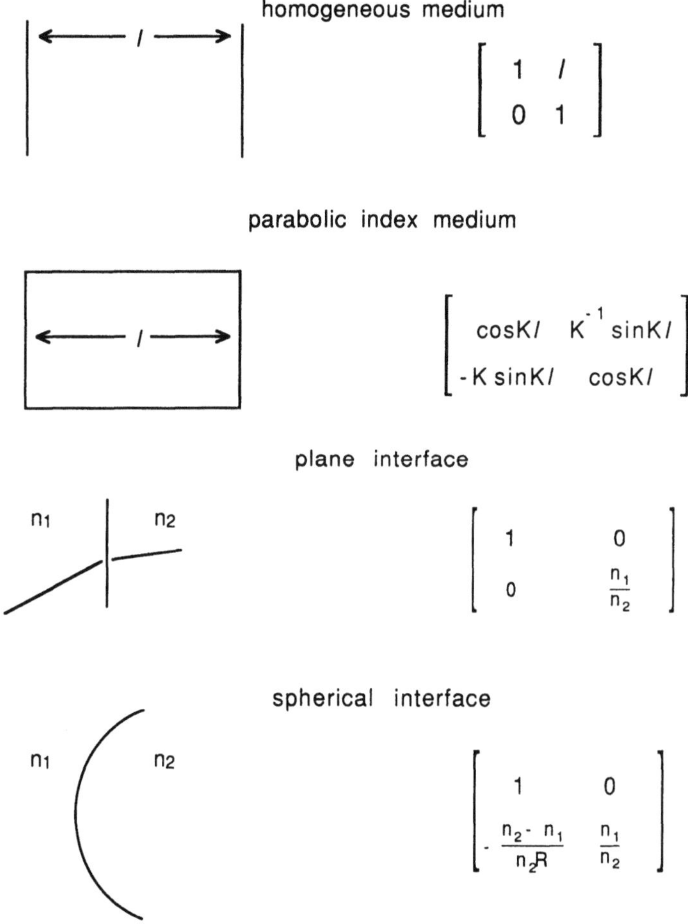

FIGURE 5.21.   Depictions of given media listed, with the *ABCD* matrix representation beside them.

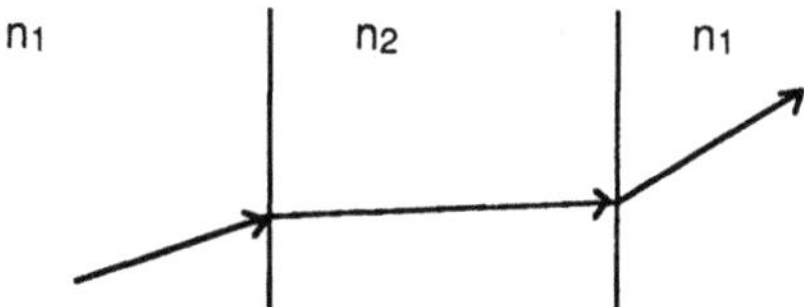

FIGURE 5.22.   Depiction of a glass plate refracting an incident ray.

5.22. The effect of the plate can be calculated as a product of three matrices, one for each interface and one for the propagation. Writing them out, we see

$$\begin{bmatrix} 1 & 0 \\ 0 & \dfrac{n_2}{n_1} \end{bmatrix} \begin{bmatrix} 1 & l \\ 0 & 1 \end{bmatrix} \begin{bmatrix} 1 & 0 \\ 0 & \dfrac{n_1}{n_2} \end{bmatrix} = \begin{bmatrix} 1 & l\dfrac{n_1}{n_2} \\ 0 & 1 \end{bmatrix} \tag{5-99}$$

and we see that the effect is much like that of propagation in a uniform medium but scaled by the ratio of the indices. A more interesting case would be that of two spherical interfaces, as depicted in Figure 5.23, which is the simplest possible model of a lens. Using the matrices from Figure 5.21, we find

$$\begin{bmatrix} 1 & 0 \\ -\dfrac{n_2 - n_1}{n_1 \kappa} & \dfrac{n_2}{n_1} \end{bmatrix} \begin{bmatrix} 1 & 0 \\ -\dfrac{n_2 - n_1}{n_2 \kappa} & \dfrac{n_1}{n_2} \end{bmatrix} = \begin{bmatrix} 1 & 0 \\ -\dfrac{1}{f} & 1 \end{bmatrix} \tag{5-100}$$

where the focal length f is defined by

$$f = \frac{n_1 \kappa}{2(n_2 - n_1)} \tag{5-101}$$

Now that we have a model matrix for a lens, we can start considering imaging systems. The simplest one we can consider is a single lens system, as is

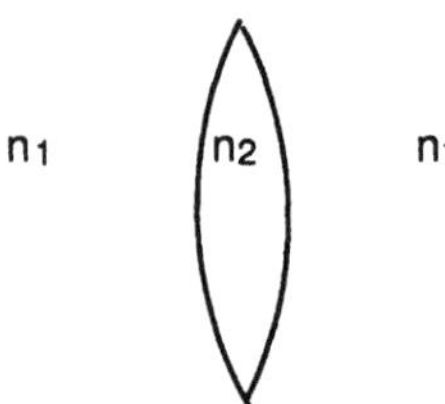

FIGURE 5.23.   Two spherical interfaces back to back.

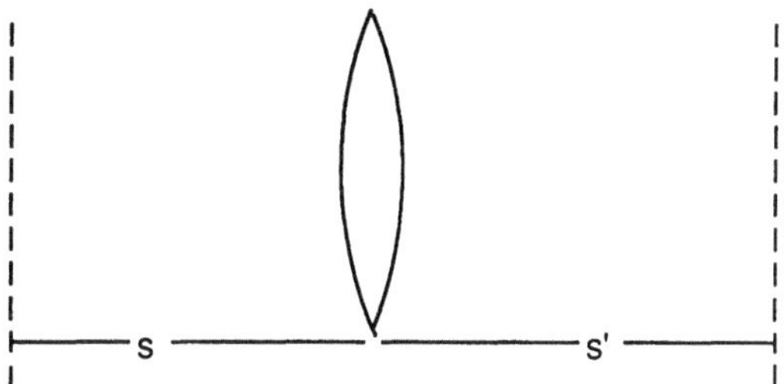

FIGURE 5.24.   Sketch of a single lens imaging system.

depicted in Figure 5.24. The matrix calculation takes the form

$$
\begin{bmatrix} 1 & s' \\ 0 & 1 \end{bmatrix}
\begin{bmatrix} 1 & 0 \\ -\dfrac{1}{f} & 1 \end{bmatrix}
\begin{bmatrix} 1 & s \\ 0 & 1 \end{bmatrix}
=
\begin{bmatrix} 1 - \dfrac{s'}{f} & s + s' - \dfrac{ss'}{f} \\ -\dfrac{1}{f} & 1 - \dfrac{s}{f} \end{bmatrix}
\tag{5-102}
$$

Equation (5-102) is indeed an interesting one, as it contains all the salient features of imaging. A schematic of an imaging system appears in Figure 5.25. The point of the figure is that if x_0 is to be imaged, then all rays emanating from x_0 must converge on the image point Mx_0. As all $B = 0$ (the 1, 2 matrix element), which gives

$$
s + s' - \frac{ss'}{f} = 0
\tag{5-103}
$$

which can be rewritten in the more familiar form

$$
\frac{1}{s} + \frac{1}{s'} = \frac{1}{f}
\tag{5-104}
$$

FIGURE 5.25.   Schematic of an imaging system in which x_0 is imaged with magnification M.

which is known as the imaging condition. Using (5-104) in (5-102), one finds

$$\begin{bmatrix} A & B \\ C & D \end{bmatrix} = \begin{bmatrix} -\dfrac{s'}{s} & 0 \\[2ex] -\dfrac{1}{f} & -\dfrac{s}{s'} \end{bmatrix} \tag{5-105}$$

and it becomes clear that the system magnification is given by $-s'/s$, where the minus sign just indicates that the system is inverting. In a microscope, one would want to maximize the ratio of s'/s in order to achieve the maximum possible signal magnification. In a telescope, one would want to minimize this ratio, as there one wishes to have the maximum possible angular magnification.

Although we have only scratched the surface of the subject of imaging, we will now present a set of examples of more complicated imaging systems that can be understood in terms of the principle enunciated above.

Examples. Refracting telescopes

(a) Keplerian astronomical telescope

This, the simplest of refracting telescopes, consists of two (originally only) convex lenses. It delivers an inverted final image. Even though telescopes are usually used to view objects at great distances ($s_0 \rightarrow \infty$), Figure 5.26 uses a finite object distance for ease of illustration.

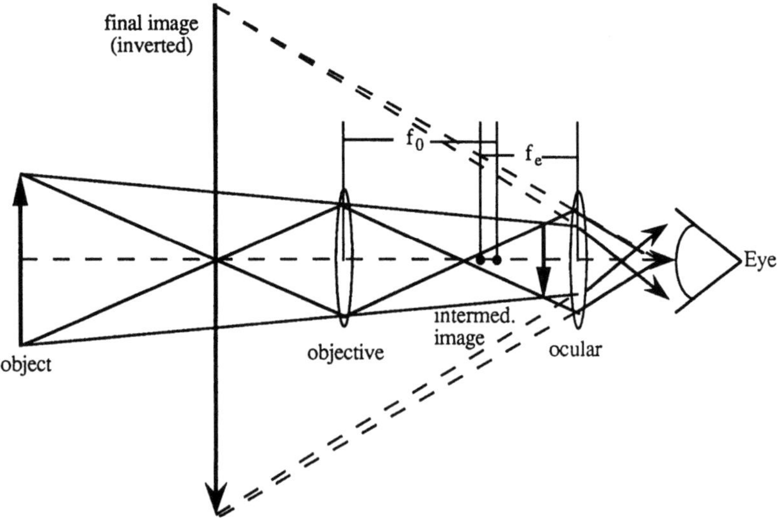

FIGURE 5.26.   Schematic of a simple refracting telescope.

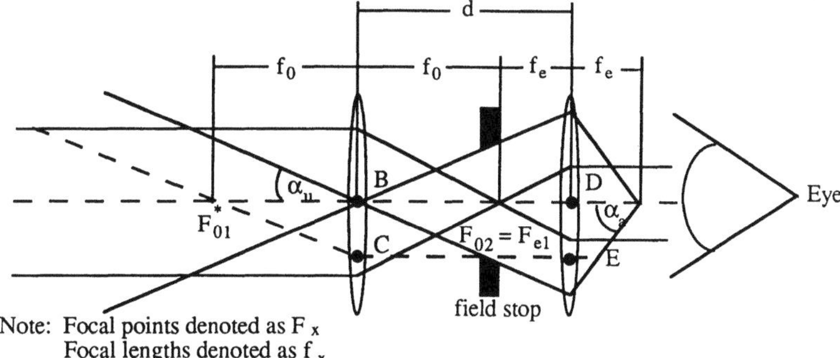

Note: Focal points denoted as F x
Focal lengths denoted as f x

FIGURE 5.27.   Illustration of the placement of an eyepiece focal point.

As previously stated, normally $s_0 = \infty$, which means that the intermediate image rests at F_0 (the focal point of the objective). For normal viewing, the focal point of the eyepiece thus must be at the point F_0 as well ($F_0 = F_e$), as illustrated in Figure 5.27. Now, α_u = half-angle to the margins of visible objects (thus α_u and α_a are measures of the field of view in the object space and the image space, respectively). As is generally known, the angular magnification is

$$M = \frac{\alpha_a}{\alpha_u} \tag{5-106}$$

Note that $F_{02} = F_{e1}$. Also, in the paraxial approximation $\alpha_u \cong \tan \alpha_u$ and $\alpha_a \cong \tan \alpha_a$. The image fills the region of the field stop, and half its extent equals $\overline{BC} = \overline{DE}$. Thus, the ratio of tangents yields

$$M = -f_0/f_e \tag{5-107}$$

Apparently, a Keplerian telescope requires large-diameter lenses for a good field of view, and, for most practical purposes will be rather long, as $d = f_0 + f_e$. Keep in mind that modern telescopes do not use simple convex lenses but generally have multielement objectives, usually doublets or triplets.

(b) Terrestrial telescopes

Terrestrial telescopes are essentially the same as Keplerian telescopes, only they have an erecting system to assure that the final image is noninverted. Figure 5.28 depicts a terrestrial telescope. Because the erecting system does not magnify the image,

$$M = \alpha_a/\alpha_u = -f_0/f_e \tag{5-108}$$

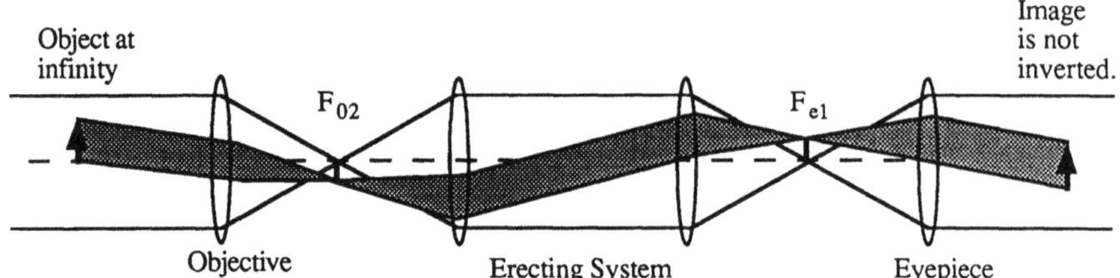

FIGURE 5.28.   Depiction of a terrestrial telescope.

The erecting system in terrestrial telescopes usually is located between the objective and the eyepiece, so the telescopes obviously will have to have long draw tubes.

(c) Binocular telescopes (binoculars)

In order to reduce the lengths of telescopes, the erecting systems from yesteryear have been replaced by erecting prisms that accomplish the same thing. Modern binoculars are very compact and have two objectives to provide stereoscopic effects. Figure 5.29 shows a very simplified schematic of binoculars. Customarily binoculars carry numerical specifications, 6 × 30, 7 × 50, 20 × 50 · · ·, where the first number is the magnification ($M = -f_0/f_e$, of course), and the second number is the clear aperture of the objective (in millimeters).

(d) Galilean telescopes

An alternative to the terrestrial telescope is the Galilean telescope. Built by Galileo, it resembles the Keplerian telescope but uses a concave eyepiece. The Galilean telescope is more compact than Kepler's design and produces a non-

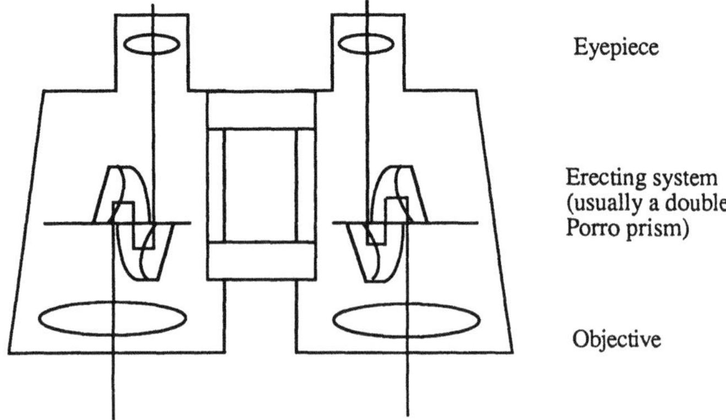

FIGURE 5.29.   Illustration of the principle of operation of binoculars.

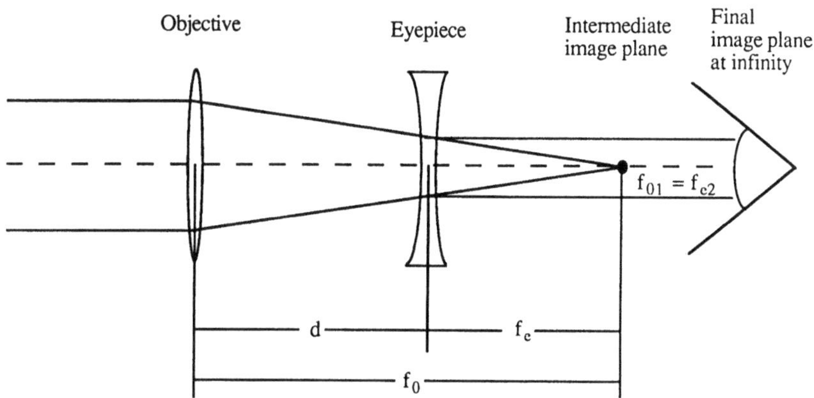

FIGURE 5.30.   Schematic of a Galilean telescope.

inverted image. A Galilean telescope is shown in Figure 5.30. As with all refracting telescopes, the principle remains unchanged. For viewing faraway objects, $d = f_0 + f_e$, only now $f_e < 0$. Also, as before, $M = \alpha_a/\alpha_u = -f_0/f_e$ because the rays can be traced in the same fashion as with the previous examples, only the eyepiece does not invert the final image.

(e) Erecting telescopes

As a numerical example to illustrate an erecting telescope, depicted in Figure 5.31, we will determine the necessary powers and spacings to produce a magnification of 10 × and a length of 20″. Assume that we have chosen $f_0 = 5″$, $f_e = 2″$, and an eye relief of 4″. The magnification of a terrestrial telescope is the magnification that the telescope would have without the erector, multiplied

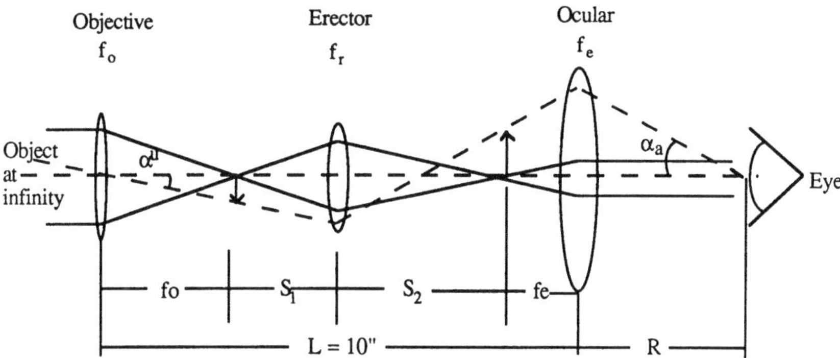

FIGURE 5.31.   Optics of a simple erecting telescope.

by the linear magnification of the erector system:

$$M = \frac{f_0}{f_e}\frac{s_2}{s_1} \tag{5-109}$$

Note that

$$L = f_0 + s_1 + s_2 + f_e \tag{5-110}$$

We can combine the expressions (5-109) and (5-110) to get

$$s_1 = \frac{f_0(L - f_0 - f_e)}{Mf_e + f_0} \tag{a}$$

$$s_2 = \frac{s_1 Mf_e}{f_0} = \frac{Mf_e(L - f_0 - f_e)}{Mf_e + f_0} \tag{b} \quad (5\text{-}111)$$

$$f_r = \frac{s_1 s_2}{s_1 + s_2} = \frac{Mf_e(L - f_0 - f_e)}{Mf_e + f_0^2} \tag{c}$$

Also,

$$f_0 = 5'' \tag{a}$$

$$f_e = 2'' \tag{b}$$

$$s_1 = \frac{5(10 - 5 - 2)}{10 \times 2 + 5} = 0.6'' \tag{c}$$

$$(5\text{-}112)$$

$$s_2 = \frac{s_1 \times 10 \times 2}{5} = 2.4'' \tag{d}$$

$$f_r = \frac{s_1 s_2}{s_1 + s_2} = 0.48'' \tag{e}$$

5.7 PHASE SPACE AND LIOUVILLE'S THEOREM

We have seen that we can make few statements about the general equations of geometrical optics other than the intensity law of geometrical optics and the associated brightness theorem. We also have seen, however, that an analogy exists between paraxial geometrical optics and mechanics. In this section, we will find that an analogy can be efficaciously exploited to yield general statements about the propagation of energy in optical systems in general and in paraxial systems in particular.

There does exist a theoretical basis for the relations between mechanics and ray optics. Perhaps the easiest way to investigate these relations is in terms of Lagrangian and Hamiltonian mechanics. We saw in the preceding section how the important parameters of a ray in one-dimensional propagation were its position, x, and the rate of change of its position, dx/dz. In mechanics it is customary to find equations of motions for particles from an action principle of the form (see, for example, Born and Wolf 1975, Appendix A; Goldstein 1980, Chapter 2; Sommerfeld 1964a, Chapter 6; Feynman and Hibbs 1965, Chapter 11; or Schiff 1968):

$$\delta \int_{t_1}^{t_2} L(x, \dot{x}, t)\, dt = 0 \quad \delta x(t_1) = \delta x(t_2) = 0 \tag{5-113}$$

where $\dot{x}$ here denotes dx/dt, the δ denotes a functional variation, and $L(x, \dot{x}, t)$ is known. It is well known that the variation of the action, that is, the integral of the Lagrangian over a path with fixed endpoints, is solved by $x(t)$, $\dot{x}(t)$, which satisfy the Euler-Lagrange equation

$$\frac{\partial}{\partial t}\left(\frac{\partial L}{\partial \dot{x}}\right) - \frac{\partial L}{\partial x} = 0 \tag{5-114}$$

Now the action principle of (5-113) essentially is a prescription for finding extrema of the action integral. This requires finding minima or maxima of the action functional. The coordinates that are necessary to maximize this functional will give the path coordinates, parameterized by time. The problem, then, is reduced to a purely mathematical core once the action principle is chosen. All the physics of the problem is contained in the action principle itself. Fermat determined such a principle around 1650, a principle known as that of least time (see, for example, Feynman, Leighton, and Sands 1963, Chapter 26 of Volume 1). Fermat's action can be written as (see, for example, Marcuse 1982, Section 3.4):

$$\int_{P_1}^{P_2} dt = \int_{P_1}^{P_2} n(x, y, z)\, ds \tag{5-115}$$

where P_1 and P_2 are the two points to be joined by the ray, and $n(x, y, z)$ and ds are, as before, the index of refraction distribution and the element of length along the ray. What we would really like, however, is a form like (5-115) with the time replaced by the coordinate z, which can be obtained from (5-115) in the form

$$\int_{P_1}^{P_2} dt = \int_{P_1}^{P_2} L(x, \dot{x}, z) \, dz \qquad (5\text{-}116)$$

where $L(x, \dot{x}, z)$ can be expressed as

$$L(x, \dot{x}, z) = n(x, z) \sqrt{1 + \dot{x}^2} \qquad (5\text{-}117)$$

where $\dot{x}$ now denotes a derivative with respect to z. Minimization of (5-116) leads to the system of (5-114), whose solution yields the rays of the equations of (5-60).

Hamilton, who first formulated the principle of least action, also developed a different formulation of mechanics. In his formulation, one defines a variable canonical to the position x by

$$p = \frac{\partial L(x, \dot{x}, z)}{\partial \dot{x}} \qquad (5\text{-}118)$$

where we continue to use z instead of the more conventionally used t, and one then defines a Hamiltonian function by

$$H(x, P, z) = p\dot{x} - L(x, \dot{x}, z) \qquad (5\text{-}119)$$

One then can show that the equations of (5-114) can be expressed by

$$\dot{x} = \frac{\partial H}{\partial p} \qquad \text{(a)}$$

$$\dot{p} = -\frac{\partial H}{\partial x} \qquad \text{(b)}$$

$$(5\text{-}120)$$

It is worthwhile to reflect on where we are going with this development. So far, all we have accomplished through defining $L(x, \dot{x}, z)$ and $H(x, p, z)$ is to find a complicated formulation to simplify the derivation preceding equation (5-60). As we already knew equation (5-60), this is not a very useful procedure. What is valuable from Hamilton's formulation is, however, the definition of the canonical variable p, and, as we will see presently, the definition of phase space. From the definition of (5-118) and from equation (5-119), one notes that

$$p = n(x, z) \frac{x}{\sqrt{1 + x^2}} = n(x, z) \sin \theta \qquad (5\text{-}121)$$

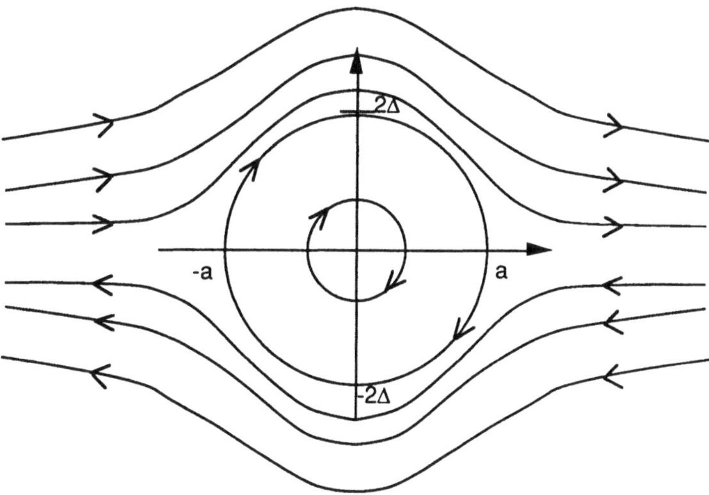

FIGURE 5.32.   Some ray trajectories for a waveguide plotted in phase space.

where θ is the usual ray tracing variable. Now p is a generalized momentum, and, indeed, in the paraxial limit we said it is closely related to $\theta = \dot{x}$, where the index of refraction plays the part that the mass plays in mechanics. Now a ray, at a given coordinate z, will be completely specified by its two coordinates x and p. With increasing z, the x and p values can change. Therefore, a ray can be specified by a curve (trajectory) on an x–p diagram, as is illustrated in Figure 5.32, where some trajectories for a waveguide, such as those discussed following equation (5-90), are selected. The idea is that trajectories with small-enough R value will be bound and therefore will exhibit a periodic behavior. Rays with R greater than 1 and non-zero momentum will eventually head off to $x = \pm\infty$. Rays with zero momentum that begin outside the core appear as points that do not move with z.

Figure 5.33 illustrates the concept of ray density and its evolution. A source can be characterized by the rays it emits. Each point on the (flat) surface of a source will emit into some solid angle with some angular distribution. This could be represented as a set of points in phase space, one for each ray, with intensity determined by density. A single point on the surface will correspond to a set of points lying on a line parallel to the x-axis, whereas all the rays emitted at a given angle would correspond to a line parallel to the p-axis. The first sketch in Figure 5.33 is that of a rectangle, drawn to enclose (exactly) the area that contains all the rays emitted from a (partially coherent) source. The second part of the figure illustrates what happens when such a distribution propagates forward. A point with a positive momentum will move along $+x$ and one with negative momentum along $-x$, causing the rectangle to rotate with

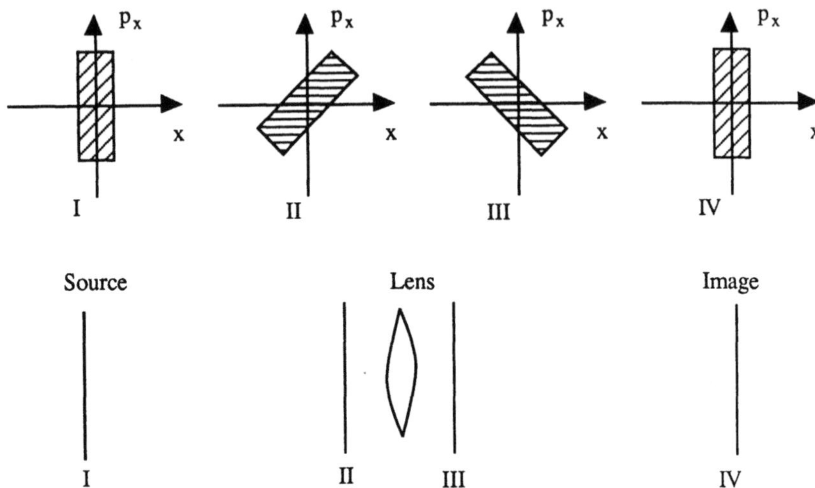

FIGURE 5.33.   Evolution of the ray density in an imaging system.

moderate deformation. An ideal imaging lens is designed to exactly reverse this pseudo rotation such that further propagation will return the rectangle to an upright position, with perhaps scaled x and p extents. It is this scaling, and, indeed, the propagation of the area itself, that will presently capture our attention.

Now to find how an element of area $\Delta A = \Delta x \, \Delta p$ transforms along a trajectory, one need recall that Θ, the relative change of area (areal dilatation) of a plane figure, is given by (Sommerfeld 1964b, Section 1; Marcuse 1982):

$$\Theta = \frac{\partial \xi}{\partial x} + \frac{\partial \eta}{\partial p} \tag{5-122}$$

where ξ and η denote the translations of a point of the area in the directions x and p, respectively. Therefore, one could write that

$$\xi = dx = \dot{x} \, dz \qquad \text{(a)}$$
$$\eta = dp = \dot{p} \, dz \qquad \text{(b)} \tag{5-123}$$

giving for Θ

$$\Theta = \frac{\partial \dot{x}}{\partial x} + \frac{\partial \dot{p}}{\partial p} \tag{5-124}$$

However, one could use (5-120) for $\dot{x}$ and $\dot{p}$ to obtain

$$\Theta = \frac{\partial}{\partial x}\frac{\partial H}{\partial p} - \frac{\partial}{\partial p}\frac{\partial H}{\partial x} = 0 \tag{5-125}$$

which shows that each elemental area is conserved, a result often referred to as Liouville's theorem (Sommerfeld 1964c, 207f; Marcuse 1982). But as a total area is made up of elemental areas, it also says that the total phase space area of any arbitrary optical excitation will be conserved in the absence of scattering or absorption. Denoting the density of rays by $\rho(x, p, z)$, one could express this invariance in the form

$$\frac{d\rho(x, p, z)}{ds} = 0 \tag{5-126}$$

where the d's denote a total derivative, and s is the arc length along a ray path. Now as ray paths can be found in any passive medium, (5-126) states that the ray density cannot be increased by any passive medium, which, in particular, must include any imaging system. This statement sounds a great deal like a statement of the earlier-discussed brightness theorem for the specific intensity. Indeed, (5-126) is a more succinct statement of the brightness theorem. The ray density is closely related but not identical to the specific intensity. This difference is both dimensional (the density is dimensionless) and due to the use of the canonical momentum p instead of the derivative of the x coordinate. The p contains the $n(x, z)$. As we saw in Chapter 2, energy densities must vary with n because of the change in the energy stored in the polarizable medium. Dimensionless quantities need not have such n dependence.

As a final point, it can be of interest at least to mention the relation of (5-126) to the brightness theorem and the equation of Liouville and Boltzmann. The Boltzmann transport equation is the one generally used to track any type of particle transport. Now, in the absence of collisions, the above-derived Liouville theorem would apply, and the particle density in phase space would remain constant along a trajectory [set of $\mathbf{r}$, $\mathbf{p}$ values that satisfy the equations of motion (5-120)]. Particles, however, interact with one another; so the collision in the Boltzmann equation is of paramount importance. If one denotes $f(\mathbf{x}, \mathbf{p})$ as the particle density in phase space, one can write (see, for example Kittel and Kroemer 1980, Chapter 14)

$$\frac{df(\mathbf{x}, \mathbf{p}, t)}{dt} = -\frac{f(\mathbf{x}, \mathbf{p}, t) - f_0(\mathbf{x}, \mathbf{p}, t)}{\tau_c} \tag{5-127}$$

where $f_0(\mathbf{x}, \mathbf{p}, t)$ is an equilibrium distribution, and τ_c is a characteristic decay time. The optical problem, that of radiometry as described by the brightness theorem, is a different problem, in that light does not collide with light. However, light does get absorbed, emitted, and scattered. These effects are not taken into account in Liouville's theorem and must be put into the equation for $\rho(\mathbf{x}, \mathbf{p}, z)$ or $I(\mathbf{r}, \Omega)$ in a manner analogous to the manner in which collisions were included in (5-127) to be the equation of transfer, which serves as a basis for the theory of radiative transfer. It should be noted in passing, that although radiometry can be generalized to sources of arbitrary coherence (Wolf 1978a, Wolf 1978b), radiative transfer only applies to incoherent sources, as scattering processes are very sensitive to the coherence of the incident light.

References

Arnaud, J. A., Mode coupling in first-order optics, *JOSA 61*, 751–758 (1971).

Babič and Buldyrev, *Asymptotic Methods in Short-Wave Diffraction Theory*, translated by E. F. Kuester, Springer-Verlag, Berlin (1991).

Bender, C. M. and S. A. Orszag, *Advanced Mathematical Methods for Scientists and Engineers*, McGraw-Hill, New York (1978).

Berry, M. V., Quantal phase factors accompanying adiabatic changes, *Proc. R. Soc. Lond. A392*, 45–47 (1984).

Berry, M. V., Classical adiabatic angles and quantal adiabatic phase, *J. Phys. A: Math. Gen. 18*, 15–27 (1985).

Born, M. and E. Wolf, *Principles of Optics*, Fifth edition, Pergamon Press, New York (1975).

Brillouin, L., *Compt. Rend. 183*, 24 (1926).

Chiao, R. T. and Y. S. Wu, Manifestations of Berry's topological phase for the photon, *Phys. Rev. Lett., 57*, 933–936 (1986).

Eckart, C., *Phys. Rev. 28*, 711–726 (1926).

Einstein, A., *Verh. Deutsch. Phys. Ges. 19*, 82 (1917).

Evans, J. and M. Rosenquist, "$F = ma$" optics, *Am. J. Phys. 54*, 876–883 (1986).

Feynman, R. P. and A. R. Hibbs, *Quantum Mechanics and Path Integrals*, McGraw-Hill, New York (1965).

Feynman, R. P., R. B. Leighton, and M. Sands, *The Feynman Lectures on Physics*, Addison-Wesley Publishing Co., New York (1963).

Goldstein, H., *Classical Mechanics*, Second edition, Addison-Wesley, Reading, MA (1980).

Grimaldi, F. M., *Physics-Mathesis de lumine, coloribus et iride*, Bologna (1665).

Hecht, E. and A. Zajac, *Optics*, Second edition, Addison-Wesley, Reading, MA (1987).

Jeffries, H., *Proc. London Math. Soc. (2)23*, 428 (1923).

Keller, J. B., Corrected Bohr-Sommerfeld quantum conditions for nonseparable systems, *Ann. Phys. 4*, 180–188 (1958).

Keller, J. B., Semiclassical mechanics, *SIAM Review 27*, 485–504 (1985).

Keller, J. B. and S. I. Rubinow, Asymptotic solution of eigenvalue problems, *Ann. Phys. 9*, 24–75 (1960).

Kittel, C. and H. Kroemer, *Thermal Physics*, Second edition, W. H. Freeman, San Francisco (1980).

Klein, M. V. and F. E. Furtak, *Optics*, Second edition, John Wiley and Sons, New York (1986).

Kramers, H. A., *Z. Physik 38*, 828 (1926).

Liouville, J., *Journal de Math. 2*, 16, 418 (1837).

Marcuse, D., *Light Transmission Optics*, Second edition, Van Nostrand Reinhold, New York (1982).

Mihalas, D., *Stellar Atmospheres*, W. H. Freeman, San Francisco (1978).

Motz, L., *Astrophysics and Stellar Structure*, Ginn, Waltham, MA (1970).

Raleigh, Lord, *Proc. Roy. Soc. Lond. A86*, 207 (1912).

Schiff, L. I., *Quantum Mechanics*, Third edition, McGraw-Hill, New York (1968).

Schrödinger, E., *Ann. Physik 79*, 734 (1926).

Sommerfeld, M., *Mechanics*, Volume 1, *Lectures on Theoretical Physics*, Academic Press, New York (1964a).

Sommerfeld, A., *Mechanics of Deformable Bodies*, Volume II, *Lectures on Theoretical Physics*, Academic Press, New York (1964b).

Sommerfeld, A., *Thermodynamics and Statistical Mechanics*, Volume V, *Lectures on Theoretical Physics*, Academic Press, New York (1964c).

Spiegel, M. R., *Vector Analysis*, Schaum's Outline Series, McGraw-Hill, New York (1959).

Tomita, A. and R. Y. Chiao, *Phys. Rev. Lett. 57*, 937–939 (1986).

Wentzel, G., *Z. Physik 38*, 518 (1926).

Wolf, E., Coherence and radiometry, *J. Opt. Soc. Am. 68*, 6–17 (1978a).

Wolf, E., The radiant intensity from planar sources in any state of coherence, *J. Opt. Soc. Am. 68*, 1597–1605 (1978b).

Yariv, A., *Optical Electronics*, Third edition, Holt Rinehart and Winston, New York (1985).

Problems

1. Consider a plane wave incident on a half-space of an inhomogeneous medium, as depicted in Figure 5.34. We wish to find the form of the wave equation satisfied in

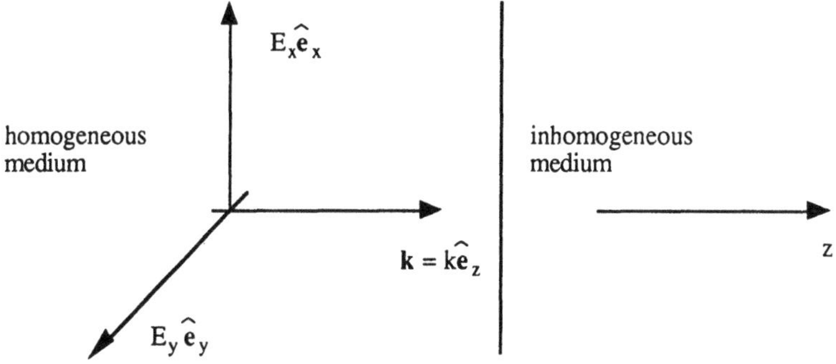

FIGURE 5.34.   Figure for problem 1.

the inhomogeneous medium, where the index in the medium has the following dependences:

(a) $n = n(x, y, z)$.

(b) $n = n(z)$.

(c) $n = n(x)$.

(d) $n = n(t)$.

2. This problem will compare various forms of solution of the propagation problem for a medium in which the index of refraction $n(x, y, z)$ is a function only of z.

(a) Find the "wave-like" equation for the E-vector in a medium with $n(z)$.

(b) Can you choose an $n(z)$ (other than the trivial one) for which you can solve the equation in (a)?

In (c) through (e), consider the medium defined by

$$n(z) = n_0 + \Delta n \sin kz$$

where one can assume $\Delta n \ll n_0$.

(c) Find the eikonal. Can one easily find the ray directions from the eikonal?

(d) Write down the paraxial ray equations. Can they be solved analytically?

(e) How would the polarization vector vary in such a medium?

3. Say that we have a semi-infinite medium ($z \geq 0$) with index of refraction $n(z)$ ($n(0) = 1$). At the plane $z = 0$ we have a ray incident from free space ($z < 0$) into the medium, with direction cosines α_1, α_2, $\sqrt{1 - \alpha_1^2 - \alpha_2^2}$ at the point $x = y = z = 0$.

(a) Find the eikonal for this medium.

(b) Find $\hat{s}$, the tangent to the ray path.

(c) Find the parametric (in s) equations of the ray path.

(d) If $n(z) = 1 + az$, what is the ray path asymptotically (at large z)?

(e) If $n(z) \cong a/(1 + bz)$, $b > 0$, what is the asymptotic ray path?

4. Consider a medium whose index of refraction is a function $n(z)$, that is, is independent of x and y. Refer to the text for an example of such a medium. Assume that the ray enters the medium at the plane $z = 0$, with direction cosines

$$\frac{1}{n(0)} \left(\alpha, \beta, \sqrt{n^2(0) - \alpha^2 - \beta^2} \right)$$

(a) Find an expression for the ray path in the medium.

(b) For $n^2(z) = n^2(0) [1 + \Delta z^2]$ and an initial condition of $\alpha = n(0)/2$, $\beta = 0$ sketch the ray path.

(c) For initial conditions as in (b) but $n^2(z) = n^2(0)/(1 + \Delta z^2)$, sketch the ray path. Is something wrong?

(d) Say that

$$n^2(z) = \begin{cases} n^2(0) & z < 0 \\ .4n^2(0) & 0 < z < a \\ n^2(0) & z > a \end{cases}$$

What are the ray paths?

(e) Say that a medium as in (b) is excited with a source that excites rays with all α's from $-.5$ to $.5$. What does the output ray congruence look like?

5. Suppose that we know the eikonal in a given medium is described by $S = Az - B(x^2 + y^2)$.
   (a) Find the index of refraction of the medium.
   (b) Find the ray paths in the medium.
   (c) Qualitatively describe what would happen to a plane wave after entering such a medium.

6. Consider a two-dimensional medium with index $n^2(x, z) = 1 - Kx^2$.
   (a) By assuming an eikonal $S(x, z) = f(x) + \alpha z$, find a form for the eikonal.
   (b) Find the ray paths (approximately) for different initial conditions on the eikonal. Interpret your answer.

7. The purpose of this problem is to compare the transport of the $\mathbf{e}$ and $\mathbf{h}$ vectors along a given curve.
   (a) Assuming that $\hat{\mathbf{s}} = \nabla S / n$, compare the differential equations obtained for the e-field components along the $\hat{\mathbf{n}}$ and $\hat{\mathbf{b}}$ directions with those for the propagation of $\hat{\mathbf{n}}$ and $\hat{\mathbf{b}}$.
   (b) Assuming a ray path given by

   $$x(t) = 3 \cos t$$
   $$y(t) = 3 \sin t$$
   $$z(t) = 4t$$

   Find the κ and τ. *Hint:* The relations

   $$\hat{\mathbf{s}} = \frac{d\mathbf{r}}{ds} = \frac{d\mathbf{r}/dt}{|d\mathbf{r}/dt|}$$

   $$\kappa = \left| \frac{d\hat{\mathbf{s}}}{ds} \right|$$

   $$\tau = \left| \frac{d\hat{\mathbf{b}}}{ds} \right|$$

   may be useful.
   (c) Discuss the evolution of an initial polarization vector along the curve given in (b).

8. Consider the imaging system depicted in Figure 5.35.
   (a) Write an *ABCD* matrix for the system.
   (b) Find the imaging condition.
   (c) What is the system magnification?
   (d) How would one take into account the effect of the finite apertures of lenses f_1 and f_2? How would this affect the calculation? How would this affect the actual imaging system?
   (e) Repeat (d) but for one opaque aperture stop between f_1 and f_2, say at a distance f_1 behind lens f_1. Model the aperture stop as a circle of radius a_s that passes no light at radius greater than a_s. Why would one use such a stop?

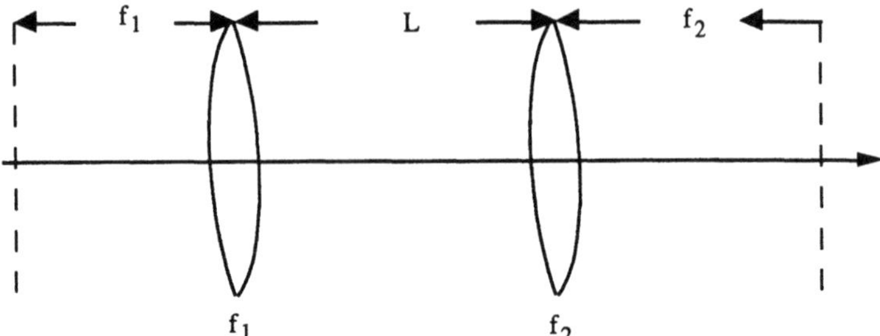

FIGURE 5.35.   Figure for problem 8.

9. Consider the two-lens system depicted in Figure 5.36.
   (a) What is the imaging condition for this system?
   (b) How would you optimize this system for microscopic or telescopic applications? Give formulae for magnifications.
   (c) What is the output if an aperture stop of radius a_s is placed at the stop plane?
   (d) Make an analytical argument that shows the dependence of the depth of field on the stop radius.

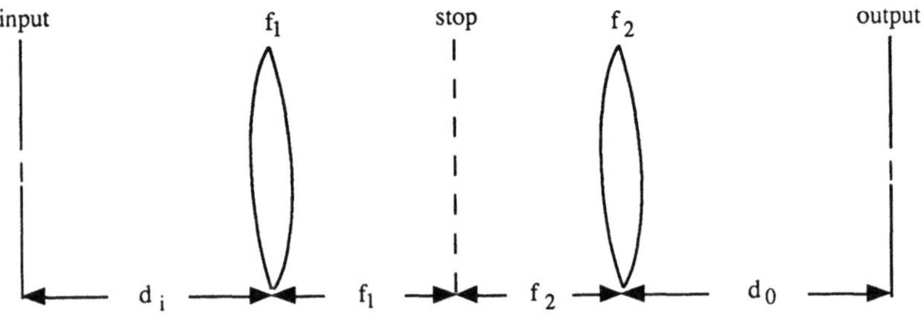

FIGURE 5.36.   Figure for problem 9.

10. As illustrated in Figure 5.37, a simple compound microscope consists of two lensing systems, known as an objective and an eyepiece. The distance from the objective back focal point to the field stop aperture where the first real image is formed is known as the tube length, and typically is given as 150 mm. The standard near point, defined as the distance of the virtual image to the eyepiece, is typically taken as 250 mm. Assume that a $10.0\times$ eyepiece with a 25 mm focal length (f_e) is used.
   (a) Given an objective with a focal length of $f_0 = 16.9$ mm, and a numerical aperture of NA $= 0.25$, what is the total magnification of the microscope?
   (b) Repeat part (a) with an objective with a focal length of $f_0 = 3.1$ mm and having a numerical aperture of NA $= 0.85$.
   (c) Assume that the object is illuminated by a monochromatic source with a wavelength of $\lambda_0 = 500$ nm. The object consists of two lines separated by one micron. Will each of the lenses given in parts (a) and (b) be able to resolve this object?

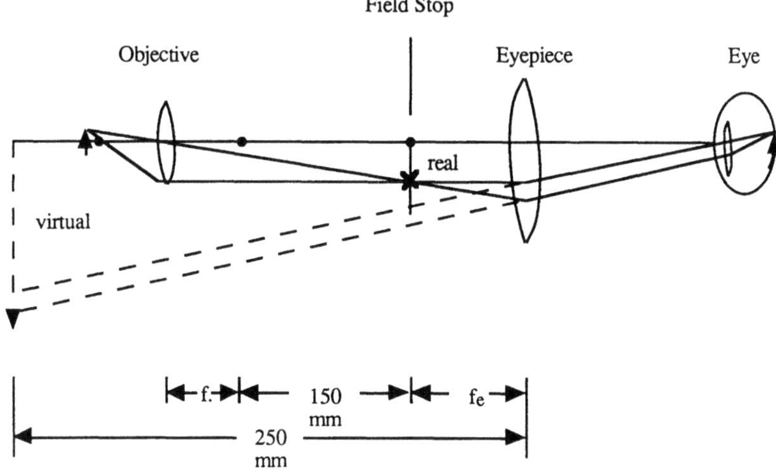

FIGURE 5.37.   Figure for problem 10.

11. Consider the imaging system depicted in Figure 5.38.
    (a) Use *ABCD* matrices to find the ray height and angle at planes IIa, IIb, III, IVa, IVb, and V for an input ray of x_1, x_1'. Say that a turbulent wave travels downward through the planar section labeled T, behind lens 1, at a velocity v. This turbulence affects the phase correlation of the incident wave such that at the output of the turbulent layer

$$\langle \epsilon^{i[(x,t)-(x,t+2)]} \rangle = \epsilon^{-\tau/\tau_\varphi}$$

    (b) Calculate the bandwidth of the wave at plane III. What can be said about the degree of polarization of the wave at this plane? The correlation $\langle a(x, t)\, a(x, t + \tau)\rangle$?

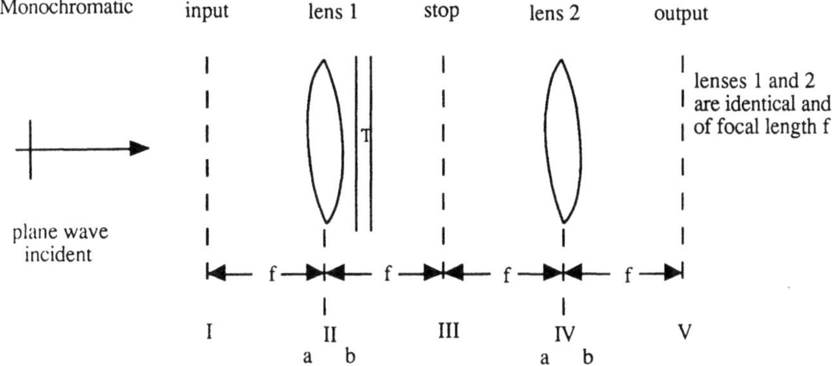

FIGURE 5.38.   Figure for problem 11.

(c) Assume that the input plane contains a tiny pinhole on the axis. Find an expression for the intensity pattern at plane III, assuming that the turbulent layer is very thin and right behind lens 1. Find the intensity at plane IV in the limits of $\tau_\varphi \rightarrow \infty$ and $\omega_0 \tau_\varphi \rightarrow 1$. Sketch the phase fronts of the waves for these two cases.

(d) What does the image (at plane V) of the point source of (c) look like for the limit $\omega_0 \tau_\varphi \rightarrow 1$? Assume that the NA of the system is roughly .1 if you need to. What effect would this have on the image of a complicated transparency? How could one correct for this problem?

12. An eye may be approximated as a simple symmetric lens of focal length F_e with a zoom range from 33 to 40 mm and a fixed image plane or retina at 40 mm. Assume that the index of refraction of the lens $n_2 = 1.44$, and that of the fluid between lens and retina $n_3 = 1.34$.

(a) Use the *ABCD* matrix. Derive the imaging condition and the magnification.

(b) In viewing a particular small object of linear dimension l_0, what range of object distances is possible, and what sizes of retinal image result?

(c) Will the vision of a near-sighted person (focus is in front of the retina) improve under water or not? The index of refraction of water is 1.34.

(d) Given a camera with an F number of 2.0, the camera is immersed in water (it is waterproof). Does the F number change? If so, to what value?

13. An eye may be approximated as a simple lens of focal length F with a zoom range from 33 to 40 mm and a fixed image plane or retina at 40 mm. In viewing a particularly small object of linear dimension l_0:

(a) What range of object distances is possible?

(b) What sizes of retinal images result?

14. This problem will consider the paraxial ray equations in a medium where the index is given by

$$n(x, z) = \frac{x^4}{a^4} + b(z) \frac{x^2}{a^2} + c$$

where it can be assumed that $c \gg 1$, and that $b(z)$ is a slowly varying function of z.

(a) Consider the limit in which $b(z) = b_0 \gg 1$ but $x < a$ such that

$$b_0 \left(\frac{x}{a}\right)^2 < c$$

What are the ray trajectories corresponding to the low ray energies in this configuration? Sketch where these trajectories lie in the effective potential.

(b) Consider a limit in which $b(z) \sim 0$. Again, consider the trajectories of lowest energy. What equation do they satisfy? Can you solve it? Sketch where these trajectories lie in the effective potential.

(c) Consider the limit in which $b(z) = b_0 < 0$. Where would the minimum energy trajectories lie if one were to plot them on an effective potential diagram? Where would they lie in phase space versus those in (a) and (b)?

(d) Consider the case in which $b(z) = b_0(1 - \alpha z)$ where b_0 is as it was in (a). What is the paraxial ray equation? Can you solve it?

(e) Consider the minimum energy ray trajectories of (a). If one of these were the initial condition in (d), what would happen to this ray when $b(z)$ passed through 0 and on to negative values? Explain your answer.

15. Consider a medium with index of refraction

$$n^2(x, z) = \begin{cases} n_c^2 \left(1 - \dfrac{2\Delta(z)}{a^2(z)} x^2 \right) & x < a(z) \\[3mm] n_c^2 (1 - 2\Delta(z)) & x > a(z) \end{cases}$$

where $\Delta(z)$ and $a(z)$ are very slow functions of z. We wish to consider rays entering this medium at $z = 0$ with direction cosines

$$\frac{1}{n(x_e, 0)} (\alpha, 0, \sqrt{n^2(x_e, 0) - \alpha^2})$$

at a height x_e above the x-z axis crossing.

In (a) and (b) take Δ and a as independent of z. In (a) through (e), sketch the ray paths:

(a) For $\alpha = 0$ and $x_e = .8a(0)$.

(b) For $x_e = 0$ and $\alpha = \sqrt{\Delta}$.

(c) For $x_e = .8a(0)$, $\alpha = 0$, a constant, and $\Delta(z) = \Delta(0) (1 + (z/z_0))$, z_0 constant.

(d) For $x_e = .8a(0)$, $\alpha = 0$, Δ constant, and $a(z) = a(0) (1 + (z/z_0))$, z_0 constant.

(e) For $x_e = .8a(0)$, $\alpha = 0$, Δ constant, and $a(z) = a(0) (1 - (z/z_0))$, z_0 constant.

16. Consider a medium described by an index of refraction distribution

$$n^2(x) = \begin{cases} n_0^2 \left(1 - 2\Delta \dfrac{x^2}{a^2} \right) & x < a \\[3mm] n_0^2 (1 - 2\Delta) & x > a \end{cases}$$

where $\Delta \ll 1$.

(a) Find the *ABCD* matrix for such a medium that extends for a length l in the z direction.

(b) What is the imaging condition for this medium? What values of l will satisfy this condition?

(c) Unfortunately, even small thermal variations will cause changes in the value of Δ. Say that an original length l of graded index medium is affected by thermal variations such that the first length l_1 is unaffected, a length of l_2 is affected, and the last length l_3 is unaffected. Call the modified Δ, Δ'. What is the modified *ABCD* matrix of the medium? Assume that $l_1 + l_2 + l_3 = l$.

(d) Two possible applications of gradient index media are imaging and temperature sensing. Say that a's from 10 μm to 5 μm are available, lengths from a couple of centimeters to 50 km are achievable, and that Δ can be affected by 1% by ambient temperature changes. Design optimal imaging and sensing systems, and estimate their sensitivity or lack thereof.

17. (a) Write the paraxial ray equations in cylindrical coordinates.

 (b) Solve these equations for the parabolic profile

$$n^2(r) = n_1^2 \left[1 - 2\Delta \left(\frac{r}{a} \right)^2 \right]$$

 The parabolic profile's ray trajectories also can be found in rectangular (x, y) coordinates.

 (c) Solve for the ray trajectories of the parabolic profile in a rectangular coordinate system.

 (d) Do the results of (c) agree with the results of (b)? Explain your results.

18. Consider a paraxial ray, with initial conditions $x(0) = x_0$, $(dx/dz)(0) = x_0'$, $y(0) = y_0$, $(dy/dz)(0) = y_0'$ in a medium for which $n = n(z)$, for $z > 0$, $n(z) = n(0)$, $z < 0$.

 (a) Solve the paraxial ray equation.

 (b) Show that these rays are perpendicular to the eikonal found earlier in this chapter, and that the rays are indeed in the direction $\hat{s} = \nabla S/n$.

 In (c) and (d) one is to plot the ray paths with the eikonals superimposed and describe what happens in the limit $z \to \infty$.

19. Say that we have a medium whose index is expressible in the form

$$n^2(x) = n_1^2 \left[1 - 2\Delta f \left(\frac{x}{a} \right) \right]$$

 where $f(x/a) = \Sigma_{n=0}^{N} a_{2n} (x/a)^{2n}$ where $f(0) = 0, f(x/a) > 0$.

 (a) Sketch all the possible shapes of the function f for $N = 2$.

 (b) For a given N, how many families (congruences) of nonintersecting rays are possible?

 (c) How would one go about exciting two different nonintersecting ray congruences?

 (d) Say that $f(s) = s^2 - 5s^4 + 4s^6$. What is the numerical aperture of this guide?

 (e) Can one selectively excite the different ray congruences of the guide in (d) by a method different from that described in (c)?

20. Consider a medium defined by the matrix transformation

$$\begin{bmatrix} x(l) \\ x'(l) \end{bmatrix} = \begin{bmatrix} \cos \kappa l & \frac{1}{\kappa} \sin \kappa l \\ -\kappa \sin \kappa l & \cos \kappa l \end{bmatrix} \begin{bmatrix} x(0) \\ x'(0) \end{bmatrix}$$

 which it performs on an incident ray at height $x(0)$ and angle $x'(0)$.

 (a) Find the imaging condition, and identify inverting and noninverting images in terms of the associated κl values.

 Consider a phase space defined by coordinates x and x'. In this phase space, illustrate the transformation affected by the above-described medium when $\kappa l = \pi/2$, by sketching the incident and transformed shapes on the phase space axes for the excitations of (b) through (d):

(b) An axial ray at the medium center.

(c) A z-directed plane wave.

(d) An expanding point source at the origin.

(e) What operation would the $\kappa l = (2n + 1) \, \pi/2$ medium perform on a spatially coherent incident waveform?

6

Interference

6.1 INTRODUCTION

The topic of optical interference is closely related to the topic of interferometric measurement, which in turn is closely tied to the topic of partial coherence. This chapter will concentrate on these operational aspects of interference. The chapter begins with a description of Michelson's interferometer and the quantities it measures, an exposition that naturally leads to a discussion of temporal coherence of light. Section 6.3 contains a discussion of some other Michelson type interferometers. There follows a discussion of a different kind of coherence (linewidth)-measuring apparatus, the Fabry-Perot interferometer. Following this is a discussion of Young's experiment and the diffractometer, which leads naturally to a consideration of the concept of spatial coherence. The last section of the chapter, that of 6.6, presents a discussion of the experiment of Hanbury-Brown and Twiss.

6.2 THE MICHELSON INTERFEROMETER

In order to carry out the analysis here, we reintroduce the analytical signal representation that was mentioned in Chapter 2 (equations 2-80 and 2-82). The idea is that one can represent an electric field as the real part of a complex disturbance (see, for example, Born and Wolf 1975, Section 10.2):

$$\mathbf{E}(t) = \mathrm{Re}\ [\mathbf{V}(t)] \tag{6-1}$$

where $\mathbf{V}(t)$ is related to the spectrum of $\mathbf{E}(t)$ by

$$\mathbf{E}(\omega) = \frac{1}{4\pi} \int_{-\infty}^{\infty} \mathbf{V}(t')e^{i\omega t'}\ dt' \tag{6-2}$$

226

where

$$\mathbf{E}(t) = 2\,\mathrm{Re}\left[\int_0^\infty \mathbf{E}(\omega)e^{-i\omega t}\,d_\omega\right] \tag{6-3}$$

The quantity of most importance, however, is the optical intensity, at least if one is considering plane wave excitations. The optical intensity is defined for $\mathbf{k}$-directed plane waves by

$$\langle \mathbf{S}(\mathbf{r}, t)\rangle = I(\mathbf{r}, t)\hat{\mathbf{e}}_k \tag{6-4}$$

where the Poynting vector is expressible in terms of the analytical signal by

$$\langle \mathbf{S}(\mathbf{r}, t)\rangle = \frac{\langle \mathbf{V}^\dagger(\mathbf{r}, t)\cdot\mathbf{V}(\mathbf{r}, t)\rangle}{2\eta}\,\hat{\mathbf{e}}_k \tag{6-5}$$

where, as in Chapter 4, the $\dagger$ denotes Hermitian transpose, and therefore one can write that

$$I(\mathbf{r}, t) = \frac{\langle \mathbf{V}^\dagger(\mathbf{r}, t)\cdot\mathbf{V}(\mathbf{r}, t)\rangle}{2\eta} \tag{6-6}$$

The basic setup of a Michelson interferometer is illustrated in Figure 6.1 [see, for example, Born and Wolf (1975, Chapter 7), or for more elementary treatments one could look at Klein and Furtak (1986) or Hecht (1987)]. In spectroscopic applications, the incident light will be polychromatic. However, only

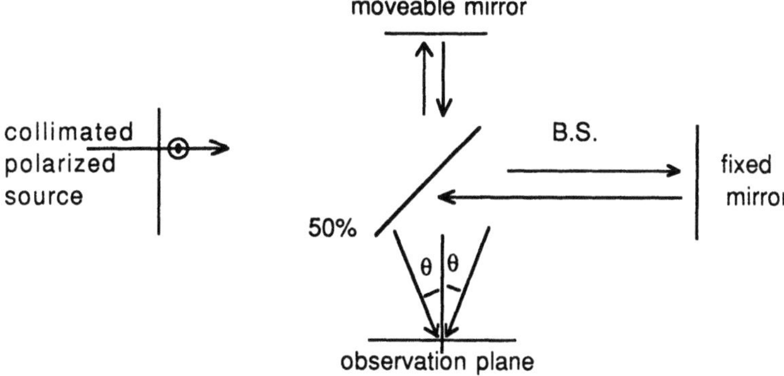

FIGURE 6.1.   Illustration of the basic configuration of a Michelson interferometer, where a 50% beam splitter splits the incident collimated polarized plane wave into two paths, which are recombined on an observation plane.

a reasonably well-collimated portion of the beam (collimation implying parallel, perhaps not evenly spaced, phase fronts) will pass through the limited numerical aperture of the interferometer. It is also somewhat important that the incident beam be polarized out of the plane of the drawing, or that the beam splitter be coated to only reflect 50% in that polarization state, as otherwise the beam splitter could not be a 50% one because the light at the observation plane would then contain both polarizations, and the interference contrast would be less than unity. Of course, if the beam splitter were coated to be exactly 50% in both polarization states and the output angle θ were zero, this problem could be alleviated. The moveable reflector (which may be a mirror or a retroreflector) generally will be mounted on some possibly automated micropositioning stage such that the pathlength increase, d, in this arm relative to the other arm can be varied with a subwavelength accuracy over distances of several (perhaps thousands) of wavelengths. The angle θ at the observation plane is due to the alignment and/or perfection of the reflector surfaces. It is very complicated to make θ exactly zero (to within one part in 10,000); therefore, for observation planes larger than 1 to 2 cm some transverse fringes may be seen. This is not a problem; in fact, this effect can be used to advantage. If one twists the fixed mirror relative to a parallel moving one, one can control the fringe size to be the detector size and thereby optimize an oscilloscope picture, but we will say more about that later. A second point also should be noted. If the mirrors were truly aligned parallel, then when the test arm was offset, a portion of the power launched into the interferometer would be reflected right back into the source. This could affect a coherent source adversely. Therefore, one can generally live with mirrors somewhat skewed relative to the incident direction.

We first wish to analyze the output of the Michelson interferometer for a monochromatic incident disturbance. One can take a coordinate system in which the optic axis, z, is the downward-pointing normal to the observation plane, with the observation plane defined by $z = 0$, and y is the coordinate pointing upward out of the plane of the paper. One can denote the y component of the delayed plane wave by V_{y2} and of the nondelayed plane wave by V_{y1}. Assuming that the delayed beam is the one with the component of motion in the $+x$ direction, one sees that the two waves are mathematically expressible as

$$V_{y1}(x, z, t) = \frac{V_{y0}}{2} e^{ikz \cos \theta} e^{-ikx \sin \theta} e^{-i\omega t} \qquad \text{(a)}$$

$$\text{(6-7)}$$

$$V_{y2}(x, z, t) = \frac{V_{y0}}{2} e^{ikz \cos \theta} e^{ikd} e^{+ikx \sin \theta} e^{-i\omega t} \qquad \text{(b)}$$

where V_{y0} is the amplitude of the wave before splitting and the splitting is assumed to be exactly 50%. The factor of 2 appears in the denominator because the forward waves have been split twice, each time with a halving of the power, which implies a field amplitude reduction of $1/\sqrt{2}$. With the identifications of (6-7), one can write that

$$
\begin{aligned}
I(x, d) &= \frac{1}{2\eta_0} \langle |V_{y1} + V_{y2}|^2 \rangle \\[2mm]
&= \frac{V_{y0}^2}{8\eta_0} |e^{+ikx \sin \theta} e^{ikd} + e^{-ikx \sin \theta}|^2 \\[2mm]
&= \frac{V_{y0}^2}{4\eta_0} \{1 + \cos (2kx \sin \theta + kd)\}
\end{aligned}
\tag{6-8}
$$

Equation (6-8) is plotted as a function of x parameterized by the delay, as in Figure 6.2. The period of the fringes is given by the expression

$$
2kX \sin \theta = 2\pi
\tag{6-9}
$$

where X is the period, which, for small angles θ is given by

$$
X = \frac{\lambda}{2\theta}
\tag{6-10}
$$

In his original work, Michelson wanted to have fringes broad enough that he could project them on a scale and read off changes of a fraction of a fringe. To do this, one would need fringes of a millimeter or more so that, for a 0.5 μm wavelength, the angle θ would have to be less than or equal to 2.5×10^{-4} radian. In some sense, this factor $1/2\theta$ is the amplification of the system, as a path length difference of d in the mirror placement would show up as a fringe displacement of $d/2\theta$. There can be cases where one could circumvent so great a need for accurate alignment. If one had a detector, for example, with a 10×10 μm active area, one could adjust the X value to something a little greater than the 10 μm value, and thereby monitor relative temporal displacements of

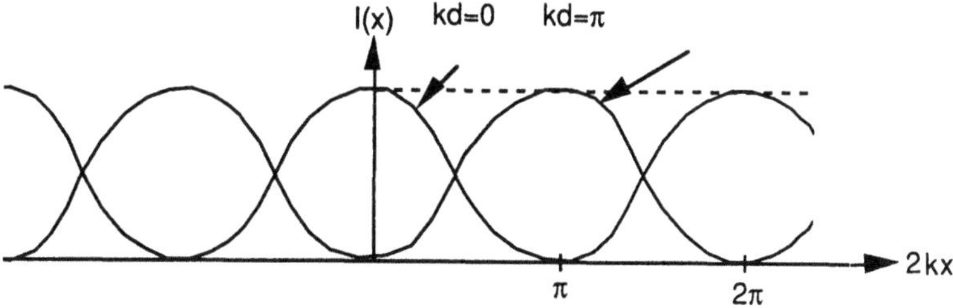

FIGURE 6.2.   A plot of the fringe function described by equation (6-8), parameterized by the delay of the moveable mirror.

the mirror to a subwavelength accuracy, as a relative change in the delay would shift the fringe pattern and thus the overlap of the detector and the fringe, changing the detector current. The configuration of such a setup is schematically depicted in Figure 6.3. In the figure, the time base is used to keep the trace sweeping at a constant rate, so that the scope face will trace out a plot of displacement versus time. Another possibility is to sweep the position of the mirror in the reference arm and use the ramp that sweeps the mirror as the second scope input. Now, under the assumption of a square detector area and perfectly plane fringes, the detector current must satisfy the equation

$$i_d(t) = K \int_{-T/4}^{T/4} (1 + \cos(\alpha + \delta(t)))\, d\alpha \qquad (6\text{-}11)$$

where K is a constant, T is the fraction of the period that the detector subtends, and, in the case of the swept mirror, $\delta(t) \sim kvt$, where v is the sweeping velocity. If the sweeping rate is sufficiently slow compared to the rise time of the detector (which is often called the fast detector limit), then the oscilloscope trace should resemble the fringe pattern (for $x = vt$). In particular, the integral from equation (6-11) could be evaluated to yield

$$i_d(t) = \frac{KT}{2}\left[1 + \cos \delta(t)\, \frac{\sin T/4}{T/4} \right] \qquad (6\text{-}12)$$

Now if one considers the detector current as a function of δ and defines the contrast ratio as the maximum value of the current minus the minimum value divided by the maximum plus the minimum, one finds that, for small fractional

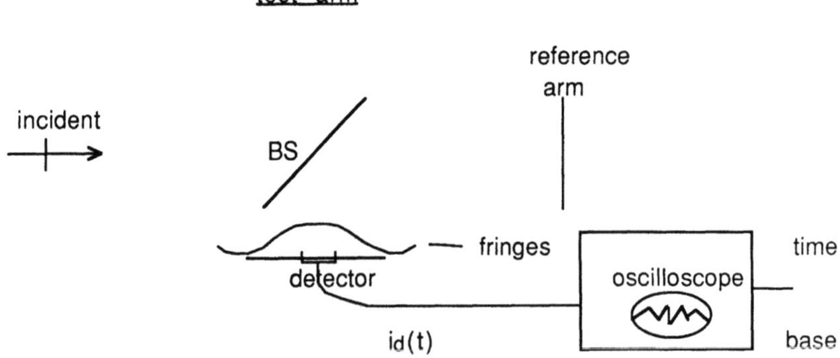

FIGURE 6.3.   Schematic depiction of the interferometer used as a detector of time-varying path-length changes.

periods T, the contrast ratio is given by

$$\frac{\max\, i_d(t) \,-\, \min\, i_d(t)}{\max\, i_d(t) \,+\, \min\, i_d(t)} \approx \left|\frac{\sin\, T/4}{T/4}\right| \tag{6-13}$$

Now this contrast ratio gives a direct indication of how accurately the minima and maxima of the oscilloscope trace generated by sweeping the mirror in the test arm could be determined. For this monochromatic case, the result of such a measurement could be a very accurate measurement of the wavelength of the source. Unfortunately, if we had a truly monochromatic source, we probably would know quite accurately what its wavelength was, and the measurement would not be very exciting. With a polychromatic source, however, we soon would see that such a sweeping procedure is a very useful measurement technique.

Now, it should be noted here that the case of perfectly collimated input radiation and well-aligned plane mirrors is not the rule, and, in fact, is somewhat the exception. For moving arms, retroreflectors are much less susceptible to vibration noise than plane mirrors. For work with low light levels, which can be due to a lack of intrinsic source collimation, one might well want to capture all light possible and not just a portion of the interference. In such a case one would want to use focusing optics in the interferometer. Fringes would no longer be plane, but would be curved. However, considerations corresponding to alignment and contrast can be handled by using the principles discussed above, except that the details of calculations become considerably more complicated and less illuminating. Here, we will stick to the simplified system and try to illuminate principles.

The next example is that of a two-tone illumination of the Michelson interferometer. This is the simplest case for which a Michelson interferometer can be used as an interference spectrograph, that is, an instrument for resolving optical spectra. In this analysis, we will start with the assumption that θ can be made equal to zero and therefore that the whole observation plane appears as a uniform gray scale. This assumption really is no restriction, as we saw in the monochromatic case that a detector placement can be found such that one reads out an intensity closely corresponding to that of the zero angle. In this case, then, the intensity that one will see in the plane $z = 0$ is given by

$$I(\tau) = \frac{V_{y0}^2}{8\eta_0} \langle |(e^{-i\omega_1 t} + e^{-i\omega_2 t}) + (e^{-i\omega_1(t+\tau)} + e^{-i\omega_2(t+\tau)})|^2 \rangle \tag{6-14}$$

where $\omega = kc$ and $\tau = d/c$ have been used, and it has been assumed, with some loss of generality, that the amplitudes of the two optical paths are equal. Something clearly must be assumed about the averaging brackets before one evaluates equation (6-14). Here, we will assume that the detector is not fast enough to register the difference frequency between the optical paths or the optical frequencies themselves. With this assumption, one can write

$$
\begin{aligned}
I(\tau) &= \frac{V_{y0}^2}{2\eta_0} \left(1 + \frac{1}{2} \cos \omega_1 \tau + \frac{1}{2} \cos \omega_2 \tau \right) \\
&= \frac{V_{y0}^2}{2\eta_0} [1 + \mathrm{Re}\, \hat{\gamma}(\tau)]
\end{aligned}
\tag{6-15}
$$

where the function $\gamma(\tau)$ is referred to as the complex coherence function.

The function $I(\tau)$ is plotted in Figure 6.4. It can readily be calculated that the function $\hat{\gamma}(\tau)$ is expressible as

$$
\hat{\gamma}(\tau) = \cos\left(\frac{\Delta\omega}{2}\tau\right) e^{-i((\omega_1 + \omega_2)/2)\tau}
\tag{6-16}
$$

and it is readily seen that this expression describes the salient features of Figure 6.4. For example, the function $\cos((\Delta\omega/2)\tau)$ has its first minimum at $\tau = 1/\Delta f$, the inverse of the difference of the two frequencies. It is in this manner that the interferometer can be used as an interference spectrograph. The larger the achievable τ offset, the finer the resolving resolution of the spectrograph. For example, consider the sodium D-lines, which are spaced apart by $\Delta\lambda = 6$ Å for a center wavelength of roughly 6000 Å, a situation that is generically illustrated in Figure 6.5. As the relative splitting, $\Delta\lambda/\lambda_0$, and 6000 Å corresponds to roughly 5×10^{14} Hz, one sees that $\Delta f = 5 \times 10^{11}$ Hz. The τ delay necessary to zero out the pattern is, therefore, about 2 psec, which corresponds in free space to a displacement of about 0.6 mm. In this case, therefore, an interferometer that can scan a millimeter or more with submillimeter controllability could easily resolve the distance between the sodium lines. If one were

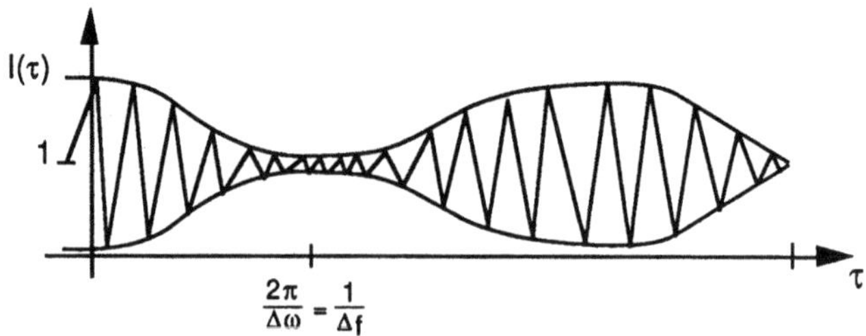

FIGURE 6.4.   A sketch of the intensity received by a ''slow'' detector as a function of the offset $\tau = d/c$.

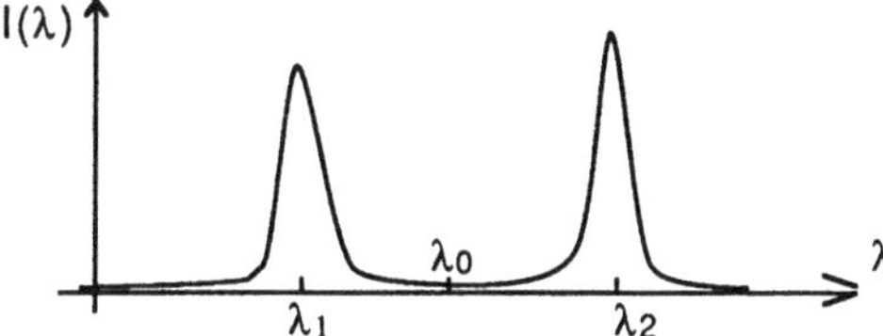

FIGURE 6.5.   The spectrum of a source exhibiting two emission lines separated by an interval $\Delta\lambda = \lambda_2 - \lambda_1$.

to consider a two-line gas laser, the situation would be quite different. In a laser, as we will see in this chapter, the spacing between adjacent longitudinal modes is given by $\Delta f = c/2l$, where l is the length of the cavity. Resolving this with the interferometer would require a displacement $d = c/\Delta f = 2l$; that is, the interferometer would have to be twice as long as the gas laser. The reason for this is essentially that a gas laser is an interferometer of sorts itself.

Before considering more general polychromatic illumination, it seems worthwhile first to consider the concept of temporal coherence length. As one might recall from the discussion of amplitude and phase jitter in Chapter 2, a possible realization for a component of the electric field of a noisy source could take the form

$$E_i(t) = a(t) \cos (\overline{\omega} t + \theta(t)) \tag{6-17}$$

where the a and θ were stochastic functions of time. A high gain source would likely have small variations of a, but, from spontaneous emission noise, would have occasional but possibly large shifts in phase. A possible realization for the electric field of such a source is illustrated in Figure 6.6. What is clear from the figure is that stretches of the field separated by random phase jumps will tend to decorrelate if interfered together. For purposes of spectral analysis,

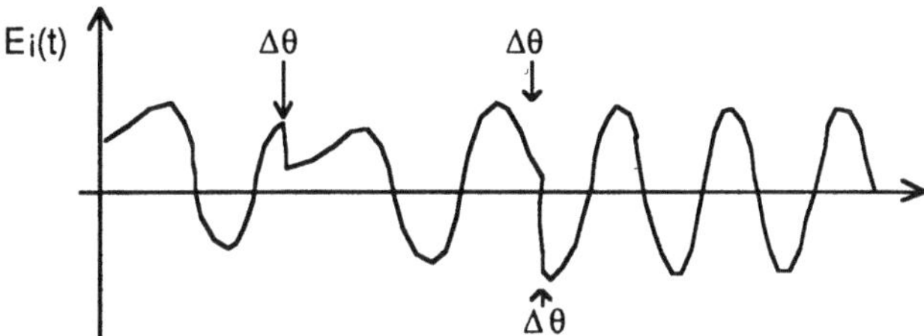

FIGURE 6.6.   Possible realization of a source that exhibits small-amplitude noise but sudden shifts in optical phase, which are pointed out in the figure.

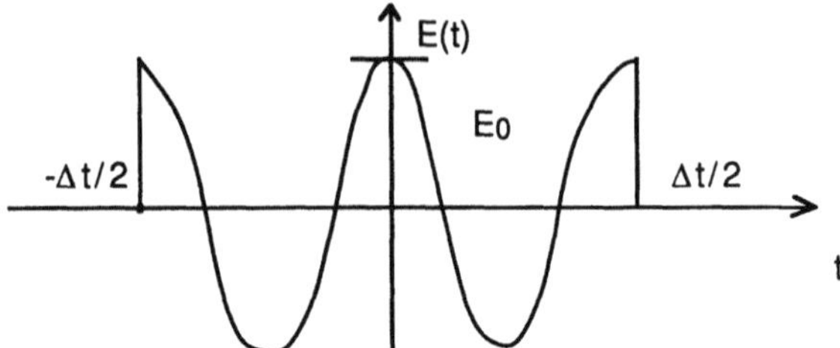

FIGURE 6.7.   Sketch of a truncated monochromatic signal.

therefore, one could consider a "gated" electric field of the form

$$E_i(t) = E_0 \cos (\omega_0 t) \operatorname{rect} \left(\frac{t}{\Delta_t} \right) \tag{6-18}$$

where the rect function is defined (as per problems 2.5, 2.25 and 3.9)

$$\operatorname{rect} (x) = \begin{cases} 1 & |x| < \tfrac{1}{2} \\ 0 & |x| > \tfrac{1}{2} \end{cases} \tag{6-19}$$

as is sketched in Figure 6.7. The spectrum of this signal can be found from equation (6-2) together with (6-18) as

$$
\begin{aligned}
E_i(\omega) &= \frac{E_0}{4\pi} \int_{-\Delta t/2}^{\Delta t/2} e^{i(\omega - \omega_0)t} \, dt \\
&= \frac{\Delta t}{4\pi} E_0 \frac{\sin \left(\dfrac{\omega - \omega_0}{2} \right) \Delta t}{\left(\dfrac{\omega - \omega_0}{2} \right) \Delta t}
\end{aligned}
\tag{6-20}
$$

which is sketched in Figure 6.8. Denoting the spectral width of the transform by $\Delta\omega$, one notes immediately that

$$\Delta t \, \Delta \omega \sim 2\pi \tag{6-21}$$

which is equivalent to

$$\Delta t \, \Delta f \sim 1 \tag{6-22}$$

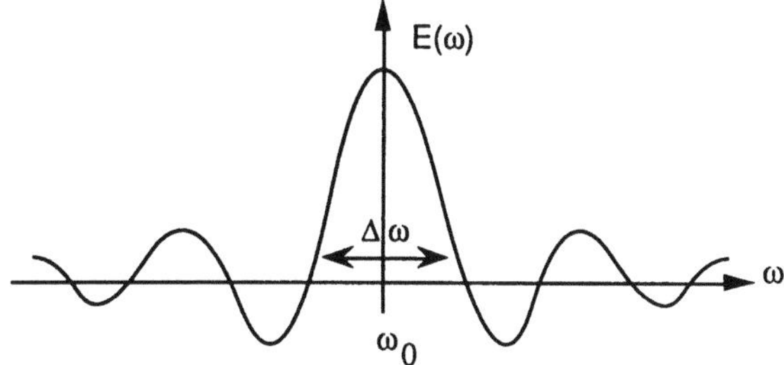

FIGURE 6.8.   Sketch of the spectral function defined by equation (6-20).

which is the same relation satisfied by the τ of the first zero of the coherence function γ of equation (6-15). The relation between these two quantities is not incidental; in fact, as we will see in more detail in the discussion of more general polychromatic sources, the entity that these quantities represent is the temporal coherence length of the source, often denoted by $\tau_c = l_c/c$, where l_c is the temporal coherence length. The physics of this entity are reasonably clear, however. If there is a uniformly distributed phase shift that occurs, on the average, each τ_c seconds, then records of the signal that are delayed and summed together will tend to decorrelate after the delay reaches τ_c seconds. This phase noise, as seen from equation (6-22), will cause a spectral broadening of roughly Δf. Figure 6.4 together with equation (6-15) shows that a spectral width of Δf for a completely different spectral shape will cause a fringe washout in a delay of τ_c also, even though the interference fringes may reappear at greater delay. This result should convince us that the "uncertainty" relation of equation (6-22) should hold independently of spectral shape, and, in particular, for other types of line broadening mechanisms than the spontaneous emission-induced phase noise discussed above.

For a general polychromatic plane wave disturbance, the analytical signal representation of the y-polarization state of the electric field is

$$V_y(x, y, z, t) = a(x, y, z, t)e^{i\bar{k}z}e^{-i\bar{\omega}t}e^{i\theta(x, y, z, t)} \qquad (6\text{-}23)$$

where $\bar{k}$ is defined by $\bar{k} = \bar{\omega}n/c$, where n is the index of refraction. The a and θ depend on x and y because of any initial phase front distortion, which would be due to the lack of coherence of different points on a finite-size source face. The z dependences of these quantities came about because of the possible effect of dispersion. If the propagation occurred in a nondispersive medium, or, better stated, if the disturbance is quasi-monochromatic enough that it does not see the (perhaps weak) dispersion of the medium, then a and θ can be taken to be independent of z, as we will assume in all future considerations. As the development proceeds, we will investigate the x and y dependences of the a and θ functions, but for the discussion of the interference spectrograph it suffices to suppress this dependence. This could practically be achieved by spatially filtering the incident disturbance; that is, the beam could be focused down, passed through a pinhole, and recollimated. The beam thus would appear as if it emanated from a pinhole and could not exhibit x–y structure. The spatial filtering process, however, must have a cost in terms of energy level transmitted, as by making the phase front uniform, one is reducing the entropy carried by the disturbance, and entropy reduction requires energy. An alternative explanation of the process could assume that only a plane wave can be focused to a point. A distorted phase front focuses to a complicated focal plane structure, the majority of which must be filtered out in the spatial filtering process.

We now consider an analytical signal representation, expressible as

$$V_y(t) = a(t)e^{i\theta(t)}e^{-i\bar{\omega}t} \tag{6-24}$$

to represent the y component of the plane wave entering the (perfectly aligned) Michelson interferometer of Figure 6.1. The intensity at the output plane will be given by

$$
\begin{aligned}
I(\tau) &= \frac{1}{2\eta_0}\langle\, |V(t) + V(t - \tau)|^2 \,\rangle \\[2mm]
&= \frac{1}{2\eta_0}\{\langle a^2(t)\rangle + \langle a^2(t - \tau)\rangle \\[2mm]
&\quad + 2\,\mathrm{Re}\,[e^{-i\bar{\omega}\tau}\langle a(t)a(t - \tau)e^{i(\theta(t) - \theta(t - \tau))}\rangle]\}
\end{aligned}
\tag{6-25}
$$

It could be noted here that the Fourier transform of the last term on the right-hand side of (6-22) is proportional to the spectral density of the field as given by equations (2-86) and (2-87).

One often defines the complex degree of the coherence function, $\hat{\gamma}(\tau)$, by

$$\hat{\gamma}(\tau) = e^{-i\bar{\omega}\tau}\,\frac{\langle a(t)a(t - \tau)e^{i(\theta(t) - \theta(t - \tau))}\rangle}{[\langle a^2(t)\rangle\langle a^2(t - \tau)\rangle]^{1/2}} \tag{6-26}$$

where $\hat{\gamma}(\tau)$ can be expressed in terms of the coherence function (also called degree of coherence) $\gamma(\tau)$ by

$$\hat{\gamma}(\tau) = \gamma(\tau)e^{i\alpha(\tau)}e^{-i\bar{\omega}\tau} \tag{6-27}$$

where $\alpha(\tau)$ is a phase that can be found from (6-25). Equations (6-25) and (6-26) enable one to express

$$I(\tau) = I_1(\tau) + I_2(\tau) + 2\sqrt{I_1(\tau)}\,\sqrt{I_2(\tau)}\,\mathrm{Re}\,[\hat{\gamma}(\tau)] \tag{6-28}$$

where the notation

$$I_1(\tau) = \langle a^2(t) \rangle / 2\eta_0 \qquad \text{(a)}$$
$$I_2(\tau) = \langle a^2(t - \tau) \rangle / 2\eta_0 \qquad \text{(b)}$$

$$(6\text{-}29)$$

has been employed. Equation (6-28) is the general form for the received intensity. If the source of the radiation is stationary in time, one immediately sees that $I_2(\tau) = I_1(\tau)$, and can write

$$I(\tau) = 2I_1(\tau)[1 + \text{Re } \hat{\gamma}(\tau)] \qquad (6\text{-}30)$$

and it becomes clear that, for $\text{Re } \hat{\gamma}(\tau) = 0$, interference effects are negligible as the two beam intensities simply add at the observation plane. From the expression (6-27), it is clear that one could make the general definition that the temporal coherence time, τ_c, is given by

$$\tau_c = \min [\tau > 0] \text{ such that } \gamma(\tau) = \epsilon \qquad (6\text{-}31)$$

where ϵ is a small constant. It is necessary to make $\epsilon > 0$ to avoid cases of exponential correlation where $\gamma(\tau)$ will decay exponentially but never quite become zero.

6.3 OTHER INTERFEROMETERS

Another example of an interferometric device is the directional coupler, as depicted in Figure 6.9. The device has much in common with the Michelson interferometer or an extended beam splitter, in that light is input at one of four ports and can exit from either of two others. A guided wave directional coupler is a device in which two single-mode waveguides are brought in close proximity to one another such that the degeneracy of the system's eigenmodes is broken. This gives rise to a symmetric and an antisymmetric eigenmode (pic-

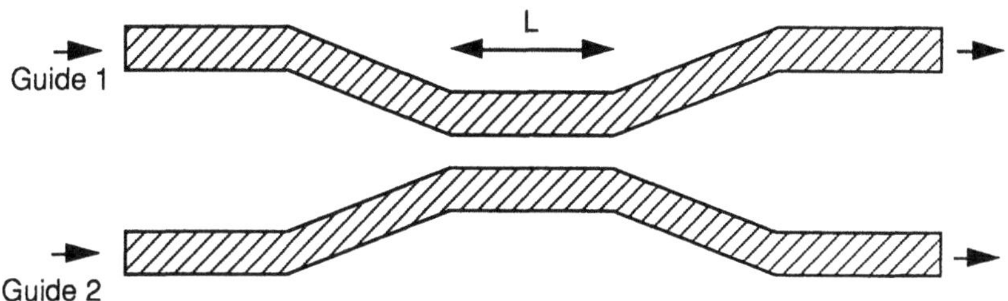

FIGURE 6.9.   Schematic depiction of an optical directional coupler.

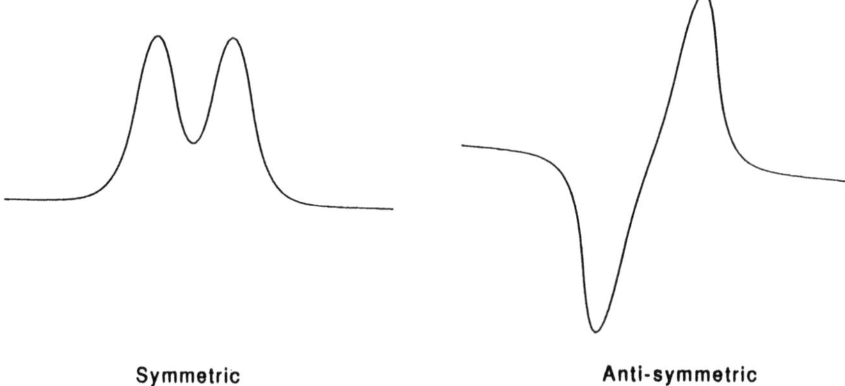

FIGURE 6.10.   Sketch of an interference pattern of two coupler modes.

tured in Figure 6.10), which propagate down the structure at different velocities. If, at a distance L, the waveguides are separated (adiabatically) such that they are no longer coupled, then the fields in the individual guides at this distance may be approximated as the sum and the difference of the fields due to the symmetric and the antisymmetric mode, or essentially the interference pattern of these two modes, as shown in Figure 6.10. The transfer matrix for this device is given as

$$
\begin{bmatrix} \text{Guide 1} \\ \text{Guide 2} \end{bmatrix}_{\text{out}} = \begin{bmatrix} \cos\left(\dfrac{\Delta\beta l}{2}\right) & -i\,\sin\left(\dfrac{\Delta\beta l}{2}\right) \\ -i\,\sin\left(\dfrac{\Delta\beta l}{2}\right) & \cos\left(\dfrac{\Delta\beta l}{2}\right) \end{bmatrix} \begin{bmatrix} \text{Guide 1} \\ \text{Guide 2} \end{bmatrix}_{\text{in}} \tag{6-32}
$$

where $\Delta\beta$ is the difference in the propagation constants of the two eigenmodes, and L is the length of the coupling region where the waveguides are in close proximity. As is evident from the transfer matrix, by initially exciting one of the input guides, any ratio of output powers of the two guides may be selected by choosing the proper L. In practice this is difficult to do, and methods to tune $\Delta\beta$ are used. The length needed to completely transfer power from one guide to another is referred to as a coupling length and is given by $L_c = \pi/2\Delta\beta$. The flow of power from one guide to another and back is shown in Figure 6.11. This length is known as the beatlength of the device.

Another important application of interferometry is the testing of optical components that have either flat surfaces (such as prisms, plane mirrors, etc.) or spherical surfaces (such as lenses, objectives, etc.). A commonly used interferometer for optical component testing is the Twyman-Green interferometer,

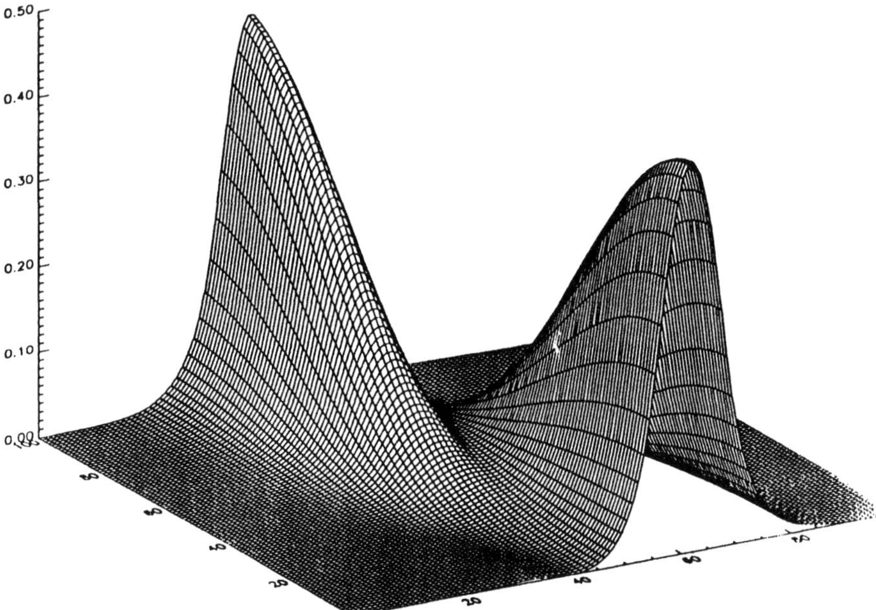

FIGURE 6.11.   A sketch of the variation of guide power as a function of length in a directional coupler.

which is a modification of the Michelson interferometer working in collimated light, as shown in Figure 6.12. The reference arm of the interferometer provides a plane wave that interferes with the wave that has been twice through the specimen in the test arm, thereby illustrating the departures from straightness of the fringes and showing the variations in optical path caused by defects in the specimen. Figure 6.13 shows different arrangements of the test arm that can be used for different test components that have spherical surfaces. In the spherical surface cases, the form of the interference pattern depends on the position of the center of curvature of the mirror in the testing arm. When the center of curvature of the mirror lies exactly at the focal point of the test object, with an ideal surface no interference fringes are observed. If the mirror is moved longitudinally, interference bands appear. From the shape of interference pattern, one may judge the quality of the specimen.

Another instrument that is often used for optical component testing is the Fizeau interferometer, shown in Figure 6.14. The flat to be tested is held below a standard flat on an adjustable table. The fringes are viewed through the standard flat. Two-beam fringes are obtained, and visual examination of the fringes gives the shape of the surface to about 0.03λ. To increase the precision, the two surfaces can be given high-reflecting coatings to produce multiple-beam

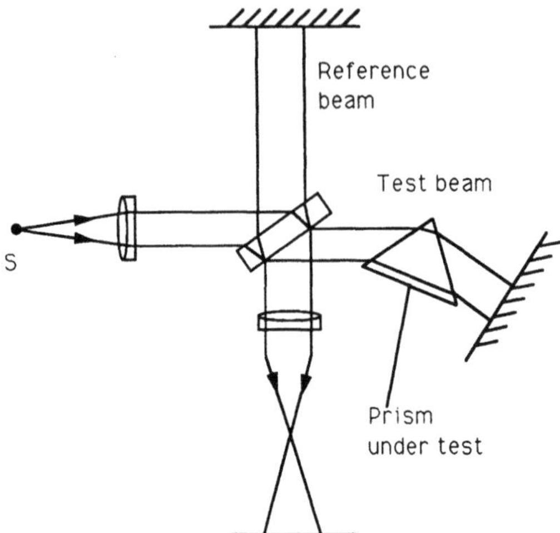

FIGURE 6.12.   The Twyman-Green interferometer, as used to test a prism.

fringes that can be judged to 0.01λ, or measured to 0.001λ by photoelectric methods.

6.4 THE FABRY-PEROT INTERFEROMETER

Before considering in greater detail the resolving power of the Michelson interferometer when used as a spectrograph, as defined by equation (6-28), we wish first to consider another type of interferometer that has a greater sensitivity as a spectrograph although a greatly reduced scanning range, the Fabry-Perot interferometer. A third spectrographic instrument, the grating spectrometer, that with the poorest resolution and the greatest spectral range, will be discussed near the end of Chapter 7. These three instruments and their multitudinous variations form the basic equipment of the field of spectroscopy and, more generally, optical spectral analysis.

A Fabry-Perot etalon is essentially an optical flat, as depicted in Figure 6.15 and considered in Chapter 2 and as is discussed in various books including Yariv (1985). Here, in one sense, we do not need the generality of Chapter 2, as, for our purposes, we can just as well assume the outside regions to be unity index. In another sense, we need greater generality than in Chapter 2, as we could just as well take the inside index to be unity and assume that the reflection comes from thin layers at the interface. We did not solve this problem in Chapter 2, but we can solve it in the same way that we could have solved the single slab problem (but did not). That is, due to the index discontinuity, there will be Fresnel coefficients at each interface, as is illustrated in Figure 6.16. For

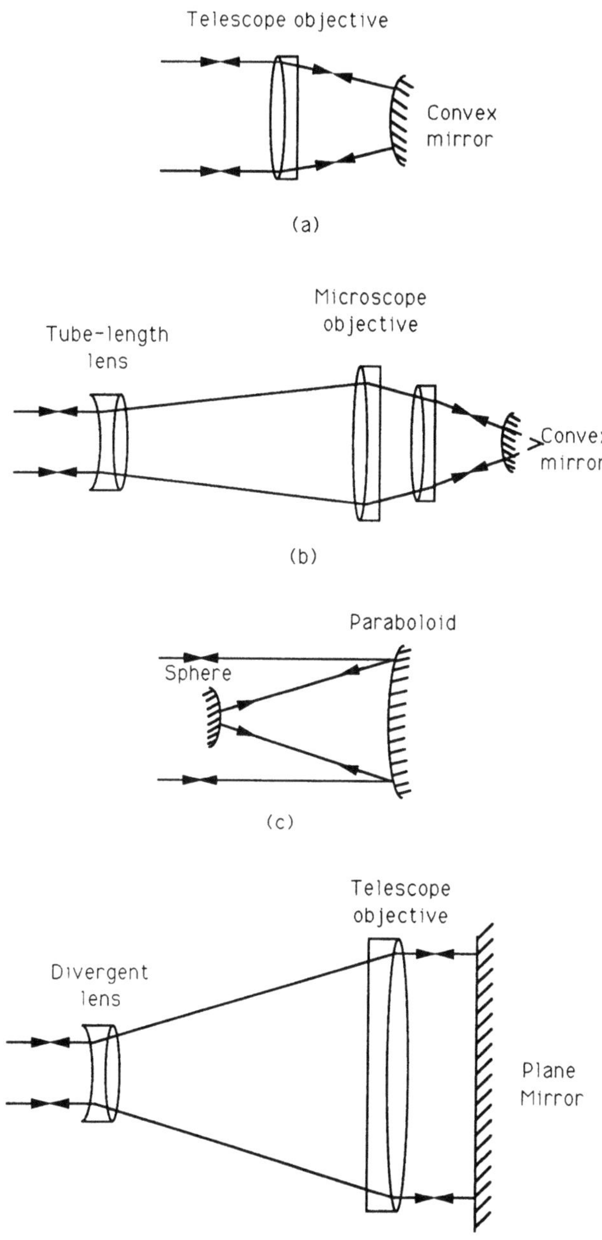

FIGURE 6.13. Test arm of a Twyman-Green or Fizeau interferometer, for testing: (a) a small telescope objective; (b) a microscope objective; (c) a paraboloidal mirror; (d) a large telescope objective.

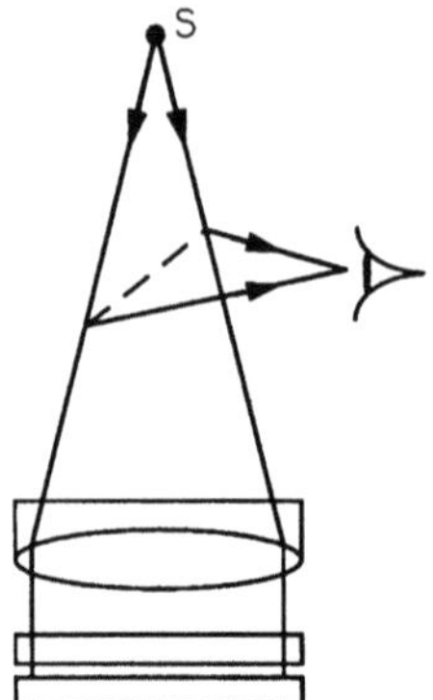

FIGURE 6.14.   Fizeau interferometer.

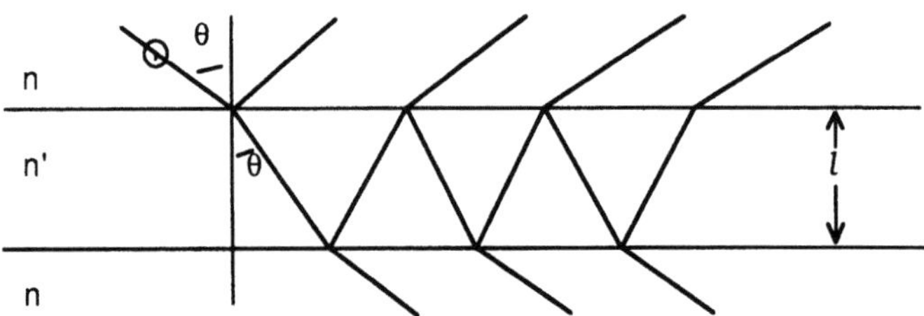

FIGURE 6.15.   Depiction of a Fabry-Perot etalon, illuminated by a plane wave at an angle θ, with the assumed multiple reflections sketched.

example, from Figure 6.10 it can be surmised that the amplitude of the reflected wave will be given by a series of the form

$$\frac{a_r}{a_i} = r + tt' r' e^{i\delta} + tt' r' e^{i\delta} r'^2 e^{i\delta} + tt' r' e^{i\delta} r'^4 e^{2i\delta} + \cdots$$

$$= r + tt' r' e^{i\delta} \sum_{n=0}^{\infty} r'^{2n} e^{in\delta}$$

(6-33)

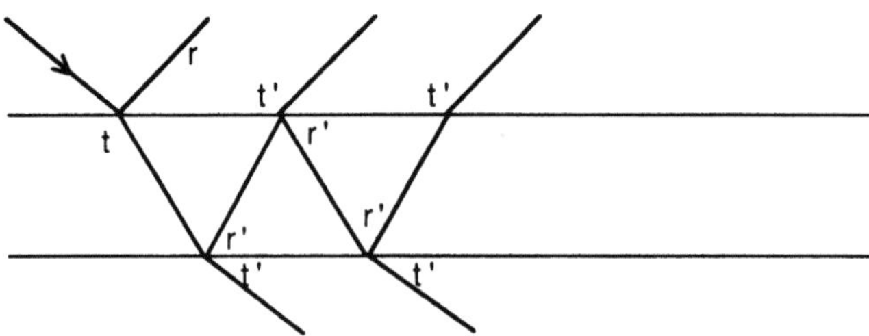

FIGURE 6.16.   Illustration of the use of Fresnel coefficients to keep track of the various contributions in the multiple bounce argument.

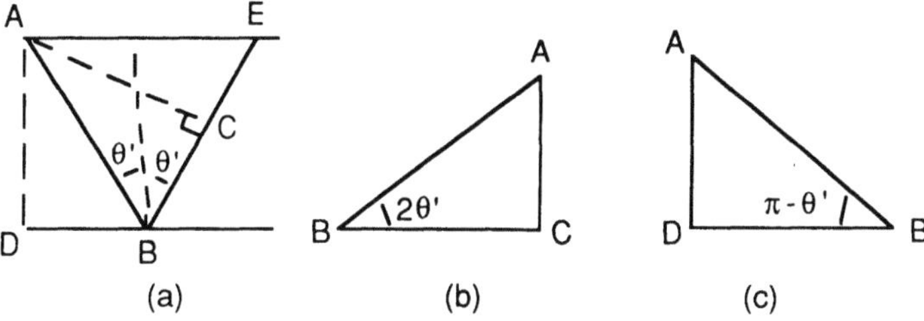

FIGURE 6.17.   Sketches of the geometrical constructions that go into deriving the phase delay δ in equation (6-33): (a) the triangle defining the propagation path ABE, (b) a breakout of ABC, and (c) a breakout of ADB.

where δ represents the delay experienced by the wave in traversing the plate one round trip. A geometrical construction such as that illustrated in Figure 6.17 rapidly shows that

$$\delta = k_0 n' (AB + BC) = k_0 n' \left(\frac{AD}{\cos \theta'} + AB \cos 2\theta' \right)$$

$$= k_0 n' \frac{l}{\cos \theta'} (1 + \cos 2\theta') = 2k_0 n' l \cos \theta' \tag{6-34}$$

Now, from the identity that

$$\sum_{n=0}^{\infty} x^n = \frac{1}{1 - x} \tag{6-35}$$

one can write that

$$\frac{a_r}{a_i} = r + \frac{tt' r' e^{i\delta}}{1 - r'^2 e^{i\delta}} \tag{6-36}$$

Now the basic idea behind what is being done here is that, as was true in solving multilayer problems in Chapter 2, one needed to assume solutions to the boundary value problem. In the multiple-bounce argument, one needs only to know interface reflections and propagation delays. If one could find a form for the intensity reflection of the slab in terms of interface intensity reflections and transmissions and propagation delays, then the solution would be more general than those obtained in Chapter 2, in the sense that one could replace the interfaces with any kind of partially reflecting mirror and find the value of δ in the limit that $n' \rightarrow 1$ to find the limit of a_r and a_t for a cavity resonator. Therefore, what is necessary is to get the magnitude of (6-36) into a form that has only R's and T's, the intensity reflection coefficients. To do this, it is useful to employ some identities valid for Fresnel coefficients. With reference to equation (2-144), one can rapidly note that

$$r = -r'$$
(6-37)

and therefore that

$$r^2 = r'^2 = R$$
(6-38)

Also, one can note from equation (2-144) that

$$t' = \frac{n' \cos \theta'}{n \cos \theta} t$$
(6-39)

and therefore that

$$tt' = T$$
(6-40)

the intensity transmission. With these identities, it can be noted that

$$
\begin{aligned}
\frac{a_r}{a_i} &= r + \frac{tt' r' e^{i\delta}}{1 - r'^2 e^{i\delta}} \\
&= \frac{-r' + r'(r'^2 + tt')e^{i\delta}}{1 - r'^2 e^{i\delta}} \\
&= -r' \left(\frac{1 + e^{i\delta}}{1 - r'^2 e^{i\delta}} \right)
\end{aligned}
$$
(6-41)

Squaring the result of (6-41), one finds

$$R(\nu) = \frac{I_r}{I_i} = \frac{2R(1 - \cos \delta)}{(1 - r')^2 + 2r'^2(1 - \cos \delta)} = \frac{F \sin^2 \delta/2}{1 + F \sin^2 \delta/2}$$
(6-42)

where the parameter F is defined by

$$F = \frac{4R}{(1 - R)^2}$$
(6-43)

The cavity (or etalon) intensity transmittance can be likewise expressed as

$$T(\nu) = \frac{I_t}{I_i} = \frac{1}{1 + F \sin^2 \delta/2}$$
(6-44)

where now the results are expressed in terms only of intensity reflections and delays and thus are valid for the general case.

The intensity transmittance as a function of δ, parameterized by F, is plotted in Figure 6.18. A first observation to make from that figure is that the intensity transmittance is a periodic function of δ of period 2π that exhibits peaks with full-width half power (FWHP) width of $4/\sqrt{F}$ for large enough F. The ratio of this FWHP width to the period (2π) is defined to be the inverse of the finesse F, which is therefore defined by

$$F = \frac{\pi\sqrt{F}}{2} \tag{6-45}$$

For example, consider an air–glass interface where the intensity reflectance is 0.04 and therefore the F is roughly 0.16, giving an FWHP width of 10, which is greater than the period. Also, for this reflectance the minimum transmission is roughly 0.84. Therefore, this reflection is sufficiently low that our picture of periodic peaks begins to break down although there is still clearly an etalon effect, in that the glass plate (optical flat) is still frequency-selective. If one considered 99% intensity reflective mirrors, then the F would be roughly 40,000, giving a finesse of roughly 310. The FWHP width here would be about 0.02, and minimum transmittance would be 0.000025, which is for all practical purposes zero. The 99% figure is achievable but a bit high in practice, where most commercial Fabry-Perot resonators have finesses in the 100 to 200 range, corresponding to reflectivities of 97 to 98%. Fabry-Perot resonators with finesses as high as 20,000 are available, however.

As this discussion began with the idea that the topic would be a second type of interferometer that would complement the action of a Michelson interferometer, it would stand to reason that the behavior of the response with δ should be translated into frequency response. Noting that maxima of the transmission occur at integral multiples of 2π gives that

$$\delta_m = 2m\pi \tag{6-46}$$

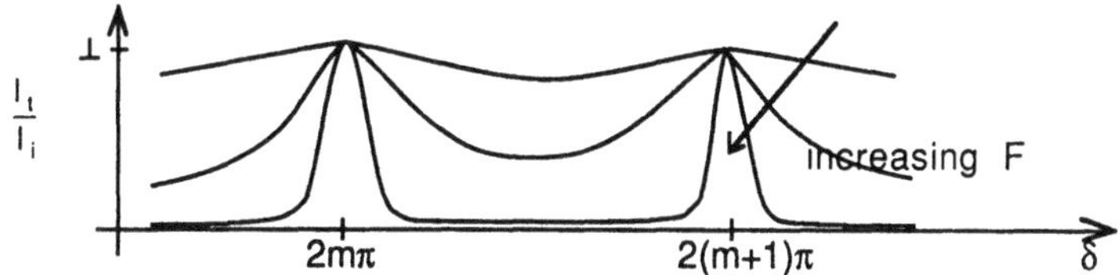

FIGURE 6.18.   Plots of the intensity transmittance as a function of δ, parameterized by $F = 4R/(1 - R)^2$.

and, from (6-30) and $k_m = 2\pi \nu_m / c$, gives that

$$\nu_m = \frac{mc}{2nl \cos \theta'} \tag{6-47}$$

From this, one can find the mode spacings $\Delta \nu$ by

$$\Delta \nu = \nu_{m+1} - \nu_m = \frac{c}{2n'l \cos \theta'} \tag{6-48}$$

which, when specialized to an empty cavity at normal incidence, gives

$$\Delta \nu = \frac{c}{2l} \tag{6-49}$$

which is exactly the well-known formula for the spacing of longitudinal modes of a gas laser of length l. This is indeed as it should be, as the discreteness of the longitudinal modes spectrum of a gas laser is due solely to the Fabry-Perot resonator in which the active medium is placed. The $\Delta \nu$ value of a Fabry-Perot that is to be used as an interferometer is called its free spectral range, as, after a free spectral range, the spectrum repeats itself. If one were to scan the peak transmission point by scanning the cavity length from $l_0 + \delta l$ to $l_0 - \Delta l$, one would scan the center frequency from $\nu_m(l_0 + \delta l)$ to $\nu_m(l_0 - \delta l)$, where

$$\nu_m(l_0 + \delta l) = \frac{mc}{2(l_0 + \delta l)} \tag{a}$$
$$\tag{6-50}$$
$$\nu_m(l_0 - \delta l) = \frac{mc}{2(l_0 - \delta l)} \tag{b}$$

Taking $\delta l / l \ll 1$ gives

$$\nu_m(l_0 - \delta l) - \nu_m(l_0 + \delta l) = \frac{mc\delta l}{l_0^2} \tag{6-51}$$

The situation is as sketched in Figure 6.19. Here it is shown that, for δl large enough, $\nu_{m-1}(l_0 - \delta l)$ will become equal to $\nu_m(l_0 + \delta l)$. The value for which this occurs is

$$\delta l = \frac{l_0}{2m} \tag{6-52}$$

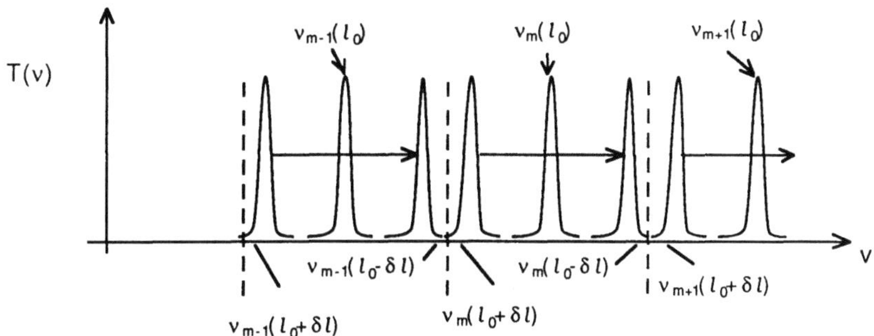

FIGURE 6.19.   The effect of a change in the cavity length from $l_0 + \delta l$ to $l_0 - \delta l$ on the intensity transmission spectrum of a Fabry-Perot cavity as a function of frequency ν.

and is the value of δl for which the free spectral range equals the actual scanning range. In this sense, the free spectral range is the maximum spectral range that one can scan without repeating oneself. As $\delta l/l_0$ is dimensionless and m is l_0/λ, it is seen that, for normal incidence, a scan of $\pm\lambda/2$ will sweep out the free spectral range.

The above discussion has now left us with all the pieces necessary to put together the workings of a Fabry-Perot spectrometer. It should be noted here that the use of plane mirrors in a Fabry-Perot is by no means universal, and, in fact, confocal configurations are probably more common in practice. However, for simplicity in what follows, only the planar mirror case will be treated. Consider the situation depicted as Figure 6.20. Here, an incident intensity spectrum $I_i(\nu)$ is collimated and shown on the input of a Fabry-Perot resonator whose back mirror is mounted on a moveable stage (possibly piezoelectric crystal) that can be driven electronically. The output of the cavity, which should still be a collimated plane wave if the mirrors are truly flat and parallel, is collected by a lens and focused onto the active region of a detector. The signal used to sweep the mirror positioner also is used as the sweep input to the oscilloscope whenever the detector output is to be used as the other scope input. Now, the intensity

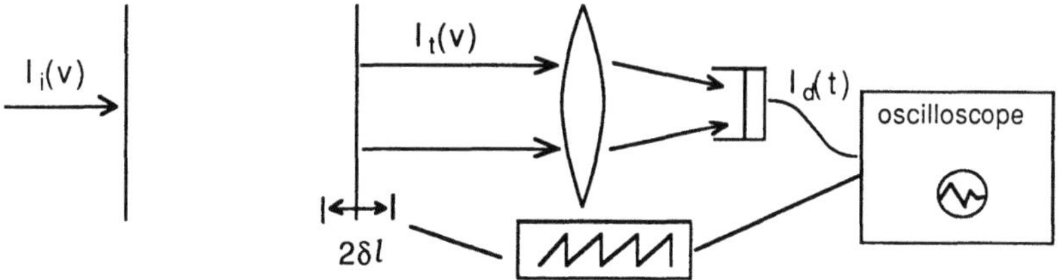

FIGURE 6.20.   Schematic illustration of the setup of a Fabry-Perot scanning interferometer.

transmitted at a frequency ν by the cavity must be given by

$$I_t(\nu) = T(\nu)I_i(\nu) \qquad (6\text{-}53)$$

where $T(\nu)$ is given by (6-44). The detector current must be proportional to the Poynting vector crossing the detector surface. Assuming that there are no frequency-dependent losses in the imaging system, and that the detector has a flat response over the spectral range of interest, then

$$i_d(t) = K\left\langle \int_0^\infty T(\nu')I_i(\nu')\,d\nu' \right\rangle \qquad (6\text{-}54)$$

where K is the proportionality constant, and the time average is necessary because $T(\nu')$ will vary with time because of the scanning of the cavity length. Now detectors do not have flat responses over very great spectral ranges, but as we soon will see, the response must be flat only over a free spectral range, which can be sufficiently small for this to be true. Clearly, if $T(\nu)$ does not change much during a detector period τ_d, then the angular brackets could be dispensed with. It is known that the total scan requires a motion of λ. For a given finesse F, the width of the transmission peak will therefore be λ/F. In this sense, F is the number of resolvable points during a scan. If the scan velocity is such that the detector time τ_d is much less than $\lambda/vF = \tau_{\text{scan}}/F$, where v is the scanning velocity and τ_{scan} is the period of the scan, (i.e., the peak does not sweep much during a detector time), then

$$i_d(t) = K \int T(\nu')\,I_i(\nu')\,d\nu' \quad \text{for } \tau_d \ll \frac{\lambda}{cF} \qquad (6\text{-}55)$$

It should be pointed out here that another time scale would be the time it takes for a photon to get out of the cavity, which is given by $F\,ln/c$. The scan must be slow enough to accommodate this period. To complete the argument for perfect Fabry-Perot operation, it can be assumed that $F \gg 1$ and $\delta L/L \ll 1$. If this is so, then one can write, approximately, that

$$T(\nu) = \sum_m \delta(\nu - \nu_m(t)) \qquad (6\text{-}56)$$

Then one can express $i_d(t)$ in the form

$$i_d(t) = K \sum_m I_i(\nu_m(t)) \qquad (6\text{-}57)$$

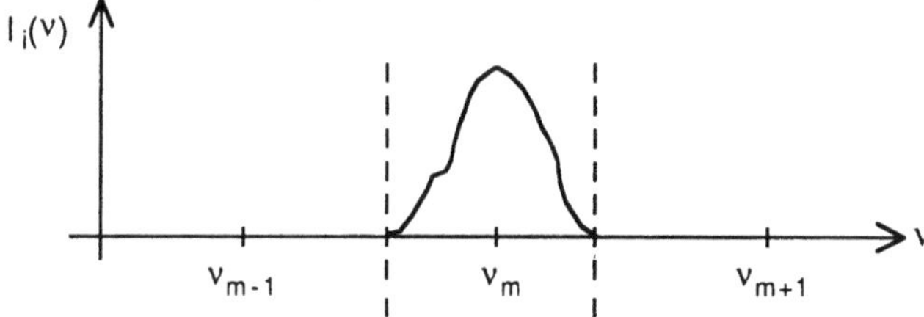

FIGURE 6.21. Sketch of an intensity spectrum that is band-limited to a single Fabry-Perot spectral range.

where

$$\nu_m(t) = \frac{mc}{2l_0}\left(1 - \frac{\delta l(t)}{l_0}\right) \tag{6-58}$$

Equation (6-57) represents the output of a long, high-finesse Fabry-Perot readout by fast electronics. This is generally the system that one designs if one wants to use a Fabry-Perot as a spectrograph. Unfortunately, the output current is still quite a complicated function, even if the scan of δl is perfectly linear. The output, however, does become simple if the intensity spectrum is band-limited in the sense of Figure 6.21. In the band-limited case, equations (6-57) and (6-58) reduce to

$$i_d(t) = KI_i\left(\nu_{m0}\left(1 - \frac{\delta l(t)}{l_0}\right)\right) \tag{6-59}$$

and, if the length is scanned linearly in time and the same signal is used to drive the oscilloscope time base, the current and the oscilloscope trace will be a scaled representation of the actual spectrum. As sawtooth waveforms such as that illustrated in Figure 6.22 are readily available from any number of commercial generators, obtaining the linear length scan is not a problem. If the signal is not

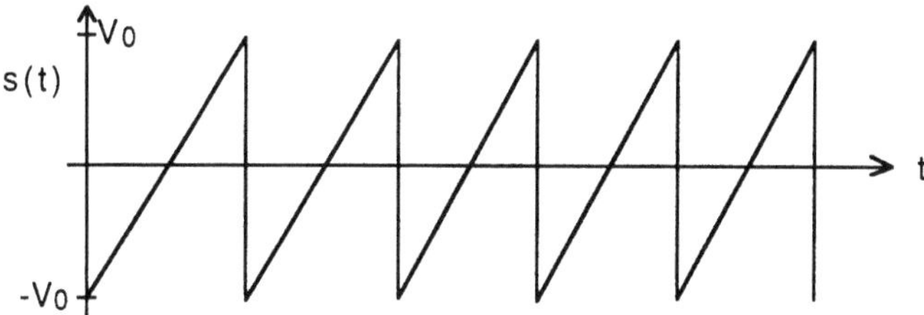

FIGURE 6.22.   Sketch of a sawtooth waveform.

bandlimited, a form of error analogous to the wrap-around error observed in taking the fast Fourier transform (FFT) of non-time-limited signals is observed [see, for example, Cochran et al. (1967) for a lucid discussion of the discrete Fourier transform]. As it turns out, various aspects of Fabry-Perot sampling are analogous to FFT analysis.

An interesting example of a Fabry-Perot would be that of a guided wave version. A guided wave Fabry-Perot (GWFP) cavity will have no diffraction losses. However, propagation losses could limit the obtainable finesse of such a cavity. We will presently calculate the intensity transmittance of a GWFP, considering both propagation losses and mirror/coupling losses. We will assume the finesses of (a) a 1 cm long cavity in $LiNbO_3$ with 1 dB/cm loss, (b) a 10 m and 100 m long cavity in an optical fiber operated in the red part of the spectrum with 10 dB/km loss, and (c) a 10 m and 100 m long cavity in a fiber operated in the infrared with 1 dB/km loss.

Assuming small losses, how does the resolving power of the interferometer vary with the length of the cavity? The transmitted amplitude will be given by a series of the form

$$\frac{a_t}{a_i} = \gamma tt' e^{i\delta}[1 + r^2 e^{i2\hat{\delta}} + r^4 e^{i4\hat{\delta}} + \cdots] \tag{a}$$

$$= \frac{\gamma tt' e^{i\delta}}{1 - r^2 e^{i2\hat{\delta}}} \tag{6-60}$$

$$\hat{\delta} = (k_r + ik_i)L = ik_iL + \delta \tag{b}$$

where γ is a coupling factor taking into account the inevitable loss when coupling into the cavity, r', t, and t' are the same as defined in Figure 6.16 and $\exp(i\hat{\delta})$ is the propagation factor, which includes propagation losses. The intensity transmittance follows as

$$\frac{I_t}{I_i} = \frac{\gamma^2 T^2 e^{-2k_iL}}{(1 - Re^{-2k_iL})^2} \frac{1}{(1 + F\sin^2\delta)} \tag{a}$$

$$F = \frac{4Re^{-2k_iL}}{(1 - Re^{-2k_iL})^2} \tag{b}$$

$$\tag{6-61}$$

If the finesse is limited by the propagation losses, we let $R \sim 1$ and find

$$F \approx \pi \left\{ \frac{e^{-2k_iL}}{(1 - e^{-2k_iL})^2} \right\}^{1/2} \tag{a}$$

$$k_i[cm^{-1}] = \frac{x[dB/cm]}{20 \log e} \tag{b}$$

$$\tag{6-62}$$

The assumed finesses are as follows:

(a) LiNBO$_3$, 1 dB/cm, $L = 1$ cm:

$$F = 13.6 \tag{6-63}$$

(b) Optical fiber 10 dB/km, $L = 10$ m, 100 m:

$$F_{10\,m} = 136 \qquad \text{(a)}$$
$$F_{100\,m} = 13.6 \qquad \text{(b)} \tag{6-64}$$

(c) Optical fiber, 1 dB/km, $L = 10, 100$ m:

$$F_{10\,m} = 1364 \qquad \text{(a)}$$
$$F_{100\,m} = 136 \qquad \text{(b)} \tag{6-65}$$

If $2k_i L << 1$, we have $F \approx \pi/2k_i L$ and the resolving power is given by

$$\Delta \nu = \frac{FSR}{F} \cong \frac{c}{2nL}\frac{2k_i L}{\pi} = \frac{ck_i}{\pi n} \tag{6-66}$$

that is, it is independent of length.

To conclude this section, we wish to discuss the relative merits of the scanning Fabry-Perot cavity and Michelson interferometer when used for spectrographic resolution. To accomplish this, first some typical figures for achievable spectral scan ranges and resolutions will be given, followed by some examples of spectroscopic tasks and how each instrument might (or might not) be able to perform them.

As was seen above, the Fabry-Perot has a free spectral range that is determined by the cavity length through the relation $c/2l$. If a cavity must be scanned over $\pm \lambda/2$, there are some practical limitations on its length that limit the length to roughly 0.1 mm $< l < 0.5$ m, which says that the free spectral range (FSR) must roughly satisfy the relation

$$300 \text{ MHz} \gtrsim FSR \gtrsim 1.5 \text{ THz} \tag{6-67}$$

The Michelson interferometer had no free spectral range limitation except the trivial one that the optics be nondispersive, or at least that the dispersion in the two arms cancel and the wavelength be small enough to allow for beam collimation in a reasonable volume. The Michelson interferometer does have a rather stringent resolution limit, however. To resolve a frequency interval Δf requires a scan length of l_c, which is λ times roughly $c/\Delta f$. Writing Δf in terms of center wavelength $\bar{\lambda}$ and wavelength spread $\Delta\lambda$, one finds that

$$l_c = \frac{\bar{\lambda}}{\Delta\lambda}\,\bar{\lambda} \tag{6-68}$$

Theoretically, one should be able to resolve a coherence length of less than a single center wavelength if the reflector were on a sufficiently accurate microstage. This would correspond to white light with a $\Delta\lambda$ of 0.5 μm (i.e., 0.25–0.75 μm) with a center wavelength of 0.5 μm. The problem would be finding the point where the two arms, each of perhaps a million wavelengths in length, would be lined up within a wavelength such that fringes, which would disappear at one wavelength displacement, would be visible. With automated micropositioning equipment and a secondary broadband source, this can indeed be done. As 0.25 μm = 2500 Å corresponds to 300 THz, one sees that the Michelson interferometer can resolve broader spectra than can the Fabry-Perot. Now, theoretically, one also should be able to scan the interferometer over infinite distances. However, a mechanical scan over meters with a submicron tolerance is not a small task. If one uses 0.5 meter as a practical limitation (recall, though, that not all experiments are reasonable in the normal industrial senses or are conducted by people who feel constrained by practical limits), then 150 MHz resolution is about the lower limit. (It should be pointed out that a Fabry-Perot with both arms traveling 30 cm for 0.6 m total travel actually flew on a 1985 space shuttle mission.) In the Fabry-Perot, the resolution limit is just the free spectral range divided by the finesse. As a finesse of 200 is very good for a commercial limit, lines down to about 1.5 MHz are resolvable. This is considerably (by a factor of F) better than the Michelson interferometer.

Examples of various spectrographic problems are in order here. One might want to resolve ''almost'' white light, for example. As was mentioned above, this would be possible in the Michelson interferometer if one were to make the initial submicron alignment, and if one had sufficiently accurate micropositioning gear. In a Fabry-Perot, the task would be hopeless. As the scan range there is fixed at λ, one cannot make the cavity length less than λ. Further, all the approximations of long cavity, large m, slow scan, and so on break down for cavity lengths less than several λ. We will later (in Chapter 7) see that the grating spectrometer is the choice for such broad spectra.

As an example of a Fabry-Perot cavity device, let us consider a semiconductor laser. If the diode is multimoded, it may well have a spectrum such as that depicted in Figure 6.23. A typical multimode gain curve for a first window laser could have a 4 nm width and a center wavelength of 0.85 μm. For a cavity length of 300 μm in a material (GaAlAs) of index $n \sim 3.5$, the intermodal spacing would be roughly 1.5 THz. Each line itself in multimode operation is of the order 10 GHz and should be roughly Lorentzian in shape. Were the laser

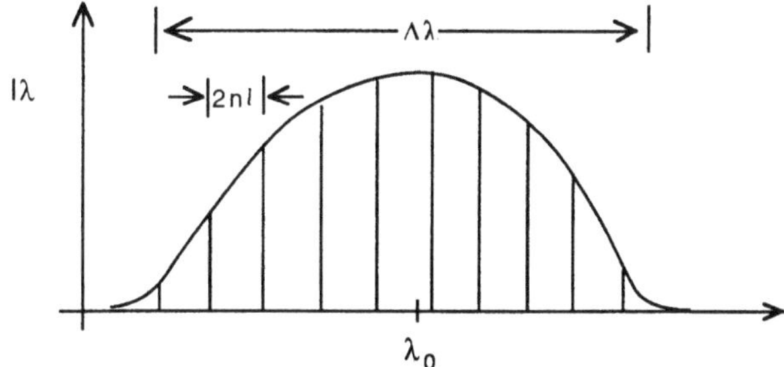

FIGURE 6.23.   Sketch of a possible laser diode spectrum centered about λ_0 with a gain curve of width $\Delta\lambda$ and an intermodal spacing of $c/2nl$.

to operate in a single mode, the linewidth could drop to a width as low as 10 MHz, indicating that the majority of the linewidth is generated by intermodal coupling mechanisms. Assuming multimode operation and using the Δf as the total width of the gain curve, one calculates the coherence length l_c to be roughly 0.2 mm. This could conveniently be resolved by the Michelson interferometer. In fact, as the spectrum is not continuous, it will ring, and this ringing also is quite resolvable on the Michelson instrument. The coherence length that would correspond to the envelope of the ringing would be the coherence length corresponding to the individual linewidths, and would be about 3 cm. Therefore, a Michelson scan of several centimeters with a μm scan step (where the stepping can be performed either by the motion or in the detection electronics) should be able to resolve both envelope and lines when Fourier-transformed. As the 4 nm gain curve width corresponds to a free spectral range of roughly 2 THz, it is just out of range for a Fabry-Perot. A modified Fabry-Perot with finesse of 20,000 thus could conceivably be used to monitor the gain curve and line placements, but without much resolution of individual lines. The Fabry-Perot could be used to monitor the 10 MHz line of a single-mode laser although one could never be sure if one were seeing the correct picture without a second monitoring instrument, as the Fabry-Perot could not monitor if a second mode started to lase, which would violate the band-limited condition of Figure 6.21, destroying the accuracy of the Fabry-Perot trace and thereby causing the earlier-discussed wrap-around error. The two-instrument solution, however, is about the only one possible, as the Michelson interferometer would need a 30-meter scan to resolve a 10 MHz linewidth.

Another example of a Fabry-Perot device is that of a gas laser. A single-mode HeNe laser may have a (short-term) linewidth as small as 1 kHz. This would correspond to a coherence length l_c of roughly 300 km. A Michelson interferometer with this scan length would indeed be hard to build. A 1-meter Fabry-Perot would need a finesse of roughly 30,000 to have this resolution, and, although some claim that this may be possible, it would be by no means trivial. Dr. John L. Hall of the Joint Institute of Laboratory Astrophysics (JILA) of the University of Colorado, Boulder, has demonstrated a HeNe laser with short-term linewidths of less than 1 Hz and a corresponding coherence length of 300,000 km, which is roughly the height of geosynchronous orbit (Salomon, Hils, and Hall 1988). The only techniques for measuring such linewidths are self-heterodyning ones, where a beam is beaten with itself in a detector and the spectrum analyzed. Such techniques are accurate only if there is no line center drift during the measurement time, but they are practically the only techniques available in this coherence length regime.

6.5 YOUNG'S INTERFEROMETER AND SPATIAL COHERENCE

In this section, we wish to consider the interference of radiation emanating from different spatial points. The degree of coherence between such spatially separated points has a dramatic effect on various optical operations such as imaging (Goodman 1965; Collier, Burckhardt and Lin 1971; Yu 1973; Cathey 1974; Marathay 1982; Goodman 1985). The archetypical geometry for this is illustrated in Figure 6.24, where two dipole antennas located in a plane $z = 0$ radiate in the forward direction (that is, have their polarizations perpendicular to the plane of the paper), and their interference pattern is observed in a second plane, $z = L$, which is located in the far field of the antennas. If one considers only a region (x, y) in $z = 0$ that is paraxial, then the antenna patterns can be considered to be uniform in this area, and the problem of calculating the interference pattern reduces to that of summing the two antenna fields and squaring. If the observation region is paraxial enough, and the antennas are driven at equal amplitude and phase, then the intensity in $z = L$ will be

$$I(x) = \frac{1}{2\eta_0} |e^{i\varphi_1(x)} + e^{i\varphi_2(x)}|^2 \tag{6-69}$$

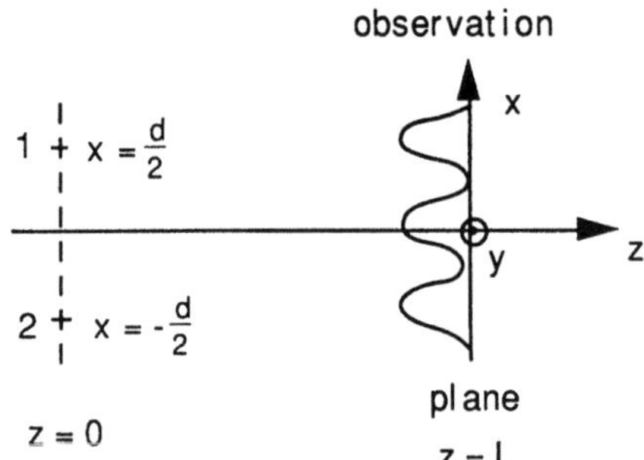

FIGURE 6.24.   Sketch of the relative locations of the radiating elements and observation plane for a simple interference problem.

where the φ's can be expressed as

$$\varphi_{\substack{1\\2}}(x) = k_0 s_{\substack{1\\2}}(x) = k_0 \sqrt{L^2 + \left(x \mp \frac{d}{2}\right)^2} \tag{6-70}$$

The intensity thus can be expressed as

$$I(x) = \frac{1}{\eta_0}\left[1 + \cos k_0(s_2 - s_1)\right] \tag{6-71}$$

The problem we wish to consider as our archetypical spatial coherence problem is an optical analog of the above-described "typical" antenna problem. The problem is often referred to as that of Young's fringes, and the apparatus for producing these fringes is illustrated in Figure 6.25. The model to be considered is essentially one-dimensional in the sense that the incident wave will be taken to be y-polarized (out of the plane of the paper), and the pinholes in the plane $z = 0$ will be considered to be in the plane $y = 0$. Therefore, at least on the line defined by $z = 0$ and $y = 0$, the problem is scalar. For this first analysis, we will consider the source S to be a temporally coherent point source. Unfortunately, such sources do not exist. However, a collimated laser beam appears as a point source at infinity in the sense that a completely coherent source at infinity could produce only coherent parallel rays at the plane $z = 0$. In this sense, we can assume that the analytical signal representation of the wave incident on $z = 0$ is

$$V_y(x, z < 0, t) = ae^{ikz}e^{-i\omega t} \tag{6-72}$$

Actually, as we soon will see, the representation of (6-72) is not crucial for the completely coherent wave because the source here is considered to be placed

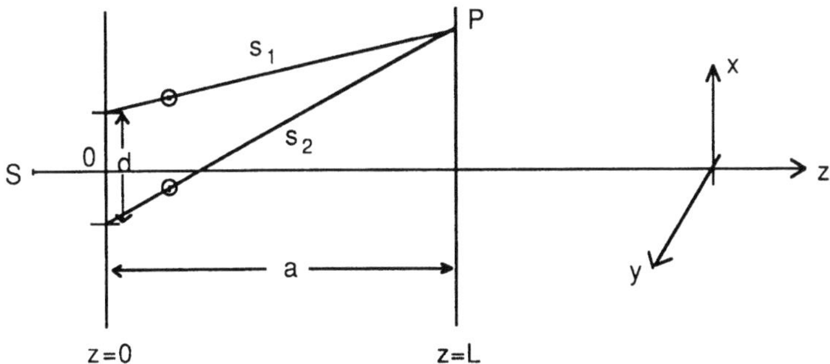

FIGURE 6.25.   A schematic depiction of the apparatus used to generate Young's fringes.

symmetrically between the pinholes, and the pinholes will remove some of the
"memory" of the source details.

Next, one must consider the effect of the pinholes on the wave. Although
we will not consider diffraction until Chapter 7, we need to use some of its
salient features here. Vector diffraction theory is quite complex and will not be
considered at all in this book. From symmetry, however, we can immediately
see that, along the line $z = L$, $y = 0$, a wave that was y-polarized in the plane
$z = 0$, $y = 0$ can only produce y polarization. Therefore, we can get by with
considering scalar diffraction. If the pinhole were truly small, less than a wavelength, the radiation, from basic physical principles, would have to be identical
to that of a small dipole antenna, that is, the far field, of the form

$$\mathbf{E}_{\text{dipole}} \sim \mathbf{E}_0 \frac{e^{i\mathbf{k}\cdot\mathbf{r}}}{r} \cos\theta \tag{6-73}$$

As a dipole becomes larger than a wavelength, the primary difference in the
radiation pattern comes in the angular factor $\cos\theta/r$. Paraxially, this factor is
going to be about constant no matter how big the pinholes are. Therefore, we
can assume that the analytical signal representation of the transmitted pinhole
must look paraxially like

$$V_y(x, z > 0, t) = K(s)a(s)e^{iks} \tag{6-74}$$

where s is the distance from the hole to the observation point, and a is a function
only of the magnitude of s. The distance to an observation point on the line
$z = L$, $y = 0$ from a point $z = 0$, $x = \pm d/2$ is given by

$$s_\pm = \sqrt{L^2 + \left(x_0 \mp \frac{d}{2}\right)^2} \tag{6-75}$$

which paraxially can be expressed as

$$s_\pm \sim L\left[1 + \frac{x_0^2 + d^2/4}{2L^2} \mp \frac{x_0 d}{2L^2}\right] \tag{6-76}$$

Certainly the last two terms in the brackets of equation (6-76) are second-order
quantities and clearly can be ignored in such terms as $1/s$ [as might show up
in $a(s)$], but they could be important in combinations such as ks, which is
measured in terms of wavelengths. Paraxially, therefore, we can take the waves
diffracted from the pinholes to be expressible in the forms

$$V_{y+}(x, z = L, t) = Kae^{iks_+}\,e^{-i\omega t} \qquad\text{(a)}$$

$$V_{y-}(x, z = L, t) = Kae^{iks_-}\,e^{-i\omega t} \qquad\text{(b)}$$

(6-77)

where the dependence of K on s has been intentionally ignored.

Now we are ready to find the (paraxial) intensity in the observation plane. It is given by

$$I(x, z = L, t) = \frac{1}{2\eta_0}\,\langle |V_{y+}(x, z = L, t) + V_{y-}(x, z = L, t)|^2 \rangle$$

$$= \frac{|K|^2\,|a|^2}{\eta_0}\left[1 + \cos k\left(\frac{xd}{L}\right)\right]$$

(6-78)

where equation (6-76) has been used to simplify the expression for $s_2 - s_1$. The result is plotted in Figure 6.26. It is immediately noticeable that the period of the fringes is

$$X = \lambda\,\frac{L}{d}$$

(6-79)

For a 1-meter L and a 0.5-mm d, the fringe period would be about 1 mm. It can be interesting to consider briefly what would be observed in the observation plane, as so many paraxial approximations have been made in the derivation of (6-78) that it is easy to lose sight of what has taken place. Two things will go wrong with the result as one moves out of the paraxial region. One is that the problem becomes a vector problem. The polarization will be exactly y on the axis, but will rotate geometrically into z as one moves in the y direction. This

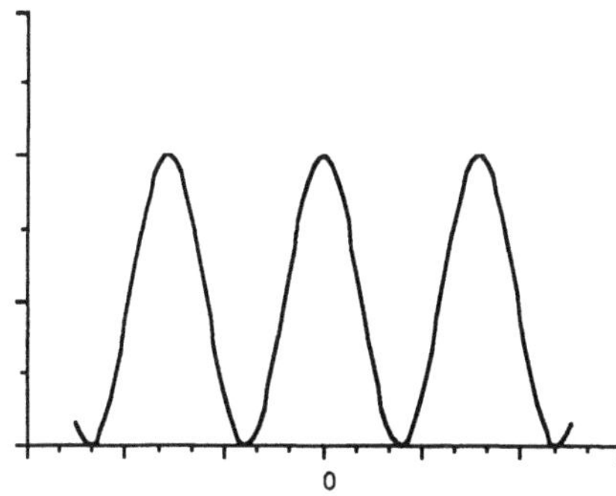

FIGURE 6.26.   Plot of the intensity fringes along the line $z = L$, $x = 0$.

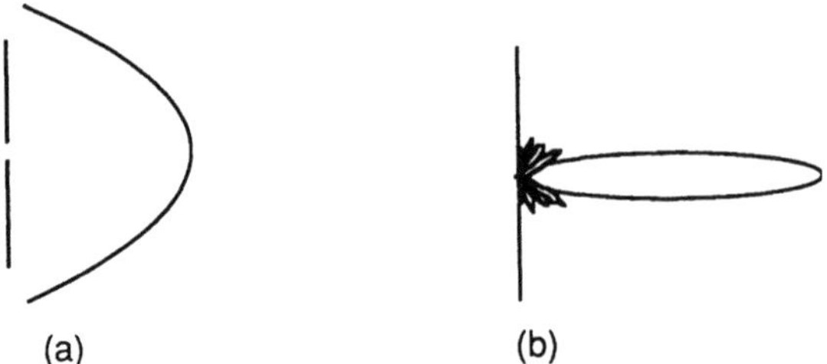

FIGURE 6.27.   Sketches of the radiation patterns from (a) a small ($<\lambda$) pinhole and (b) a large ($>\lambda$) pinhole.

effect, at least for multiwavelength pinholes, takes second place to the amplitude decay. In Figure 6.27, sketches of the radiation patterns ($I(\theta)$) are given for small and large pinholes. For smaller-than-wavelength slits, the radiation intensity will be uniform; for smaller-than-wavelength pinholes, the pattern will be uniform in the x direction but fall off as the cosine along the direction of polarization, assumed to be y. For a larger hole, the radiation intensity can fall off to its first zero in a few degrees. It is this falloff that in this problem defines the paraxial region. Therefore, one now realizes that the actual pattern will be an illuminated circle (roughly) that exhibits reasonably straight fringes surrounded by much weaker regions of probably quite distorted fringes. For our $L = 1$ m, $d = 0.5$ mm geometry, assuming roughly 10 μm pinholes (3° paraxial region for 0.5 mm illumination), the illumination region would be a 5-cm-radius circle. This would mean that about 100 fringes would be visible.

Before going to the most general case, it can be instructive to see what happens to Young's fringes when one illuminates the pinholes with two-tone illumination. Here the wave incident on the screen is taken to be

$$V_y(x, z < 0, t) = a(e^{ik_1 z - i\omega_1 t} + e^{ik_2 z - i\omega_2 t}) \tag{6-80}$$

The intensity in $z = L$, $y = 0$ is therefore given by

$$I(x, z = L, t) = \frac{|K|^2 |a|^2}{2\eta_0} \langle |e^{ik_1 s+} e^{-i\omega_1 t} + e^{ik_2 s+} e^{-i\omega_2 t}$$

$$+ e^{ik_1 s-} e^{-i\omega_1 t} + e^{ik_2 s-} e^{-i\omega_2 t}|^2 \rangle \tag{6-81}$$

and, assuming that the detector time τ_d (the response time of the eye) is much larger than $2\pi/(\omega_1 - \omega_2)$, the intensity can be expressed as

$$I(x, z = L, t) = \frac{|K|^2 |a|^2}{2\eta_0} \left[2 + \cos \frac{k_1 xd}{L} + \cos \frac{k_2 xd}{L} \right] \tag{6-82}$$

and is illustrated in Figure 6.28. It is quite interesting to note that the intensity pattern is just the sum of the intensity patterns for the two colors taken individually because of the time averaging out of the cross terms in (6-81). The composite pattern would appear quite striking. If, for example, blue and red were the colors, the pattern would be bright purple at the center and would have interestingly mixed bands of purple, red, and blue extending out from the center. Equally interesting and perhaps more important would be the pattern observed for a point source of (possibly white) spectrum $I(\lambda)$, which we now can easily work out. As the intensities add, we see that the composite pattern would be given by

$$I(x, z = L, t) = c \int_0^\infty d\lambda\, I(\lambda) \left[1 + \cos\left(\frac{k(\lambda)xd}{L}\right) \right] \qquad (6\text{-}83)$$

This pattern, if the light were white, would have a bright white maximum at $x = 0$ ringed by rainbows that would fade into a gray level, and thus is quite colorful.

Now we are ready to consider an even more general case, that in which the incident wave can be represented by

$$V_y(x, z < 0, t) = a(x, t)e^{i\bar{k}z}e^{-i\bar{\omega}t}e^{i\varphi(x,t)} \qquad (6\text{-}84)$$

where V_y is sufficiently general to represent most any quasi-monochromatic source at any placement behind the plane $z = 0$. The intensity in the observation plane for this incident disturbance is represented by

$$I(x, z = L,t) = \frac{|K|^2}{2\eta_0}\left[\left\langle a^2\left(\frac{d}{2}, t\right)\right\rangle + \left\langle a^2\left(\frac{-d}{2}, t\right)\right\rangle + 2\,\mathrm{Re}\left(\left\langle a\left(\frac{d}{2}, t\right)\right.\right.\right.$$
$$\left.\left.\left. \cdot\, a\left(\frac{-d}{2}, t\right) e^{i(\varphi(d/2,\,t) - \varphi(-d/2,\,t))} e^{ik(xd/L)} \right\rangle\right)\right]$$

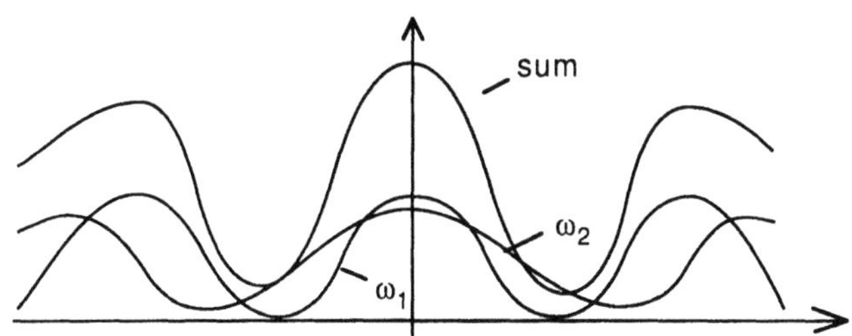

FIGURE 6.28.   Plots of the intensity patterns for ω_1, ω_2, and their sums.

which is often expressed as

$$I(x, z = L, t) = \frac{|K|^2}{2\eta_0} \left[\Gamma_{11}\left(\frac{d}{2}\right) + \Gamma_{22}\left(\frac{-d}{2}\right) + 2\,\text{Re}\,[\Gamma_{12}(d)] \right] \quad (6\text{-}86)$$

where (evidently)

$$\Gamma_{11}\left(\frac{d}{2}\right) = \frac{|K|^2}{2\eta_0} \left\langle a^2\left(\frac{d}{2}, t\right)\right\rangle \qquad\qquad\qquad (a)$$

$$\Gamma_{22}\left(\frac{-d}{2}\right) = \frac{|K|^2}{2\eta_0} \left\langle a^2\left(\frac{-d}{2}, t\right)\right\rangle \qquad\qquad (b) \quad (6\text{-}87)$$

$$\Gamma_{12}(d) = \frac{|K|^2}{2\eta_0} \left\langle a\left(\frac{d}{2}, t\right) a\left(\frac{-d}{2}, t\right)\right\rangle \qquad (c)$$

$$\cdot\; e^{i(\varphi(d/2,\,t) - \varphi(-d/2,\,t))} e^{ik(xd/L)}$$

where Γ_{11} represents the intensity that would be measured in the plane $z = L$ if pinhole 2 were blocked, Γ_{22} represents the intensity that would be measured in the plane $z = L$ if pinhole 1 were blocked, and Γ_{12} is the interference term, which is generally called the mutual coherence function. One often makes the definition that

$$\Gamma_{12}(d) = \sqrt{\Gamma_{11}\left(\frac{d}{2}\right)} \sqrt{\Gamma_{22}\left(\frac{-d}{2}\right)} \hat{\gamma}_{12}(d) \qquad (6\text{-}88)$$

where the complex degree of coherence $\hat{\gamma}_{12}(d)$ can be expressed in the form

$$\hat{\gamma}_{12}(d) = \gamma_{12}(d)e^{i\alpha(d)} e^{ik(xd/L)} \qquad (6\text{-}89)$$

Now that we have performed a number of calculations and made numerous definitions, perhaps it is time to consider the meaning of these manipulations. Generally, one is interested in such quantities as the degree of coherence as defined by (6-89) or (6-27). In (6-27), the temporal degree of coherence is a function of only the time delay between the two waves, and its first zero determines the temporal coherence time. In (6-89), because of the quasi-monochromatic approximation that is implicit in the representation of (6-84), the degree of coherence is a function of only the distance between the two pinholes. By analogy with temporal coherence length, one can therefore define the spatial coherence length by the separation at which the function γ_{12} has its first zero or at least its first value below a small, predetermined constant. One must bear in mind, however, the meaning of this distance. It is the pinhole separation for which the fringes in the observation plane completely (or almost completely) disappear. In the temporal case, the coherence length was the pathlength difference between the two paths that would cause the envelope function of the ringing with separation function to have its first zero. In an ideally aligned Michelson interferometer, the observation plane should never exhibit fringes. In a slightly misaligned one, however, the coherence length would be the misalignment length for which the fringes would exhibit minimal contrast across the observation plane. In this sense, the temporal and the spatial coherence have similar operational meaning. The temporal coherence, however, is due to time variation of the a and φ, and the spatial coherence is due to time variation of the spatial dependence of a and φ. Clearly, finite-sized polychromatic sources will have limited coherences of both types. With the realization that the spatial degree of coherence depends only on pinhole separation, the x dependence of such quantities as Γ_{12} becomes superfluous, and one often defines auxiliary quantities such as

$$J_{12}(d) = \Gamma_{12}(d, x = 0) \qquad \text{(a)}$$

$$\hat{\mu}_{12}(d) = \hat{\gamma}_{12}(d, x = 0) \qquad \text{(b)}$$

$$(6\text{-}90)$$

where J_{12} is called the mutual intensity, and $\hat{\mu}_{12}$ is called the complex degree of coherence, as was $\hat{\gamma}_{12}$ (unfortunately). It is clearly seen that

$$\hat{\mu}_{12}(d) = \gamma_{12}(d)e^{i\alpha_{12}(d)} \qquad (6\text{-}91)$$

In both the two-tone and the general cases, the disturbance directly behind the apertures was given. In general, the problem that one wishes to solve is that of what the effect of the pinholes will be on the radiation emitted by a source at some distance. The temporal variation of the spatial dependence of the a and φ of (6-84), however, certainly must vary with propagation distance. This should be clear from earlier discussions of what happens when one tries to image incoherent (Lambertian) sources with finite aperture lenses. Essentially the only problem that we have solved at this point, therefore, is the problem of setting the pinhole plane right up against the surface of the finite-sized source whose specific intensity is known. Here, we could consider the effects of a finite-sized source at a finite distance from our pinholes by doing integrals over the source. In the next chapter, however, we will learn how to propagate forward disturbances of arbitrary coherence and thereby how to propagate a degree of coherence. In light of this, it is best to defer consideration of these special cases.

A final point should be made here concerning the relation between spatial and temporal coherence. If a signal is truly monochromatic, then it is both temporally and spatially totally coherent, as the a and φ functions become time-independent. The a and φ functions could still be even rapidly varying functions of space, but the point is that the averaging brackets in (6-87) disappear, and it becomes clear that there is a value of x such that the total phase goes to zero and the value of $\hat{\gamma}_{12}(d) = 1$, giving that $\gamma_{12}(d) = 1$. In this sense, temporal coherence implies spatial coherence. The converse is not true. Consider the source to be a polychromatic point source at infinity (very large compared to a wavelength) distance. Even through the phase fronts will not be evenly spaced at $z = 0$, they all will be plane there. Delaying the wave with respect to itself will lead to a finite temporal coherence time, but spatially separated samples will always be 100% correlated regardless of the separation.

6.6 HANBURY-BROWN AND TWISS INTERFEROMETER

The conventional interference experiments (Michelson, Young, etc.), measuring the coherence length of light, cannot distinguish between states of the same spectral distribution with quite dissimilar statistical properties. An example of such states would be coherent laser light and thermal light from a lamp passed through a filter to give the same spectrum. The Hanbury-Brown and Twiss experiment measures the correlation of two optical intensities and can distinguish between such sources.

The principle of the experiment is shown in Figure 6.29, with the correlator principle depicted in Figure 6.30. The incident intensity $I(f)$ is divided into two identical beams with intensity $I_1(t) = I_2(t) = 1/2\ I(t)$. The correlator introduces a controlled time delay to one of the signals and mixes the signals in a linear mixer and integrates them to measure the long-time average. In the Hanbury-Brown and Twiss (H-B and T) experiment, the correlator further sub-

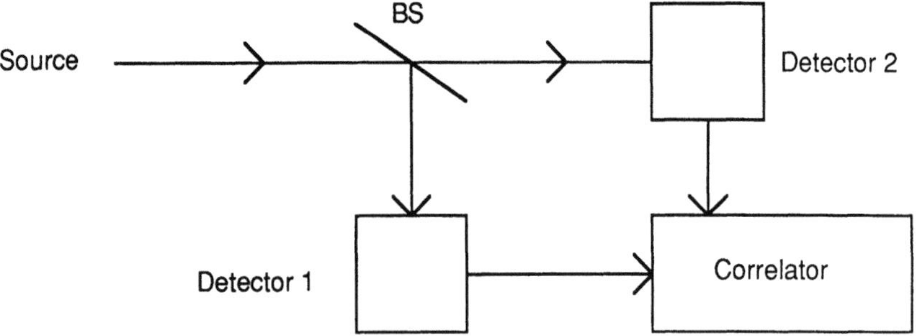

FIGURE 6.29.   Schematic depiction of the experiment of Hanbury-Brown and Twiss.

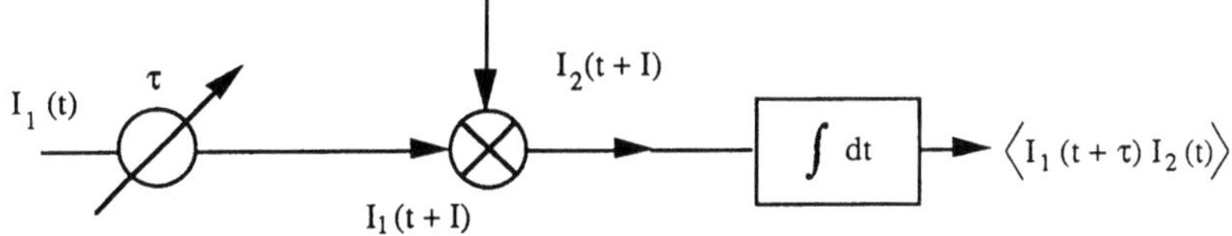

FIGURE 6.30.   Schematic depiction of the contents of the correlator block of the Hanbury-Brown and Twiss experiment of Figure 6.29.

tracts a long-term average $\langle I_1 \rangle \langle I_2 \rangle$ and normalizes to give

$$\frac{\langle I_1(t + \tau)\, I_2(t)\rangle - \langle I_1(t)\rangle \langle I_2(t)\rangle}{\langle I_1(t)\rangle \langle I_2(t)\rangle} = g^{(2)}(\tau) - 1 \tag{6-92}$$

where $g^{(2)}(\tau)$ is the degree of second-order temporal coherence. For a well-stabilized laser source high above threshold, the spectral width is solely due to phase noise. Hence, the intensity $I(t)$ is constant, and $g^{(2)}(\tau) = 1$. In this case, the H-B and T experiment results in zero signal.

Thermal light, on the other hand, is made up of independent contributions from the different radiating atoms:

$$E(t) = \sum_i E_i(t) \tag{6-93}$$

The intensity-correlation function follows as

$$\langle I(t + \tau)\, I(t)\rangle = \sum_{ijkl} \langle E_i^*(t)\, E_j^*(t + \tau)\, E_k(t + \tau)\, E_l(t)\rangle \tag{6-94}$$

Since each contribution is independent, only terms in which the ith field is multiplied by its complex conjugate are used; that is, we have contributions from

$$i = l, \quad j = k, \quad i = k, \quad \text{and } j = l \tag{6-95}$$

such that

$$\begin{aligned}
\langle I(t + \tau)\, I(t)\rangle &= \sum_i \langle E_i^*(t)\, E_i(t)\rangle \sum_j \langle E_j^*(t + \tau)\, E_j(t + \tau)\rangle \\
&+ \sum_i \langle E_i^*(t)\, E_i(t + \tau)\rangle \sum_j \langle E_j^*(t + \tau)\, E_j(t)\rangle
\end{aligned} \tag{6-96}$$

which can be written as

$$\langle I(t + \tau)\, I(t)\rangle = |\langle E^*(t)\, E(t)\rangle|^2 + |\langle E^*(t + \tau)\, E(t)\rangle|^2 \tag{6-97}$$

Normalizing to $\langle I \rangle^2$, we find $g^{(2)}(\tau)$ as

$$g^2(\tau) = 1 + |g^{(1)}(\tau)|^2 \qquad (6\text{-}98)$$

and the output from H-B and T experiment is given by $|g^{(1)}(\tau)|^2$, the absolute square of the first-order temporal coherence. For a collision-broadened light source $g^{(1)}(\tau) = e^{-\gamma t}$, resulting in

$$g^2(\tau) = 1 + e^{-2\gamma\tau} \qquad (6\text{-}99)$$

References

Born, M. and E. Wolf, *Principles of Optics*, Fifth edition, Pergamon Press, New York (1975).

Cathey, W. T., *Optical Information Processing and Holography*, John Wiley and Sons, New York (1974).

Cochran, W. T., J. W. Cooley, D. L. Favin, H. D. Helms, R. A. Kaenel, W. W. Lang, G. C. Maling, D. E. Nelson, C. M. Rader, and P. D. Welch, What is the Fast Fourier Transform, *Proc. IEEE 55*, 1664–1674 (1967).

Collier, R. J., C. B. Burckhardt, and L. H. Lin, *Optical Holography*. Academic Press, New York (1971).

Goodman, J. W., *Introduction to Fourier Optics*. McGraw-Hill, San Francisco (1965).

Goodman, J. W., *Statistical Optics*, Wiley, New York (1985).

Hecht, Optics, Second Edition, Addison-Wesley, Reading MA (1987).

Klein, M. V. and F. F. Furtak, *Optics*, Second edition, John Wiley and Sons, New York (1986).

Marathay, A. S., *Elements of Optical Coherence Theory*, John Wiley and Sons, New York (1982).

Salomon, Ch., D. Hils, and J. L. Hall, Laser stabilization at the millihertz level, *J. Opt. Soc. Am. 5*, 1576–1587 (1988).

Yariv, A., *Optical Electronics*, Third edition, Holt, Rinehart and Winston, New York (1985).

Yu, F. T. S., *Introduction to Diffraction, Information Processing and Holography*, MIT Press, Cambridge, MA (1973).

Problems

1. Suppose one were to collimate a polychromatic beam of light with amplitude spectrum $f(\omega)$ and analyze it in a Michelson interferometer as shown in Figure 6.31. Assume that the two beams interfere at a small angle θ.

   (a) What is the observed fringe pattern for equal arm lengths of the interferometer?

   (b) What is the interference pattern if there is a length mismatch of d between the interferometer arms?

   (c) Find and plot the expression for the fringe visibility as a function offset d.

   (d) What is the result in (c) for $f(\omega) = A \, \text{rect}\,((\omega - \omega_0)/\Delta\omega)$, with rect (x) defined

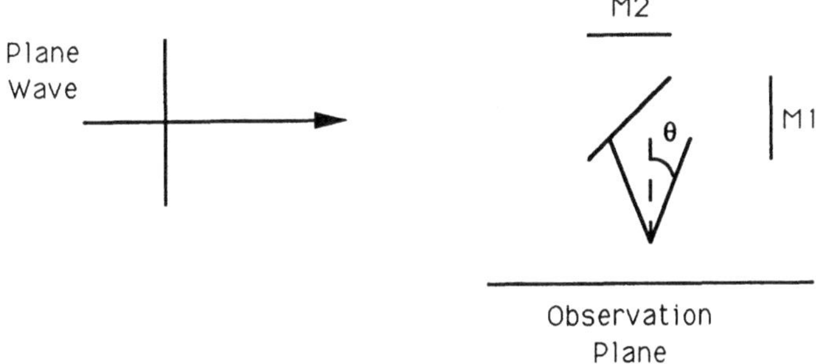

FIGURE 6.31.   Figure for problem 1.

as per equation 6-19 with A determined by the condition $\int f(\omega)\, d\omega = 1$? Find the limits of your result for $\Delta\omega \to 0$ and $\Delta\omega \to \infty$.

2. Consider a perfectly aligned Michelson interferometer, as discussed in the text. Calculate δ_{12}, α_{12}, and τ_c for the following:

   (a) An input spectrum of $E(\omega) = 1/\Delta\omega\ \text{rect}\ (\omega - \omega)/\Delta\omega$ with rect (x) defined as per equation (6.19). Consider in particular the limits where $\Delta\omega \to 0$ and $\Delta\omega \to \infty$.

   (b) A constant phase field whose amplitude decorrelates as

   $$\langle a(t)a(t + \tau)\rangle \approx e^{-\tau/\tau_a}$$

   (c) A constant-amplitude field whose phase decorrelates as

   $$\langle e^{i\varphi(t)} e^{-i\varphi(t + \tau)}\rangle \approx e^{i(q + i/\tau_\varphi)\tau}$$

3. Consider a plane wave, with power spectrum $W(\omega)$ given by

   $$W(\omega) = A_1 e^{-(\omega - \bar\omega_1)^2/2\sigma_1^2} + A_2 e^{-(\omega - \bar\omega_2)^2/2\sigma_2^2}$$

   where A_1 and A_2 are suitable normalization factors, incident on the input of a Michelson interferometer.

   (a) Find the function $\hat\gamma(\tau)$ from the power spectrum.

   (b) In the limit where $A_1 = A_2 = A$ and $\sigma_1 = \sigma_2 = \sigma$, find and sketch the function $I(\tau)$, the output of the Michelson interferometer.

   (c) Define a coherence time for the incident wave. Indicate this coherence time on a sketch of Re $[\hat\gamma(\tau)]$.

4. In (a) and (b), find the $\hat\gamma(\tau)$ that corresponds to the output of a Michelson interferometer with the following conditions:

   (a) The input

   $$E(\omega) = \frac{1}{\sqrt{2\pi}\sigma_\omega} e^{-(\omega - \omega)^2/2\sigma_\omega}$$

   Check whether your answer agrees with your intuition in the limits $\sigma_\omega \to 0$, $\sigma_\omega \to \infty$.

(b) The interferometer has diffraction losses such that the interfering beams at the output are given by

$$V_1(t) = c_1 a(t + \tau)e^{i\varphi(t+\tau)}$$

$$V_2(t) = c_2 a(t)e^{i\varphi(t)}$$

where $c_1^2 + c_2^2 = 1/2$.

In (c) and (d), find the $E^2(\omega)$ that corresponds to the following correlations:

(c) $\langle e^{i\varphi(t+\tau)} e^{-i\varphi(t)} \rangle = e^{-|\tau|/\sigma_2}$ where $a(t)$ is constant in time.

(d)

$$\frac{\langle a(t+\tau)a(t) \rangle}{\langle a^2(t) \rangle} = \frac{\sin \pi\tau/\sigma_r}{\pi\tau/\sigma_r}$$

5. Consider a Michelson interferometer. Find, from the statistics of a and θ, the intensity $I(\tau)$. Sketch the function $\gamma(\tau)$, and invert the $\hat{\gamma}(\tau)$ to find the spectral density of the incoming disturbance. Assume that the incident wave is time-stationary to allow for substitution of time averages for ensemble averages if necessary. In (a) and (b) assume that a is constant and take

(a) $\langle e^{i\theta(t)} e^{-i\theta(t-\tau)} \rangle = Ce^{-|\tau|/\tau_c}$.

(b) $p_{\Delta\theta}(\Delta\theta) = 1/\sqrt{2\pi\langle(\delta\theta)^2\rangle}\, e^{(\Delta\theta^2/2\langle(\delta\theta)^2\rangle)}$ with $\Delta\theta = \theta(t) - \theta(t-\tau)$.

In (c) and (d) assume θ is constant, but

(c) $\langle a(t)a(t-\tau) \rangle = Ce^{-a|\tau|} \sin \omega_r\tau$.

(d) $p_a(a) = a/\sqrt{2\pi\langle a^2 \rangle}\, e^{-a^2/2\langle a^2 \rangle}$ independently of time.

6. Say that the beam incident into a Michelson interferometer is not plane but actually Gaussian in cross section such that the field distribution is given by

$$E(x) = E_0\, e^{-(x/w_0)^2}$$

Find the intensity in the output plane of the interferometer.

7. Consider the system of Figure 6.32. The system is a rather unique modification of

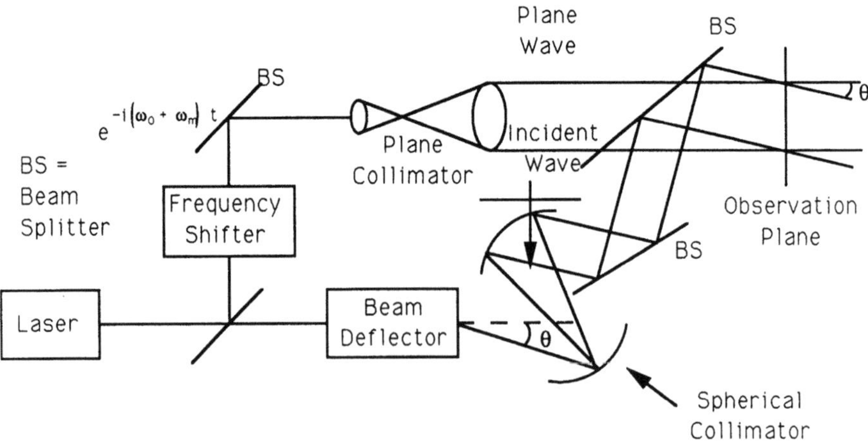

FIGURE 6.32.   Figure for problem 7.

the Michelson interferometer. The source is coherent and radiates at a frequency $f_0 = \omega_0/2\pi$. In one arm, this frequency is upshifted to $f_0 + f_m$ and then collimated to be interfered with the other arm. The other arm is beam-deflected such that the angle θ with which it will meet the other beam can be closely and rapidly controlled. Assume that the angle of inclination θ, caused by the beam deflector, is $\ll 1$ rad. Find the temporo-spatial distribution of the light in the observation plane. How could one use this system to advantage?

8. We wish to test lenses with an interferometer. The interferometer we choose to use interferes two beams together at an angle 2θ. One arm of the interferometer has a calibrated lens in it, such that the beam in this arm gets a spherical curvature and can be expressed as

$$E_1 = ae^{ikz\cos\theta}e^{-ikx\sin\theta}e^{-(ik/2f)(x^2\cos^2\theta + y^2 + z^2\sin^2\theta)}$$

The second arm contains the lens under test, which, nominally, should be identical to the calibration lens. The two arms are identical except for the lenses.

(a) If the test lens is identical to the calibration lens, what pattern will be seen in the observation plane, $z = 0$?

(b) Let us say the test lens is identical to the calibration lens except that its focal length is slightly shorter. Say that $f_{\text{test}} = f = \epsilon$. Sketch the interference patterns for several values of ϵ.

(c) Let us say the test lens is aberrated, that is, its surface is not spherical. How will this show up in the test pattern?

9. Say that we have a beam of polychromatic light of spectrum $f(\omega)$ incident at normal incidence on an electro-optic slab of index n_1 from free space (index n_0) as shown in Figure 6.33.

(a) Find and plot n_0, the transmitted intensity as a function of ω for $n_1 = 2.2$.

(b) Find an expression for the total transmission for the polychromatic beam.

(c) Say that the facets of the plate are silvered such that the reflectivity goes up to a value r_s. What happens to the plots in (a)?

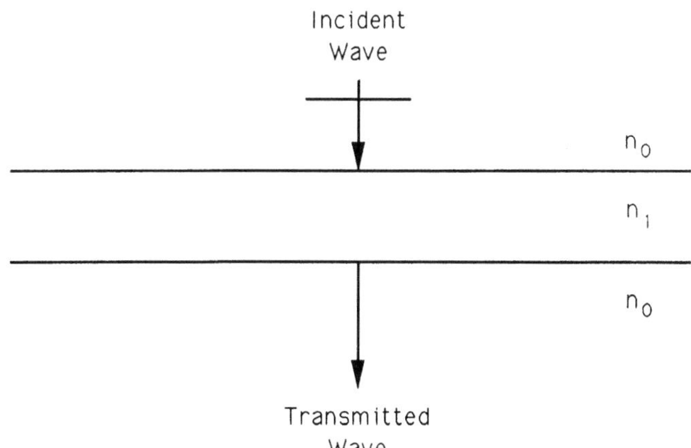

FIGURE 6.33.   Figure for problem 9.

(d) Say that a sinusoidal voltage is applied to the crystal such that the index is varied in the manner $n(t) = n_0 + \Delta n \sin \omega_m t$. What is the transmitted field as a function of time?

10. Let us say that we wish to use a Fabry-Perot etalon as a spectrum analyzer. Say that we have a signal with spectrum $I(\lambda) = I_0/\Delta\lambda \; e^{(\lambda - \lambda_0)^2/\Delta\lambda}$. Say that the mirror separation is being modulated through the free spectral range at a frequency f. In (a) and (b), assume a very high finesse. In (a) through (d) find the transmitted intensity as a function of time.

(a) If the frequency range corresponding to $\Delta\lambda$ is small compared to the *FSR*.

(b) If the frequency range exceeds the *FSR*.

(c) and (d) repeat (a) and (b) but for a case where the finesse is poor, let us say on the order of magnitude 10.

11. Generalize the multiple-bounce argument as depicted in Figure 6.34 to solve for the transmission of Fabry-Perot cavities satisfying the following conditions:

(a) The mirrors are lossless but are of two different values R_1 and R_2.

(b) The mirrors are the same, but they are both lossy such that $R + T < 1$.

(c) The mirrors are confocal such that it takes four bounces to complete a round trip.

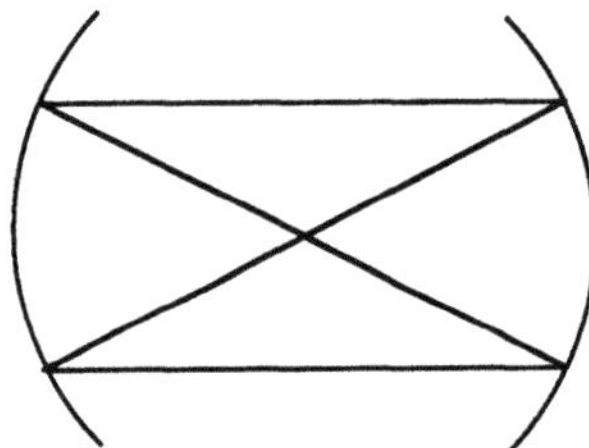

FIGURE 6.34.   Figure for problem 11.

(d) The Fabry-Perot is a guided wave version, as depicted in Figure 6.35, with a channel loss of α dB/cm. What is the maximum resolution of this device, and for what length does it occur?

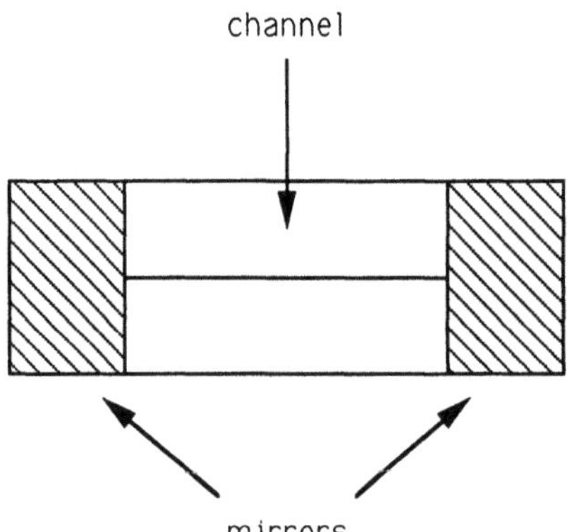

FIGURE 6.35.   Figure for problem 11.

12. Consider the multiple-bounce argument given in the text for the transmission of Fabry-Perot cavities. The reflectivities are the same, $R_1 = R_2 = 0.93$, but they are both lossy because of the absorbance A ($A = 0.05$) by the transparent metal films that are used to increase the reflectance, such that $R + T + A = 1$. Calculate the finesse and the loss of the transmitted intensity.

13. Consider a Fabry-Perot with a "fast" detector (slow compared to optical frequency and optical beat frequency) that has a transmission intensity spectrum given approximately by

$$T(\omega) = \frac{1}{\Delta\omega} \sum_{n=1}^{\infty} \mathrm{rect}\left(\frac{\omega - n\omega_B}{\Delta\omega}\right)$$

with rect (x) defined as per equation (6-19). Say that an intensity spectrum

$$I_i(\omega) = \frac{1}{\Delta\omega_a} \mathrm{rect}\left(\frac{\omega - \omega_a}{\Delta\omega_a}\right) + \frac{1}{\Delta\omega_b} \mathrm{rect}\left(\frac{\omega - \omega_b}{\Delta\omega_b}\right)$$

is incident on it.

(a)   Find the detector current $i_d(t)$.

(b–d) Take three "reasonable" limits of your results in (a) for bandlimited and non-bandlimited responses.

14. If a slow detector instead of a fast one is used, what will the output current be for the Michelson interferometer and the Fabry-Perot interferometer?

15. When the sweep period of a Fabry-Perot interferometer approaches the rise time of the detector electronics, it is necessary to take the time dependence of the Fabry-Perot transmission function ($T(V)$) and the detector circuit impulse response into account.

(a) Assuming high cavity finesse ($F \gg 1$) and $\delta l/l \ll 1$, write a general expression for the observed detector current given an optical intensity of $I_i(V)$ and a detector with impulse response $h(t)$.

(b) Evaluate the function obtained in (a) for a system with $h(t) = e^{-t/\tau_d}$ and an optical spectrum centered within a free spectral range whose upper frequency is ν_0 [i.e., $I_i(\nu) = f(\nu - (\nu_0 - (\Delta l/2l)\nu_0))$], where Δl is the maximum scan distance used,

$$\delta l(t) = \Delta l \sum_{n} r_1(t - nT), \text{ and } r(t) = \begin{cases} t & j^0 < t < 1. \\ 0 & \text{otherwise} \end{cases}$$

(c) Find an expression for the detector current if $T \ll \tau_{\text{scan}}, \tau_{\text{detector}}$ if $f(V') = e^{-|V'|/\omega}$.

16. Consider a Fabry-Perot interferometer with spherical mirrors. The electric field inside the cavity exists in discrete modes abbreviated as TEM_{lm}. The phase characteristic of each mode is described by

$$\varphi(z) = kz - (l + m + 1) \tan^{-1} \frac{z}{z_0}$$

where $z_0 = (\pi \omega_0^2 n / \lambda)$, and ω_0 is the beam waist size.

(a) For a given mode, find ΔV for this type of Fabry-Perot.

(b) Let $z_z - z_1 = d$ (the distance between the mirrors). Consider two different modes, but leave q constant.

$$k_1 d - (l + m - 1)_1 \left(\tan^{-1} \frac{z_z}{z_0} = \tan^{-1} \frac{z_1}{z_0} \right) = q\pi$$

$$k_2 d - (l + m - 1)_2 \left(\tan^{-1} \frac{z_z}{z_d} = \tan^{-1} \frac{z_1}{z_0} \right) = q\pi$$

Find ΔV caused by a change in $(l + m)$, $\Delta(l + m)$.

(c) In the confocal case $R = d$, $z_z = -z_1 = z_0$. For this situation find ΔV due to $\Delta(l + m)$

17. Consider a two-beam interference pattern as would be observed on the screen depicted in Figure 6.36 where the waves are polarized in the plane of the paper.

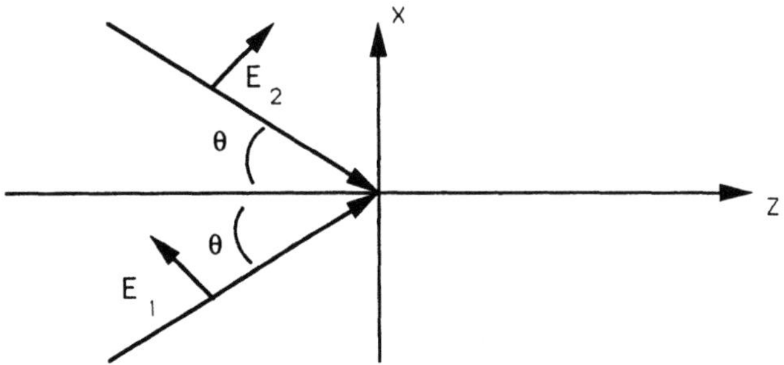

FIGURE 6.36.   Figure for problem 17.

(a) Write out expressions for E_1 and E_2.

(b) Find the interference pattern.

(c) Find the visibility as a function of θ.

(d) Repeat (c), but this time assuming that the incident wave is circularly polarized.

18. Consider two beams:

$$E_1 = e^{ikz \cos\theta} e^{-ikx \sin\theta} e^{-(ik/2f_-)(x^2 \cos^2\theta + y^2 + z^2 \sin^2\theta)}$$

$$E_2 = e^{ikz \cos\theta} e^{ikx \sin\theta} e^{(ik/2f_+)(x^2 \cos^2\theta + y^2 + z^2 \sin^2\theta)}$$

incident from $-z$ on an observation plane at $z = 0$.

(a) Find an expression for the intensity pattern seen on the screen.

(b) Sketch the intensity pattern in the limit where $f_+, f_- \to \infty$.

(c) How does the pattern of (b) change when f_- becomes finite and positive? Negative? Sketch the results.

(d) What happens when $f_+ = f_-$? Explain your answer.

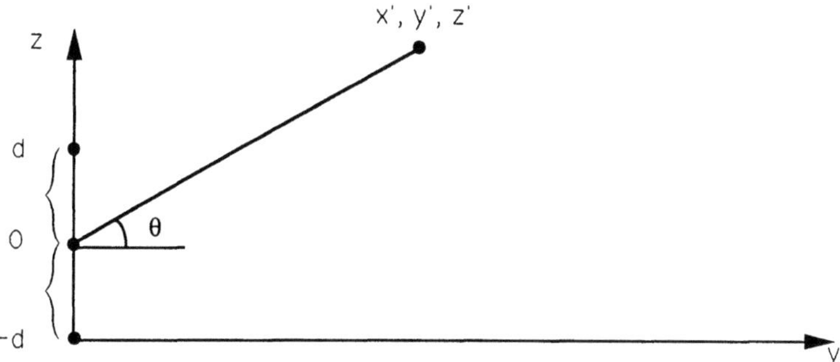

FIGURE 6.37.   Figure for problem 19.

19. Find the radiation pattern due to three dipoles, each separated by a distance d, each with an excitation of 3 radians with respect to the previous one, as is depicted in Figure 6.37.

20. Consider a diffractometer arrangement with an incoherent source of transverse length as located at a distance l_s in front of the input plane at $z = 0$.

    (a) Calculate and plot the intensity at the origin of the plane $z = L$ as a function of the split spacing d for the following intensity distribution:

$$I(x_s) = \frac{1}{a_s} \, \text{rect} \left(\frac{x_s}{a_s} \right)$$

    with $\text{rect}(x)$ defined as per equation (6-19).

    (b) Find an expression for the coherence length observing the limit $a_s \to \infty$ such that $a_s/l_s = .2$, and also when $a_s \to 0$.

21. Suppose one were to carry out a Young's interference experiment, but with a finite length source of narrow bandwidth but spatially incoherent, as in Figure 6.38. Assume $z_0 \gg d$ and $z_0 \gg L$.

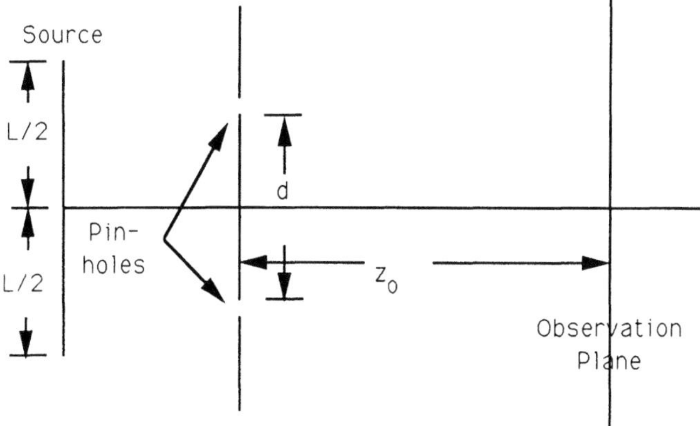

FIGURE 6.38.   Figure for problem 21.

(a) What is the intensity pattern on the observation screen for an arbitrary point on the source?

(b) What is the general expression for the intensity pattern at the output?

(c) What is the limit of (b) for $L \to 0$? $L \to \infty$?

22. Consider a pulsed point source at a location z_s in front of the input to a Young's interferometer, as depicted in Figure 6.39. Find the pattern in the image plane if the source intensity is given by

$$I(t) = I_0 e^{-(t/\Delta t)^2}$$

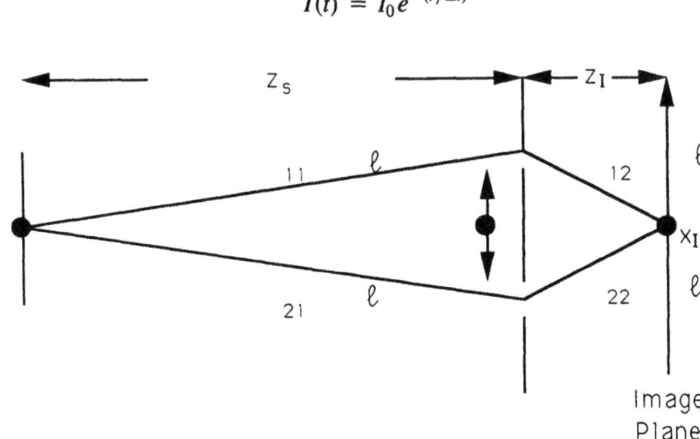

FIGURE 6.39.   Figure for problem 22.

23. Consider the Mach-Zehnder interferometer shown in Figure 6.40. Assume that the input wavelength is 0.63 μm, and the index of the modulator at 0 volts is 2.2. Also, ignore reflections from the modulator, and assume that the mirrors are perfectly aligned. At $V = 0$ the modulator arm of the interferometer is adjusted until maxi-

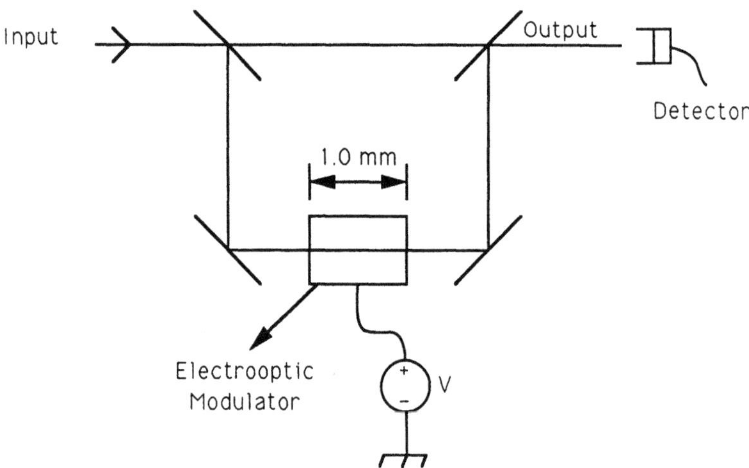

FIGURE 6.40.   Figure for problem 23.

mum power is achieved at the detector. Next the voltage is increased, causing the index of the modulator to increase. This process causes the output power of the interferometer to drop. The voltage is increased until the output power reaches the "first" minimum. What is the index of the modulator at I_{min}?

24. Consider a diffractometer arrangement with a finite incoherent source at a distance l_s from the diffracting plane, as depicted in Figure 6.41, where the intensity of the source is given by

$$I_s = \frac{1}{d_s} \, \text{rect}\left(\frac{x}{d_s}\right)$$

with rect (x) defined as per equation (6-19).

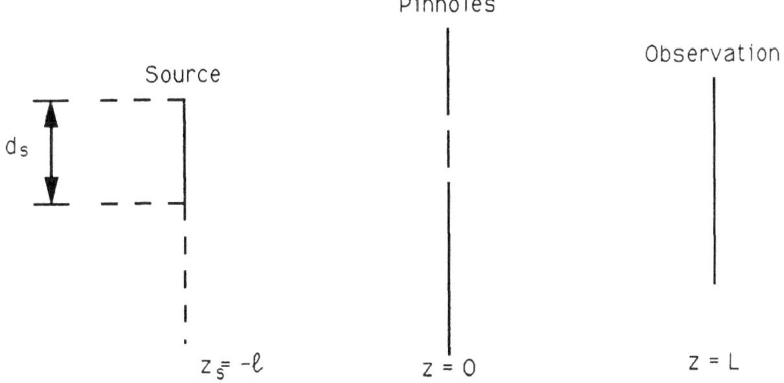

FIGURE 6.41.   Figure for problem 24.

(a) Find the fringe pattern.
(b) Find an expression for the coherence length.
(c) Find the limit of (b) when $d_s \to \infty$ such that $d_s/l_s = 0.1$.
(d) Find the limits of (a) and (b) when $d_s \to 0$.

25. Consider a diffractometer much like that of Figure 6.40 but with an incoherent source of transverse length, as located at a distance L_s in front of the input plane at $z = 0$. One wants to calculate and plot the intensity at the origin of the plane $z = L$ as a function of the slit spacing d for the following intensity distributions of the incoherent source.

(a) $I(x_s) = I_0/a_s \, \text{rect}\,(x_s/a_s)$.
(b) $I(x_s) = (10I_0/a_s)\, e^{-10 x_s/a_s}$.
(c) $I(x_s) = I_0 \displaystyle\sum_{n=-N/2}^{N/2} \delta(x_s = ns)\, \text{rect}\,(x_s/a_s)$, where $S = a_s/N$.
(d) $I(x_s) = I_0 \sin \omega x_s \, \text{rect}\,(x_s/a_s)$

with rect (x) defined as per equation (6-19).

26. Consider the diffractometer depicted in Figure 6.42. Say that the diffractometer slits are illuminated by a finite source at a finite distance. Say that the source is com-

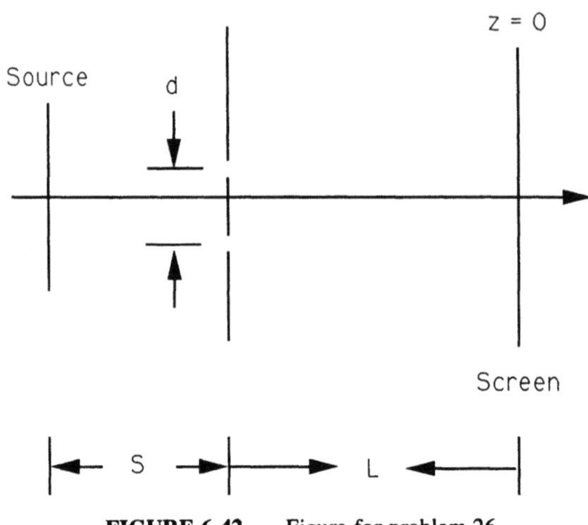

FIGURE 6.42.   Figure for problem 26.

pletely spatially incoherent but radiates according to the rules

$$I(x, \omega) = \frac{I_0}{2\pi \sqrt{\sigma_x \sigma_\omega}} \, e^{-x^2/\sigma_x} e^{-(\omega - \omega_0)^2/\sigma_\omega}$$

such that $\int I(x, \omega)\, dx\, d\omega = I_0$. In (a) through (c) find the intensity on the observation screen for:

(a) $\sigma_x \to 0$, σ_ω arbitrary.

(b) $\sigma_\omega \to 0$, σ_x arbitrary.

(c) σ_x, σ_ω both arbitrary.

(d) Take the limits of (c) for $\sigma_x \to \infty$ and for $\sigma_\omega \to \infty$, and describe what happens. Does it agree with your intuition?

27. Say that we wish to analyze an incoherent source with a diffractometer, as depicted in Figure 6.43. Say the intensity of the source is $I_0 e^{-|x|/a_s}$.

(a) Find the intensity pattern on the screen.

(b) Sketch the intensity pattern in the limits of $a_s \to 0$ and $a_s \to \infty$, for some fixed $P_1 - P_2$.

(c) Sketch the fringe visibility as a function of $\tau - |P_1 - P_2|/c$.

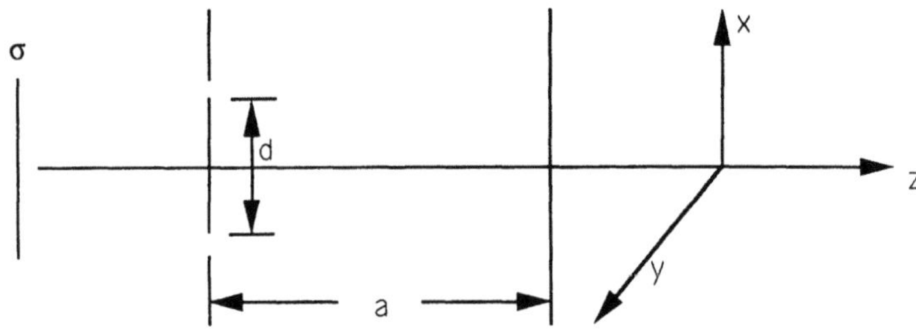

FIGURE 6.43.   Figure for problem 27.

7

Diffraction

7.1 INTRODUCTION

This chapter reviews the standard formulation of scalar diffraction theory to the depth necessary for formulation of the Van Cittert-Zernicke theorem and analysis of the grating spectrometer. Various problems in partially coherent propagation and diffraction are considered, including an analysis of Michelson's stellar interferometer. At the end of the chapter, a more accurate analysis of propagation in gratings is outlined for the purpose of describing some of the design limitations of grating-based instruments and devices.

7.2 GREEN'S THEOREM AND SCALAR DIFFRACTION

Consider the scalar wave equation

$$\nabla^2\psi + k^2\psi = 0 \tag{7-1}$$

and the archetypical diffraction problem as depicted in Figure 7.1. One might well wonder how a scalar equation such as (7-1) could be applied to an inherently vector problem such as that of Figure 7.1. Clearly, each Cartesian component of the field satisfies an equation like (7-1). However, the tangential and normal components of the **E** and **H** vectors will satisfy different boundary conditions on the surface. Even in the lossless Fresnel case, we found that we needed two kinds of waves to describe the propagation phenomena. Here, however with two simplifications, we will find that a single scalar function can suffice. The first of these is the paraxial approximation, that is, that the θ of

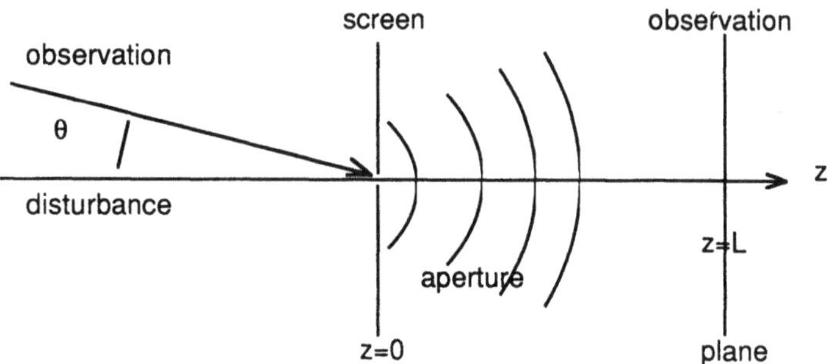

FIGURE 7.1.   Archetypical diffraction problem where an aperture diffracts an incident plane wave.

Figure 7.1 is small. With this, one needs only consider transverse field components. The second is a condition about the screen. Essentially, the screen is taken to be infinitely conductive, and therefore the condition that tangential **E** is zero on the screen is sufficient to determine the field everywhere. Vector formulations of diffraction theory, however, do exist and, for example, can be found in the books by Jackson (1975, Chapter 9) and Born and Wolf (1975, Chapter 11). These rigorous theories are generally based on a formulation by Smythe (1947; 1969, Section 12.8), which was later adapted for variational calculation by Levine and Schwinger (1948, 1949, 1950). It is very hard to get numbers out of these formulations, however, as they generally require a variational approach with trial functions. Someone who truly wanted to check the validity of the scalar theory could check it against the exact solution for the straight edge of Sommerfeld (1896; 1964, Chapter 5 of Volume 3) or that of Meixner and Andrejewski (1950) for the circular aperture. So few exact solutions exist due to the fact that these problems are quite complex.

Having accepted the scalar approximation, we can continue to formulate scalar diffraction theory [see, for example, Born and Wolf (1975, Chapter 7) or Goodman (1968, Chapters 1–4) or the undergraduate texts of Hecht (1987) or Klein and Furtak (1986).] Consider the surface S with unit normal $\hat{n}$, surrounding a volume V with a point P, as is depicted in Figure 7.2. Now, we can use Green's scalar theorem

$$\int_V (U\nabla^2 U' - U'\nabla^2 U)\, dV = \int_S \left(U\frac{dU'}{dn} - U'\frac{dU}{dn}\right) dS \qquad (7\text{-}2)$$

to relate values of functions on the surface of the volume to values of functions inside the volume. Equation (7-2) is an amazing relation when one considers that we have said nothing about U and U'. Perhaps they need to be continuous but nothing more. However, if one takes U and U' to be solutions of the wave

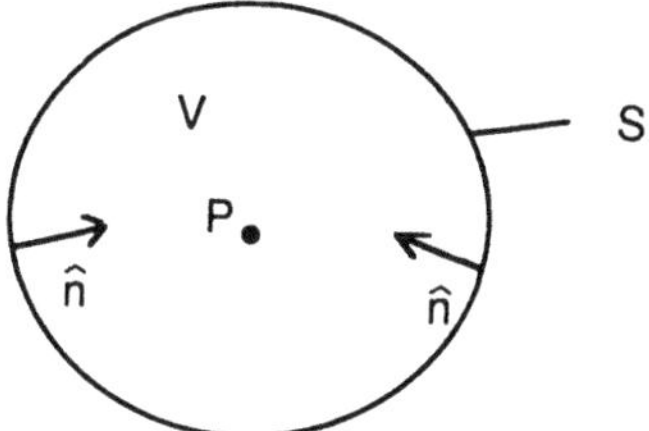

FIGURE 7.2.   A volume V, containing a point P, surrounded by a surface S with unit normal $\hat{n}$.

equation within V, that is,

$$\nabla^2 U + k^2 U = 0 \qquad \text{(a)}$$
$$\nabla^2 U' + k^2 U' = 0 \qquad \text{(b)}$$

(7-3)

one soon discovers that something truly amazing occurs. Plugging (7-3) back into (7-2), one finds that

$$\int_S \left(U \frac{dU'}{dn} - U' \frac{dU}{dn} \right) dS = 0 \qquad (7\text{-}4)$$

That is, the volume integral has disappeared. However, this is not yet the desired result. If one now takes a solution of (7-3) (b) of the form

$$U' = \frac{e^{iks}}{s} \qquad (7\text{-}5)$$

where

$$s = \sqrt{(x_p - x)^2 + (y_p - y)^2 + (z_p - z)^2} \qquad (7\text{-}6)$$

where (x_p, y_p, z_p) are the Cartesian coordinates of the point P, and (x, y, z) are the Cartesian components of a generic point in V, one sees that there is a problem. The function U' does not satisfy (7-3) (b) at the point P. This will not be a concern, however, as one can make an infinitesimal spherical surface S' around P, thereby excluding it from V, as depicted in Figure 7.3. One can now write that

$$\int_S \left(U \frac{d}{dn} \left(\frac{e^{iks}}{s} \right) - \frac{e^{iks}}{S} \frac{dU}{dn} \right) dS$$
$$= - \int_{S'} \left(U \frac{e^{iks}}{s} \left(ik - \frac{1}{s} \right) - \frac{e^{iks}}{s} \frac{dU}{dn} \right) dS'$$

(7-7)

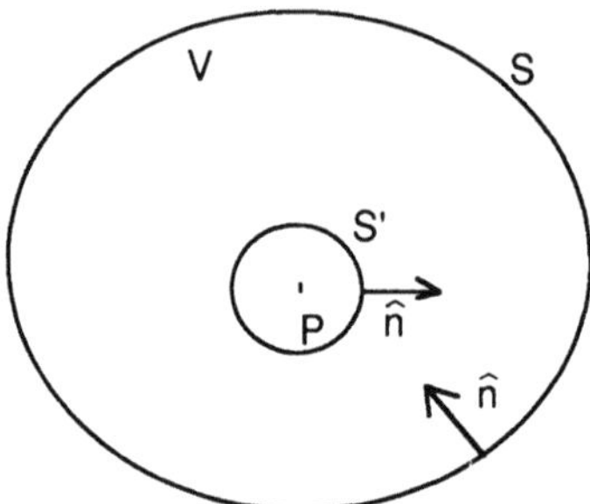

FIGURE 7.3.   A modification to the surface of Figure 7.2.

where the normal derivatives on S' have been carried out explicitly. Now if one is to define the radius of the sphere whose surface is S' to be ϵ and then takes the limit as $\epsilon \to 0$, one finds that

$$-\lim_{\epsilon \to 0} \int_{S'} \left(U \frac{e^{ik\epsilon}}{\epsilon} \left(ik - \frac{1}{\epsilon} \right) - \frac{e^{ik\epsilon}}{\epsilon} \frac{dU}{dn} \right) \epsilon^2 \, d\Omega = \int U \, d\Omega = 4\pi U(P) \quad (7\text{-}8)$$

where U and dU/dn are assumed to be bounded, and $d\Omega$ is the element of the solid angle. Combining (7-8) with (7-7), one obtains the rather surprising result that

$$U(P) = \frac{1}{4\pi} \int \left(U \frac{d}{dn} \left(\frac{e^{iks}}{s} \right) - \frac{e^{iks}}{s} \frac{dU}{dn} \right) dS \quad (7\text{-}9)$$

The result is surprising in the sense that the value of the function $U(P)$ at any point P inside the medium is determined solely by the values of U on the (arbitrary) surface of the volume. This result is naturally the basis of inverse scattering theory as well as diffraction theory, as it seems to say that one can do various forms of remote sensing. The result is closely tied to the fact that we used functions that were solutions of the free space wave equation in Green's theorem. The free space wave equation admits constant-velocity solutions such that, by measuring the fields over a closed surface, the relative phasing of the interfering fields on this surface will give determining information about the internal field structure. If the material internal to the surface were inhomogeneous or contained sources, however, the information would not be complete, and further information would be necessary to determine the internal structure. Perusal of (7-2) indicates that the volume integral may also be removed in the case of an inhomogeneous medium, but it may be quite difficult to determine a function that gives a relation as simple as (7-9).

Consider a half-space, as is depicted in Figure 7.4. Kirchhoff's boundary

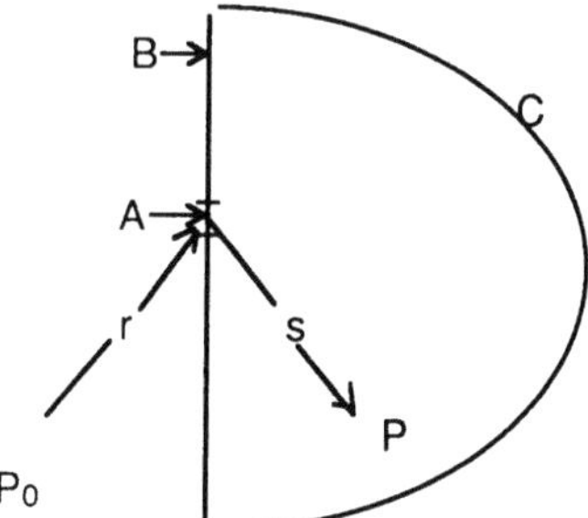

FIGURE 7.4.   A half space containing an infinitely long, infinitely conducting screen, labeled *B*, with an aperture, labeled *A*, and a semicircular boundary *C* that will be allowed to recede to infinity.

conditions essentially state that the field in the aperture *A* is the same field as would be measured in the aperture were there no screen *B*. This approximation obviously cannot be good within a wavelength or so of the screen boundary, but for apertures much larger than a wavelength it could be valid throughout the majority of the aperture region. Further, as will be discussed later, were the aperture to become wavelength-sized, it still could be treated as uniformly illuminated as far as paraxial fields are concerned. The second part of the Kirchhoff boundary condition is harder to justify in the sense that it states that the field and its normal derivative are taken to be zero on the screen. This cannot be the case. For example, suppose we were to consider a plane wave (ray) normally incident on the wall. If the wall is highly conductive, the electric field will be approximately zero on the wall. However, the normal derivative of the E-field will be an H-field that will achieve a maximal value on the wall. This indicates that the electric field energy incident goes into generating a current on the wall. In fact, a well-known law of complex analysis states that if a function and its normal derivative are zero over a finite segment of a surface, the function is zero everywhere. This is not a surprising result, as a material that absorbed all electromagnetic energy incident and could not either radiate or conduct the energy away could not exist, as the material itself would spontaneously convect away. This means that the Kirchhoff boundary conditions are inconsistent. As we will see in the next section, however, a fix-up can be applied to this Kirchhoff–Fresnel formulation that makes the theory consistent without greatly altering its numerical predictions; so we will continue the formulation while duly noting its deficiencies. The Kirchhoff boundary conditions mathematically take the form on *A* that

$$U = U^{(i)} \qquad \text{(a)}$$

$$\frac{dU}{dn} = \frac{dU^{(i)}}{dn} \qquad \text{(b)} \qquad (7\text{-}10)$$

and on B that

$$U = 0 \qquad\qquad\text{(a)}$$

$$\frac{dU}{dn} = 0 \qquad\qquad\text{(b)} \qquad\text{(7-11)}$$

It only remains for us to determine the form of the field on C. As we will presently learn, if C is allowed to recede to infinity, we will not actually need the form at all. Clearly, one can write (7-9) in the form

$$U(P) = \frac{1}{4\pi} \int_{A+B+C} \left(U \frac{d}{dn}\left(\frac{e^{iks}}{s}\right) - \frac{e^{iks}}{s}\frac{dU}{dn} \right) dS \qquad (7\text{-}12)$$

Now C is a semicircle whose radius is R. In the limit as R becomes large, the radius vector s that points from the aperture to the surface of C will become closer and closer to R. Further, as R increases without bound, the inward-pointing unit vector $\hat{n}$ will just become the unit vector inward along R, $-\hat{e}_R$. With these simplifications, one can write

$$\lim_{R\to\infty} \frac{1}{4\pi} \int_C \left(U \frac{d}{dn}\left(\frac{e^{iks}}{s}\right) - \frac{e^{iks}}{s}\frac{dU}{dn} \right) dS$$

$$= -\lim_{R\to\infty} \frac{1}{4\pi} \int_C \left(U \frac{d}{dR}\left(\frac{e^{ikR}}{R}\right) - \frac{e^{ikR}}{R}\frac{dU}{dn} \right) R^2 \, d\Omega \qquad (7\text{-}13)$$

where $d\Omega$ is the differential of the solid angle. Clearly, the contribution from along C will vanish if the condition

$$\lim_{R\to\infty} R\left(\frac{dU}{dn} + ikU \right) = 0 \qquad (7\text{-}14)$$

is satisfied on C. Equation (7-14) is often referred to as the Sommerfeld radiation condition. One should be able to rapidly convince oneself that a spherical wave emanating from the aperture A will satisfy (7-14). Indeed, as the aperture A is finite in size, for an observer who is at near-infinite distance from the aperture, A will appear infinitesimal, and, at least within some paraxial solid angle, its field will appear as a spherical wave. Therefore, by invoking Kirchhoff boundary conditions and using the Sommerfeld radiation condition, we have reduced the integral in (7-12) to one only over the aperture A. By further assuming that the source for the problem is a point source at P_0 at a distance r from the aperture (see Figure 7.4), one can write the incident field in the form

$$U^{(i)} = A \, \frac{e^{ikr}}{r} \qquad \text{(a)}$$

$$\frac{dU^{(i)}}{dn} = A \, \frac{e^{ikr}}{r} \left(ik - \frac{1}{r} \right) \cos \, (\hat{\mathbf{n}}, \mathbf{r}) \qquad \text{(b)}$$

$$\text{(7-15)}$$

and therefore (7-12) reduces to

$$U(P) = \frac{A}{4\pi} \int_A \left(\frac{e^{ikr}}{r} \frac{e^{iks}}{s} \left(ik - \frac{1}{s} \right) \cos \, (\hat{\mathbf{n}}, \mathbf{s}) \right.$$

$$\left. - \frac{e^{iks}}{s} \frac{e^{ikr}}{r} \left(ik - \frac{1}{r} \right) \cos \, (\hat{\mathbf{n}}, \mathbf{r}) \right) ds \qquad \text{(7-16)}$$

Equation (7-16) is often referred to as the result of the Fresnel–Kirchhoff formulation.

Perhaps the most interesting thing about (7-16) is its complete symmetry between s and r. This is often known as Helmholtz reciprocity, and it indicates that if the source location and the observer's position are interchanged, the same field will be observed. That this should be the case should be obvious from (7-1), the original statement of the scalar Green's theorem, which is also symmetric under the interchange of U and U'. Even in the vector case, a source-observer symmetry will exist although it will require some interchange of $\mathbf{E}$ and $\mathbf{H}$ fields.

A second symmetry is built into the relation (7-16). If one says that the plane surfaces A and B of (7-12) lie in the surface $z = 0$, one can write abstractly that

$$\int_A + \int_B = \int_{z=0} \qquad \text{(7-17)}$$

which says that the field transmitted by the aperture plus the field transmitted by an aperture stop the same size as the aperture must equal the field that would be transmitted in the absence of any screen at all. This symmetry is often referred to as Babinet's principle, as described in either Jackson (1975, Chapter 9) or Born and Wolf (1975, Chapter 11). It has some interesting consequences. For example, let us consider the optical intensity on the axis behind an aperture stop for a plane wave incident on the aperture, as illustrated in Figure 7.5. Equation (7-17) tells us that the intensity measured there will be the time-averaged square of the difference of the field observed in the absence of the screen and that which would be observed behind the open aperture. Simple-minded geometrical optics would tell us that no field is transmitted. This could

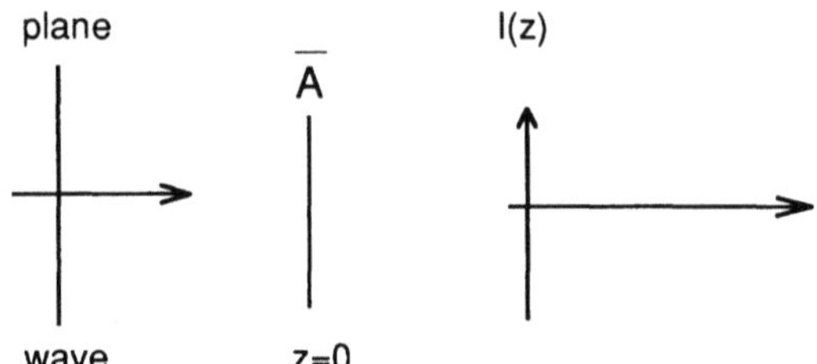

FIGURE 7.5.   An aperture stop $\overline{A}$ in the plane $z = 0$, where $I(z)$ is the optical intensity that would be measured on the z axis behind the stop.

only be the case for a truly large aperture, however, as the effect of the edges will cause a change in the total transmission of the aperture as well as a ringing with distance due to the scattered rays going in and out of phase with those rays that are transmitted straight through. The situation can be illustrated schematically as in Figure 7.6. Here, it is clear that the intensity on the axis behind the stop will oscillate with the z-coordinate, in contradistinction to both geometrical optics and common sense. This is one of the basic results of scalar diffraction theory. It should be noted that full vector diffraction theory also has a Babinet symmetry, albeit one more complicated to state [see, as were cited above, either Jackson (1975, Chapter 9) or Born and Wolf (1975, Chapter 11)]. In the paraxial limit, however, the two forms of Babinet's principle yield identical predictions that have been well verified over the years.

Although equation (7-16) is greatly simplified from equation (7-12), it can be simplified much more because of an implicit paraxial approximation made from the outset of the argument. For the purpose of the present discussion, we only want to make the approximations that kr and ks are large compared to unity. If they were not, the Kirchhoff boundary condition would be especially bad. With this assumption, one sees that

$$U(P) = -\frac{iA}{2\lambda} \int_A \frac{e^{ik(s+r)}}{sr} \left[\cos\,(\hat{\mathbf{n}},\,\mathbf{r}) - \cos\,(\hat{\mathbf{n}},\,\mathbf{s})\right]\,ds \qquad (7\text{-}18)$$

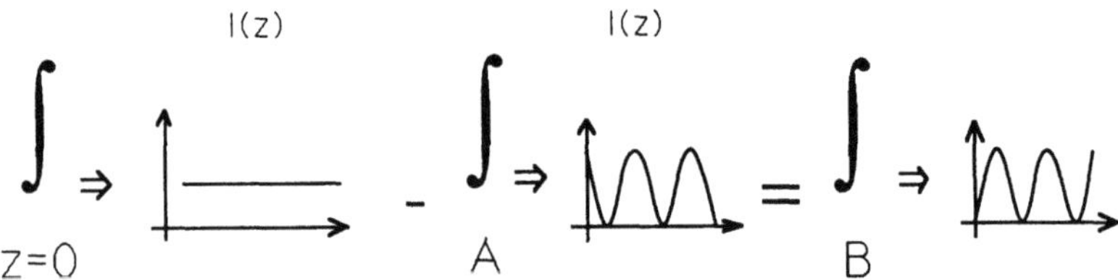

FIGURE 7.6.   Schematic illustration of the calculation of the intensity on the axis behind a (small) aperture stop, using Babinet's principle.

7.3 RAYLEIGH–SOMMERFELD THEORY

In this section, we will discover a way to modify the mechanics of the Fresnel–Kirchhoff relation so that the inconsistency is removed, but the numerical predictions of the theory are not greatly affected. Formulation of the problem was made independently by Rayleigh (1945) and Sommerfeld (1896, Chapter 5), and this is generally referred to as the Rayleigh–Sommerfeld theory. For a comparison of Fresnel, Kirchhoff, and Rayleigh–Sommerfeld theories, see Wolf and Marchand (1964).

The result that we found in equation (7-9) can be expressed in a more general form as

$$U(P) = \frac{1}{4\pi} \int_{S+S'} \left(\frac{dU}{dn} G - U \frac{dG}{dn} \right) dS \qquad (7\text{-}19)$$

where U and G are solutions of

$$\nabla^2 U + k^2 U = 0 \qquad \text{(a)}$$
$$\nabla^2 G + k^2 G = \delta(\mathbf{r} - \mathbf{r}_p) \qquad \text{(b)}$$

$$(7\text{-}20)$$

respectively. Clearly here, S and S' refer to the surfaces of Figure 7.3, where S' already has been allowed to shrink to zero. Previously in the derivation of (7-9), the function G had been chosen to be e^{iks}/s specifically. The function G, in general, is known as a Green's function and needs only to satisfy (7-20) (b). Solutions of this equation clearly are not unique, as, for one reason, boundary conditions are not specified, and, second, the volume V is finite, and G could have as many singular (source) points outside of V as one might like. Of course, these two conditions are intimately related because the only way to force G to satisfy given boundary conditions in free space is to add "image" sources to the function outside of the boundary region. Now the basic idea of Rayleigh and Sommerfeld was that, if G (or dG/dn) were made to be zero on B, then dU/dn (or U) would not need to be specified on B. The inconsistency in the Fresnel–Kirchhoff formulation came solely from this double specification. If one takes G to be

$$G = \frac{e^{iks}}{s} - \frac{e^{ik\tilde{s}}}{\tilde{s}} \qquad (7\text{-}21)$$

where $\tilde{P}$ is the conjugate point of P, defined by

$$P = (x_p, y_p, z_p) \qquad \text{(a)}$$
$$\tilde{P} = (x_p, y_p, -z_p) \qquad \text{(b)}$$

$$(7\text{-}22)$$

and s and $\tilde{s}$, which are given by

$$s = \sqrt{(x - x_p)^2 + (y - y_p)^2 + (z - z_p)^2} \qquad \text{(a)}$$

$$\tilde{s} = \sqrt{(x - x_p)^2 + (y - y_p)^2 + (z + z_p)^2} \qquad \text{(b)}$$

$$(7\text{-}23)$$

are the associated distances to the integration point (x, y, z), as is illustrated in Figure 7.7, then clearly $G(x, y, z = 0) = 0$, as $s = \tilde{s}$ on $z = 0$. Therefore, one can write equation (7-19) in the form

$$U(P) = -\frac{1}{4\pi} \int_{S+S'} U \frac{dG}{dn} dS \qquad (7\text{-}24)$$

Recalling the derivation that led from equation (7-9) to equation (7-17) except that this time, instead of assuming conditions as in (7-10) and (7-11), assuming that on A

$$U = U^{(i)} = \frac{e^{ikr}}{r} \qquad (7\text{-}25)$$

and on B

$$U = 0 \qquad (7\text{-}26)$$

then one finds the consistent equation

$$U(P) = \frac{A}{i\lambda} \int_A \frac{e^{ik(r+s)}}{rs} \cos(\hat{n}, s) \, dS \qquad (7\text{-}27)$$

which can be compared with a rewritten form of equation (7-17)

$$U(P) = \frac{A}{i\lambda} \int_A \frac{e^{ik(r+s)}}{rs} \left(\frac{\cos(\hat{n}, r) - \cos(\hat{n}, s)}{2} \right) dS \qquad (7\text{-}28)$$

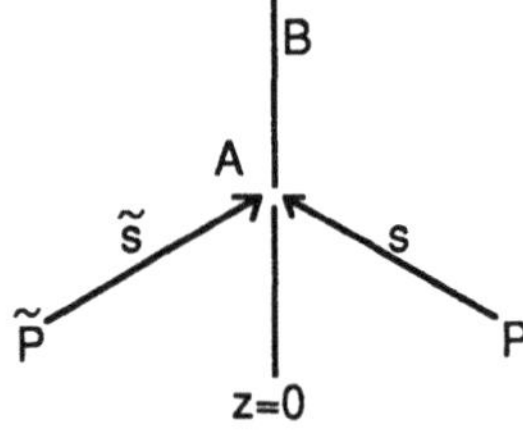

FIGURE 7.7.   Figure illustrating the field point P and its conjugate point $\tilde{P}$, and the associated distances S and $\tilde{S}$ from P and $\tilde{P}$ to a generic integration point $(x_s, y_s, 0)$.

It should further be noted here that if one were to take a plus sign in the expression for G in equation (7-21) instead of the minus sign and applied the conditions of (7-25) and (7-26) to the derivatives of U and $U^{(i)}$, then one would find the expression

$$U(P) = \frac{A}{i\lambda} \int_A \frac{e^{ik(r+s)}}{rs} \cos(\hat{\mathbf{n}}, \mathbf{r})\, dS \qquad (7\text{-}29)$$

Interestingly, although the "more correct" expressions of (7-27) and (7-29) share the "Babinet" symmetry with (7-28), they no longer have the Helmholtz reciprocity because of the lack of symmetry of their so-called obliquity factors, that is, the terms involving the cosines. The difference of (7-27) and (7-29) does exhibit reciprocity and, further, is equal to (7-28) [see, for example, Wolf and Marchand, (1964)]. But, in some sense, these distinctions are specious, as the paraxial approximation has long since been invoked in the derivation, and in the paraxial approximation the cosine factors all reduce to ± 1. Therefore, all three expressions become equal apart from a minus sign. We therefore see that, to within our paraxial approximation, we may as well set

$$U(P) = \frac{1}{i\lambda} \int \frac{e^{ik(r+s)}}{rs}\, dS \qquad (7\text{-}30)$$

where a unity-amplitude incident wave has been assumed.

We now would like to further simplify equation (7-30), at least within the confines of the paraxial approximation that we have already made. Before doing this, we will change the way we write equation (7-30) in two ways. Although we took the incident wave to be of the form of a diverging spherical wave, we could have taken the incident electric field to have been anything we wanted in the Rayleigh–Sommerfeld formulation. We thus can write that

$$U(P) = \frac{1}{i\lambda} \int U^{(i)} \frac{e^{iks}}{s}\, dS \qquad (7\text{-}31)$$

where the integral is over the aperture A. A second change we wish to make is notational but will simplify further calculations. What we wish to do here is to label the field point P by the coordinates (x, y, L) and the aperture point by the coordinates (x', y') and denote the distance between these points by r_{12}, a situation depicted in Figure 7.8. With these identifications, the diffraction integral can be written in the form

$$U(x, y, L) = \frac{1}{i\lambda} \int U^{(i)}(x', y') \frac{e^{ikr_{12}}}{r_{12}}\, dx'\, dy' \qquad (7\text{-}32)$$

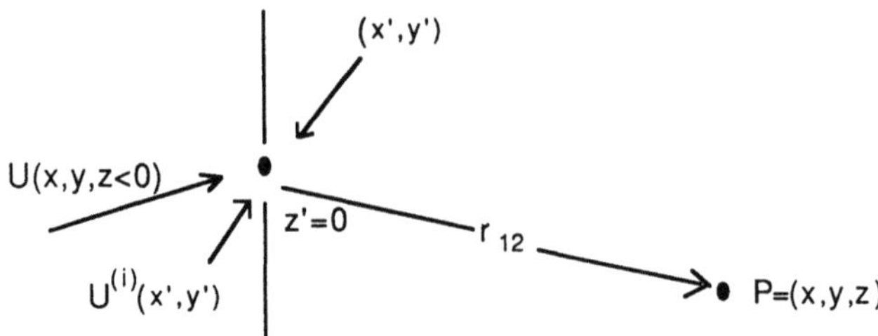

FIGURE 7.8.   A new labeling for the quantities of the diffraction problem.

Now we are ready to complete the paraxial approximation. The paraxial approximation requires that the longitudinal distance L greatly exceed the maximum values of the quantities x, y, x', and y'. The r_{12} in the denominator of (7-32) can be written in the form

$$r_{12} = L\left[1 + \frac{(x - x')^2}{L^2} + \frac{(y - y')^2}{L^2}\right]^{1/2} \tag{7-33}$$

In the exponent in (7-32), small changes in $x - x'$ or $y - y'$ compared to L could still be large compared to λ, and great care must be used in expanding the exponent. In the denominator, however, such care is unnecessary, and one can expand (7-33) in the form

$$r_{12} \sim L\left(1 + \frac{(x - x')^2}{2L^2} + \frac{(y - y')^2}{2L^2}\right) \sim L \tag{7-34}$$

and therefore write (7-32) in the form

$$U(x, y, L) = \frac{1}{i\lambda L}\int U^{(i)}(x', y')\, e^{ikr_{12}}\, dx'\, dy' \tag{7-35}$$

Equation (7-35) is the ultimate paraxial simplification of the diffraction integral. Further simplification will require further approximation. Obviously, the next thing to approximate away is the exponential, as everything else is in its simplest form already. One could do this by the so-called method of stationary phase [see, for example, Born and Wolf (1975, Appendix 3) or Carrier, Krook, and Pearson (1966, Chapter 6)], which would amount to Taylor-expanding the square root of equation (7-33) about the point $x = x'$, $y = y'$. It should be noted that this expansion could contain only even powers of $x - x'$ and $y - y'$ and therefore would appear in the form

$$r_{12} = L \left[1 + K_2 \frac{(x - x')^2 + (y - y')^2}{L^2} \right.$$
$$\left. + K_4 \frac{((x - x')^2 + (y - y')^2)^2}{L^4} + \cdots \right] \tag{7-36}$$

where the K_{2n}'s, n integer, are constant coefficients, independent of coordinate. The coefficients are well known from the binomial expansion, and r_{12} can be written explicitly in the form

$$r_{12} = L \left[1 + \frac{(x - x')^2 + (y - y')^2}{2L^2} \right.$$
$$\left. + \frac{((x - x')^2 + (y - y')^2)^2}{8L^4} + \cdots \right] \tag{7-37}$$

The condition for the first two terms to be a good enough approximation in the exponent would be that

$$kL \frac{((x - x')^2 + (y - y')^2)^2}{8L^4} \ll 2\pi \tag{7-38}$$

Defining d^4 to be the maximum value of $((x - x')^2 + (y - y')^2)^2$, one notes that the maximum numerical aperture, for which the first two terms give a good approximation to the square root, is given by

$$NA_m \sim \frac{d}{2L} \tag{7-39}$$

where the situation is schematically depicted in Figure 7.9. Using (7-35) and

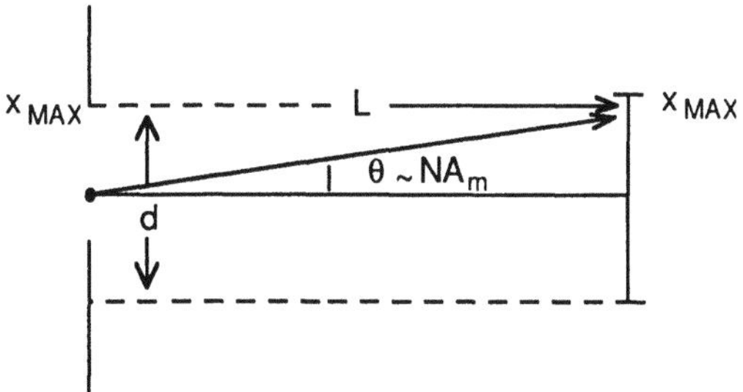

FIGURE 7.9.   A figure that exhibits the definitions of NA_m and d.

(7-39), one finds that

$$NA_m \ll \left(\frac{\lambda}{2L}\right)^{1/4} \tag{7-40}$$

The result of (7-40) is rather stringent in that, even for a 2.5 mm L, the NA_m must be less than roughly .1. As L increases, the NA_m gets smaller in proportion to the fourth root of L. It has been argued that (7-40) is not a necessary condition, as the real condition would be that the sum of all the terms excluded should be less than 2π, and the series does have alternating signs. It is hard to imagine that this adds too much range to (7-40), but the two-term expansion, better known as the Fresnel approximation, does find applicability in practice. The Fresnel approximation is much like the paraxial geometrical optics limit that was discussed in Chapter 4. In this limit, phase fronts and lenses have spherical surfaces. The higher-order terms of r_{12} have, therefore, the meaning of higher-order aberrations and certainly become important in cases where lens correction and so on become important, that is, in high numerical aperture systems. Systems for generation of wavelength-sized spots need numerical apertures approaching unity and therefore require a diffraction analysis different from the one applied here. To the best of my knowledge, no general diffraction analysis exists for such high-NA cases, where even vector effects could become important. For the present exposition, we will stick to the Fresnel approximation.

Plugging the Fresnel approximation for equation (7-33) back into the diffraction integral of equation (7-35), one finds that

$$U(x, y, L) = \frac{e^{ikL}}{i\lambda L} \int U^{(i)}(x', y') \, e^{(ik/2L)[(x-x')^2 + (y-y')^2]} \, dx' \, dy' \tag{7-41}$$

Interestingly, the integral in equation (7-41) has the form of a convolution as was defined in equation (3-85). That is, it is of the form

$$f(x) * g(x) = \int f(x - x')g(x') \, dx' = \int f(x')g(x - x') \, dx' \tag{7-42}$$

The convolution has a graphical interpretation of taking one of the functions, turning it backward, and sliding it across the other function as depicted in Figure 7.10. Here the effect of the convolution operation is a smoothing, which is done in the time domain by "slow" detectors receiving "fast" signals. Although we cannot sketch two-dimensional complex functions very easily, we can get a feeling for what the solution of (7-41) looks like by the following

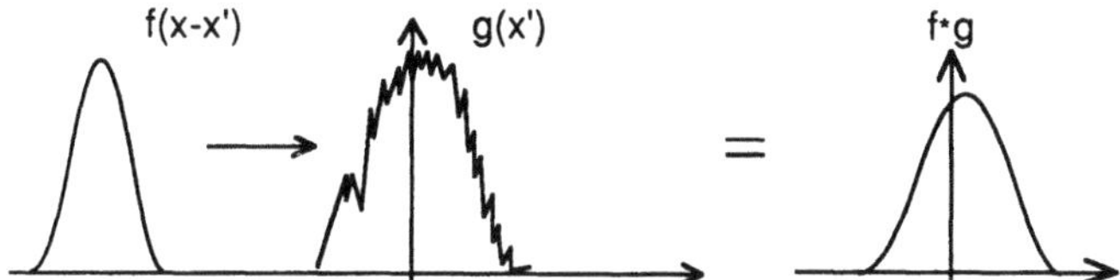

FIGURE 7.10.   Schematic illustration of the convolution operation.

simplified argument. Let us say that $U(x, y, z < 0)$ is given by a uniform plane wave traveling along z. Then the wave will be uniform in the aperture, and $U^{(i)}(x, y)$ will be given by

$$U^{(i)}(x', y') = P(x', y') \qquad (7\text{-}43)$$

where P is a pupil function that is equal to one in the aperture and zero on the screen. Now the exponential is a rapidly varying phase function except near its stationary point $x = x'$ and $y = y'$. If we were to sketch the real part of it in one dimension and slide it across the pupil function, we would get an idea of what the Fresnel zone integral looks like. This operation is schematically depicted in Figure 7.11. It is seen that the "kernel" of equation (7-41) (that is, the exponential) does not in any way smooth the pupil function but rather causes noise-like ringing all around it. This is the effect of near-zone diffraction.

The result of equation (7-41) can be written in another interesting and somewhat telling form. Multiplying out the term in the exponent, one can write that

$$U(x, y, L) = \frac{e^{ikL}}{i\lambda L}\, e^{(ik/2L)(x^2 + y^2)} \int U^{(i)}(x', y')\, e^{(ik/2L)(x'^2 + y'^2)}$$
$$\cdot\, e^{(-i2\pi/\lambda L)(xx' + yy')}\, dx'\, dy' \qquad (7\text{-}44)$$

Recalling that the Fourier transform is defined by

$$\mathcal{F}[f(x, y)] = \int e^{-i2\pi(f_x x + f_y y)}\, f(x, y)\, dx\, dy \qquad (7\text{-}45)$$

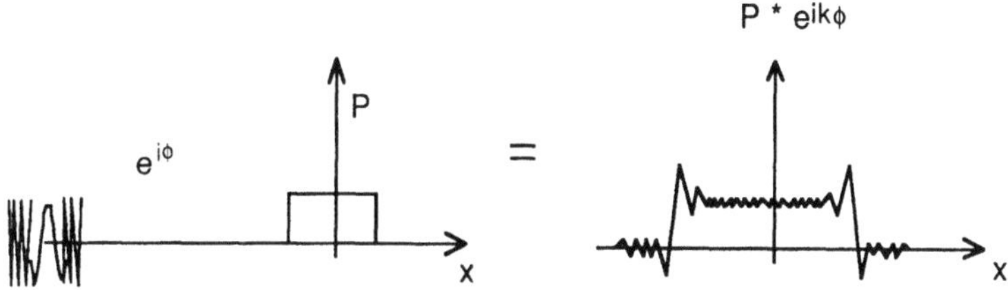

FIGURE 7.11.   Schematic depiction of the convolution operation of equation (7-41).

one sees that the expression of (7-44) can be rewritten as

$$U(x, y, L) = \frac{e^{ikL}}{i\lambda L}\, e^{(ik/2L)(x^2 + y^2)}\, \mathscr{F}[U^{(i)}(x, y)e^{(ik/2L)(x^2 + y^2)}]{\genfrac{}{}{0pt}{}{f_x = x/\lambda L}{f_y = y/\lambda L}} \qquad (7\text{-}46)$$

The interpretation of equation (7-46) is rather clear and quite interesting. Inside the transform argument, we see that the function $U^{(i)}$ is multiplied by a quadratic phase factor. It is as if each point on the phase front is radiating out a spherical wave. The Fourier transform operation is one that essentially takes an angular spectrum of the function to be transformed. This is equivalent to getting in the far field of an object, that is, getting so far away from an object that the object appears as a point radiating into some angular distribution. The phase factor outside the transform is another spherical one, which makes it seem as if each point on the output plane is also radiating a spherical wave. This concept of each phase front point being a secondary source for the wave is often referred to as the Huygens and Fresnel principle, and was used by Huygens in the seventeenth century to predict diffraction results.

Although the forms of equations (7-41) and (7-46) are simpler ones to evaluate than that of equation (7-35), they still are not trivial to solve. However, if the argument of the complex exponential in the transform argument of equation (7-46) is smaller than 2π, then one can find cases where the indicated operations are easy to perform. Denoting the maximal value of $x^2 + y^2$ in the aperture by d', one sees that the condition that allows the complex exponential to be dropped is

$$\frac{d'}{L} \ll \left(\frac{\lambda}{L}\right)^{1/2} \qquad (7\text{-}47)$$

which is a stringent requirement indeed. It defines a region known as the Fraunhofer region, and it is a condition that holds well for astronomical applications, where numerical apertures are small and distances great. One could well rewrite (7-47) in the form

$$L \gg \left(\frac{d'^2}{\lambda}\right) \qquad (7\text{-}48)$$

to get some feeling for the length scales involved. For a wavelength of 0.5 μm and an aperture of 0.5 mm, L would have to exceed a half meter for Fraunhofer to apply. For a 5 mm aperture, L would have to exceed 50 meters. For a half meter telescope aperture, L would have to exceed 0.5 million meters or 500 kilometers. Clearly, this far-field approximation will not be achieved in the

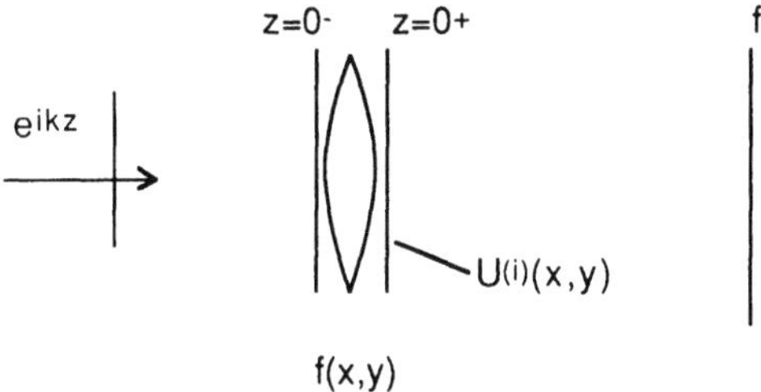

FIGURE 7.12.   Schematic illustration of a system for obtaining the Fourier transform of a transparency $f(x, y)$.

laboratory. There is another way, however, to obtain the Fraunhofer diffraction pattern. Consider the situation depicted in Figure 7.12. Here a lens is placed between a transparency $f(x, y)$ and the plane where we define $U^{(i)}(x, y)$. If we assume that the lens has a simple phase transformation effect on the incident light, then we can write

$$U^{(i)}(x, y) \cong f(x, y)e^{(-ik/2f)(x^2 + y^2)} \tag{7-49}$$

which should hold reasonably well in the absence of any strong scattering. If the observation plane were the plane $z = f$, we see that the observed pattern would be

$$U(x, y, f) = \frac{e^{ikf}}{i\lambda f} e^{(ik/2f)(x^2 + y^2)} \, \mathfrak{F}[f(x, y)]\Big|_{f_y = y/\lambda f}^{f_x = x/\lambda f} \tag{7-50}$$

which is indeed a Fourier transform relation that illustrates the so-called Fourier-transforming property of lenses.

Further illumination of the Fourier transform concept can be obtained by consideration of a 4-f optical system. A 4-f optical system is illustrated in Figure 7.13, including a "pupil function," which can be either the effective pupil of the system or an intentionally inserted transparency to be used in image processing. All of the lenses are assumed to be thin. The U's denote the optical field at the designated planes. The field, U_p^+, can be calculated by considering the Fourier transform (F.T.) module of Figure 7.14, where f is the focal length of the lens. Rewriting equation (7-41) for the free space propagation from 1 to 2,

$$U_2(x_2, y_2) = \frac{e^{ikf}}{i\lambda f} U_1(x_1, y_1) \cdot e^{(i(k/2f))(x_1^2 + y_1^2)} \tag{7-51}$$

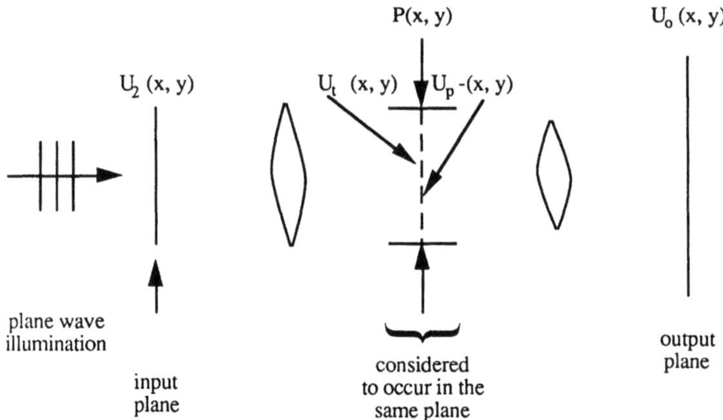

FIGURE 7.13.   Schematic of a 4-f system.

Taking the F.T., we get:

$$\tilde{U}_i = \mathcal{F}(U_i) \tag{7-52}$$

$$\tilde{U}_2(x_2, y_2) = \frac{e^{ikf}}{i\lambda f} \tilde{U}_1 \mathcal{F}\left\{e^{(i(k/2f))(x_1^2 + y_1^2)}\right\} \text{ evaluated at } f_x = \frac{x_2}{\lambda f}, f_y = \frac{y_2}{\lambda f} \tag{7-53}$$

and, dropping the constant phase,

$$\mathcal{F}\left\{e^{(i(k/2f))(x_1^2 + y_1^2)}\right\}\Big|_{f_x, f_y} = e^{-i\pi f \lambda (f_x^2 + f_y^2)} \tag{7-54}$$

we get

$$\tilde{U}_2(x_2, y_2) = \frac{1}{i\lambda f} \tilde{U}_1 e^{-i\pi f \lambda (f_x^2 + f_y^2)} \tag{7-55}$$

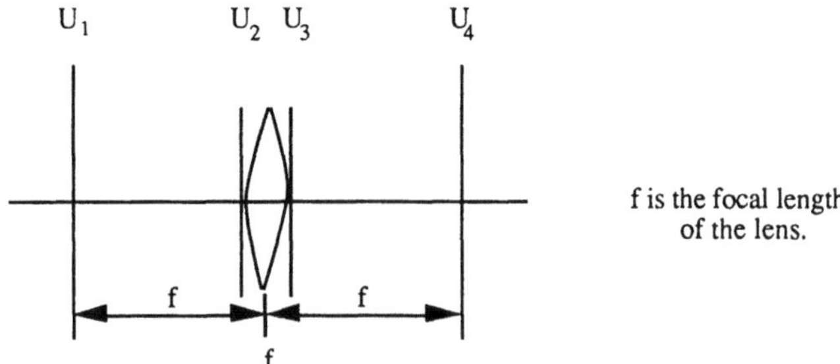

f is the focal length
of the lens.

FIGURE 7.14.   Illustration of a Fourier transform module.

Using equation (7-50) (again dropping constant phases):

$$U_4 \cong \frac{1}{i\lambda f} e^{(ik/2f)(x_4^2 + y_4^2)} \mathcal{F}\{U_2(x_2, y_2)\}$$

$$= \frac{-1}{(\lambda f)^2} e^{(ik/2f)(x_4^2 + y_4^2)} [e^{-i\pi f \lambda(f_x^2 + f_y^2)} \tilde{U}_2]_{\substack{f_x = x_4/\lambda f \\ f_y = y_4/\lambda f}}$$

$$= \frac{-1}{(\lambda f)^2} e^{(ik/2f)(x_4^2 + y_4^2)} e^{-i\pi f \lambda(x_4^2/\lambda^2 f) + (y_4^2/\lambda^2 f^2)} \tilde{U}_2\left(\frac{x_4}{\lambda f_1}, \frac{y_4}{\lambda f_1}\right)$$

$$= \frac{-1}{(\lambda f)^2} \tilde{U}_2\left(\frac{x_4}{\lambda f_1}, \frac{y_4}{\lambda f_1}\right)$$

(7-56)

Thus, it is evident that the field U_4 in Figure 7.14 is an amplitude-scaled Fourier transform of the field $U_i(x, y)$. Note also that this only works out so neatly when the distances on either side of the lens are exactly one focal length. Using

$$U_{P_-} = P(x, y)U_{t_-}(x, y) = P(x_4, y_4)\tilde{U}_i\left(\frac{x_4}{\lambda f}, \frac{y_4}{\lambda f}\right)$$

(7-57)

and using the results derived above

$$U_0(x_0, y_0) = \frac{1}{(\lambda f)^2} \mathcal{F}\{U_{P_-}\}\Big|_{\substack{f_x = x_0/\lambda f \\ f_y = y_0/\lambda f}}$$

$$= \frac{1}{(\lambda f)^4} \mathcal{F}\left\{P(x, y)\tilde{U}_i\left(\frac{x}{\lambda f}, \frac{y}{\lambda f}\right)\right\}_?^?$$

$$= \frac{1}{(\lambda f)^4} \tilde{P}\left(\frac{x_0}{\lambda f}, \frac{y_0}{\lambda f}\right) \cdot (\lambda f)^2 U_i(-\lambda f x_{01} - \lambda f y_{01})$$

$$= \frac{1}{(\lambda f)^4} \tilde{P}\left(\frac{x_0}{\lambda f}, \frac{y_0}{\lambda f}\right) \cdot U_i(-\lambda f x_{0_1} - \lambda f y_{01})$$

(7-58)

Thus, the net effect of the system is to image the input.

An interesting application of the above 4-f system is that of spatial frequency filtering. As has been shown, a simple lens can be used as a Fourier transform device. The optical spatial distributions in the front and back focal planes of a lens are Fourier transform pairs. If an object is placed at the front focal plane of a lens, its spatial frequency spectrum can be obtained in the back focal plane.

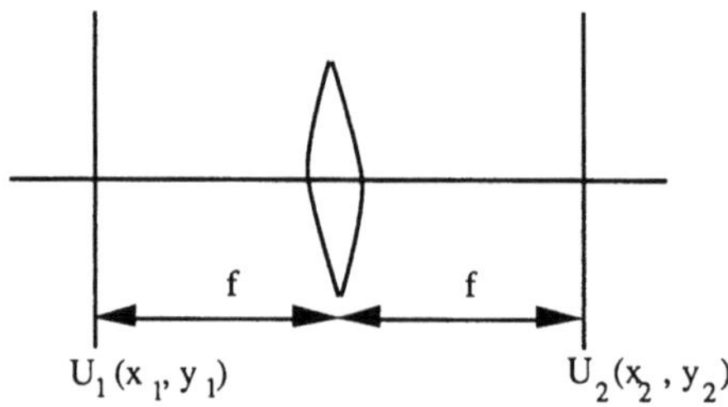

FIGURE 7.15.   Depiction of the forward propagation of a field distribution.

As shown in Figure 7.15, we have

$$U_2(x_2, y_2) = \mathcal{F}\{U_1(x_1, y_1)\}$$

$$= \frac{1}{i\lambda f} \iint U_1(x_1, y_1) \exp\left\{-i2\pi\left[x_1\left(\frac{x_2}{\lambda f}\right) + y_1\left(\frac{y_2}{\lambda f}\right)\right]\right\} dx_1\, dy_1$$

$$(7\text{-}59)$$

where $x_2/\lambda f$ and $y_2/\lambda f$ are spatial frequencies of $U_1(x_1, y_1)$. It thus is possible to change the spatial frequency content, using spatial filtering at the back focal plane. For example, an opaque circle placed at the center of the back focal plane acts as a high-pass spatial filter, and a stop or an iris acts as a low-pass filter (Figure 7.16).

Using a two-lens system with a spatial frequency filter arranged as in Figure 7.17, a modified image can be obtained in the back focal plane of the second lens.

Some of the applications of spatial frequency filtering are as follows:

1. *Edge enhancement.* The edges are regions of rapid change and are characterized by high spatial frequency. A high-pass filter will allow the edge region to be imaged and block the low frequency regions. Thus the edges are enhanced in the image formed.

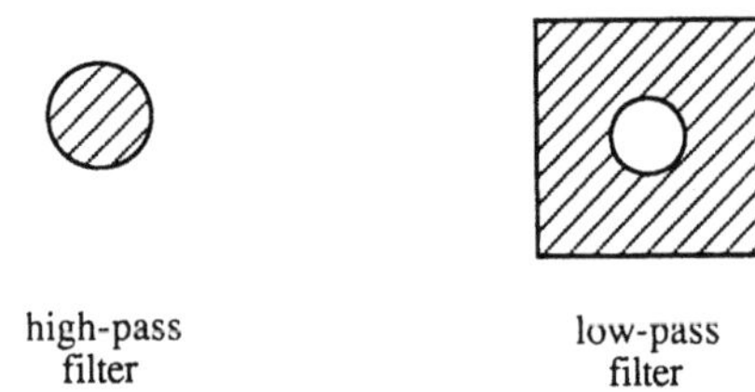

FIGURE 7.16.   Illustration of the form of spatial filters.

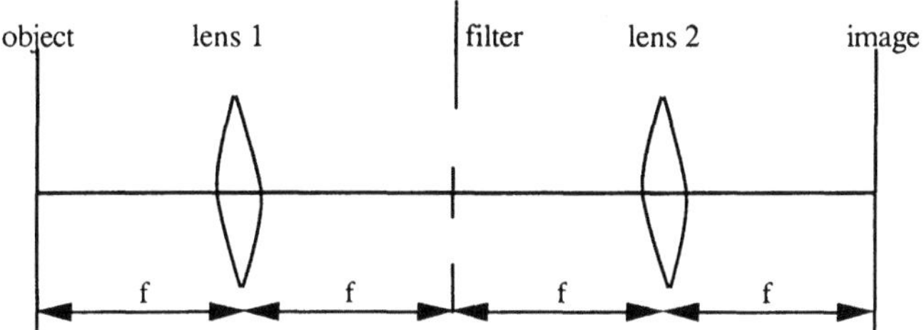

FIGURE 7.17.   Depiction of a simple two-lens spatial filtering system.

2. *Signal/noise ratio improvement.* If the signal spectrum has a known limited band, the S/N ratio of the image can be improved by blocking bands other than the signal spectral band. Because the noise spectrum spreads over the entire spectral plane, the noise is reduced.

3. *Differentiation and integration.* Placing a transparency having a transmittance

$$T(x_2) = i2\pi \frac{x_2}{\lambda f} \tag{7-60}$$

causes a differentiation of the image to be obtained in the image plane because

$$\mathcal{F}\left[\frac{\partial}{\partial x_1} g(x_1)\right] = i2\pi \frac{x_1}{\lambda f} G(x_1) \tag{7-61}$$

Similarly, a transparency having a transmittance

$$T(x_2) = i2\pi \frac{\lambda f}{x_2} \tag{7-62}$$

will result in an integration of the image distribution. The 180° phase shift required by the differentiation and integration can be achieved by providing a transparency with a π radians shift in optical thickness at the origin.

Before we go on to propagation of partially coherent fields, it is quite useful to present an important example of Fraunhofer diffraction. The case was originally worked out by Lord Rayleigh, and its result is generally known as the Rayleigh criterion or the diffraction limit. The problem is illustrated in Figure 7.18. In any optical system, there will be a limitation on the radius of the beam caused by an aperture. Hopefully this limiting aperture will be much smaller than the radius of any lens in the system, as the many paraxial approximations made in the theory of imaging tend to become invalid if one uses the spherically

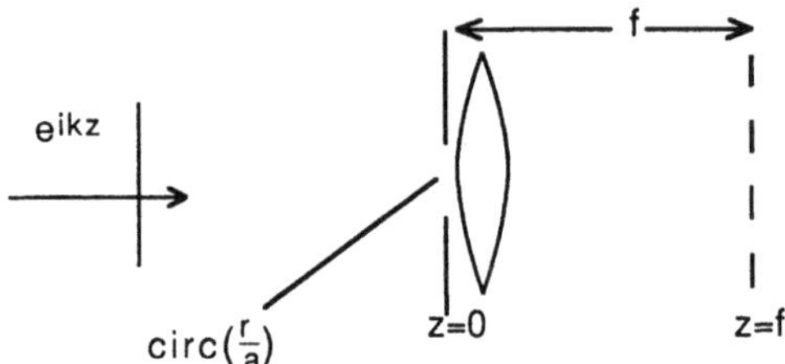

FIGURE 7.18.   The archetypical problem of diffraction theory, that of the diffraction of a circular aperture of radius a.

aberrated nonparaxial portions of noncorrected lenses. Microscope objectives are corrected. The problem illustrated in Figure 7.18 is the simplest one of diffraction by an aperture, that is, the problem where the Fraunhofer diffraction of an aperture is observed directly. Although this problem is simple, it is also generic in that any optical system can be analyzed in terms of a diffraction-limited response, convolved with an ideal geometrical optics response. It is beyond the scope of this book to show this, but it is discussed in some detail in the book by Goodman (1968, Chapter 7). In any case, to analyze the problem at hand, we note that the pattern that will be observed in the plane $z = f$ is given by

$$U(x, y, z) = \frac{e^{ikf}}{i\lambda f} e^{(ik/2f)(x^2 + y^2)} \, \mathscr{F}\left[\text{circ}\left(\frac{r}{a}\right)\right]_{f_\rho = \sqrt{(x^2 + y^2)}/\lambda F} \tag{7-63}$$

where the circ function is unity for its argument less than unity and zero otherwise, in a manner analogous to the definition of rect function of equation (6-19) and where the use of polar coordinates has already been anticipated. The Fourier transform integral can be written out in the form

$$\mathscr{F}\left[\text{circ}\left(\frac{r}{a}\right)\right] = \int\int dx' \, dy' \, \text{circ}\left(\frac{r}{a}\right) e^{-i2\pi(f_x x + f_y y)} \tag{7-64}$$

A transformation of the input and output planes to polar coordinates can be carried out by making the following substitutions:

$$x = r \cos\theta \tag{a}$$

$$y = r \sin\theta \tag{b}$$

$$f_x = f_\rho \cos\varphi \tag{c}$$

$$f_y = f_\rho \sin\varphi \tag{d}$$

$$\tag{7-65}$$

thereby obtaining the integral in the form

$$\mathcal{F}\left[\text{circ}\left(\frac{r}{a}\right)\right] = \int_0^{2\pi} d\theta \int_0^a r\,dr\,e^{-i2\pi r\rho \cos(\theta - \varphi)} \tag{7-66}$$

where the circ function has now been substituted into the integral limits. Recalling the identity

$$J_0(a) = \frac{1}{2\pi}\int_0^{2\pi} d\theta e^{-ia \cos(\theta - \varphi)} \tag{7-67}$$

where $J_0(a)$ is the zeroth-order Bessel function, one can simplify the integral of (7-66) to

$$\mathcal{F}\left[\text{circ}\left(\frac{r}{a}\right)\right] = 2\pi \int_0^a r\,dr\,J_0(2\pi r f_\rho) \tag{7-68}$$

A second identity

$$\int_0^x \zeta J_0(\zeta)\,d\zeta = x J_1(x) \tag{7-69}$$

together with the substitution

$$R = 2\pi r\rho \tag{7-70}$$

can be used to evaluate the integral in (7-68) and obtain the result

$$\mathcal{F}\left[\text{circ}\left(\frac{r}{a}\right)\right] = \pi a^2 \frac{J_1(2\pi f_\rho a)}{\pi f_\rho a} \tag{7-71}$$

and thereby obtain the field in the form

$$U(x, y, f) = \pi a^2 \frac{e^{ikf}}{i\lambda f} e^{(ik/2f)(x^2 + y^2)} \frac{J_1(2\pi f_\rho a)}{\pi f_\rho a} \tag{7-72}$$

and thereby an intensity of the form

$$I(x, y, f) = \frac{1}{2\eta_0}\left(\frac{\pi a^2}{\lambda f}\right)^2 \left(\frac{J_1(2\pi f_\rho a)}{\pi f_\rho a}\right)^2 \tag{7-73}$$

The factor with the Bessel function in it was written in this form because its value at $f_\rho a = 0$ is 1. The factor in front of the Bessel function is the ratio of the aperture area to the Fraunhofer distance, λf, squared. Because of the normalization, this gives the relative amount of the unity intensity wave that gets transmitted by the aperture.

The Bessel function term is plotted as a function of af_ρ in Figure 7.19. The first order Bessel function divided by its argument is sometimes called the Jinc function. The half width of the main lobe of this function is given by the relation

$$af_\rho \cong .61 \tag{7-74}$$

Writing out that $f_\rho \sim r_{HW}/\lambda f$ and approximating $.61 \sim 1/2$, one finds that

$$r_{HW} = \frac{1}{2}\frac{f}{a}\lambda \tag{7-75}$$

Noting that the numerical aperture of the system is roughly given by (paraxial limit)

$$NA \sim \frac{a}{f} \tag{7-76}$$

one can write that

$$r_{HW} = \frac{\lambda}{2NA} \tag{7-77}$$

a relation that gives the diffraction limit of an optical system. A standard optical system cannot produce a spot less than half a wavelength wide, and can achieve half a wavelength only if the numerical aperture of the system is near unity. If one steps down the numerical aperture to obtain depth of field, the diffraction

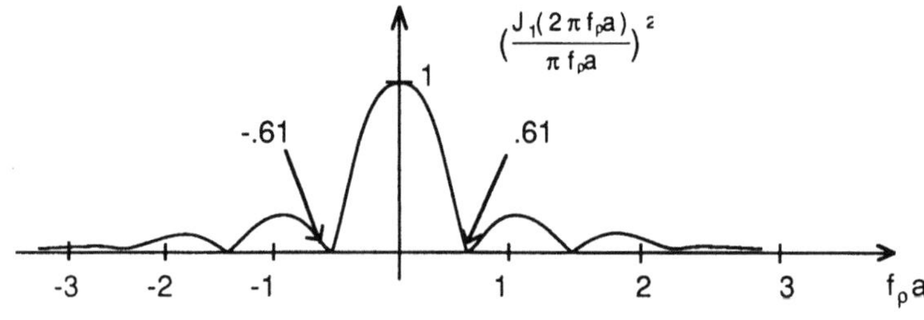

FIGURE 7.19.   A plot of the JINC function squared of equation (7-73).

spreading increases. This is not much of a consideration in regular photography, as the transfer functions of regular ASA films cut off at less than about 100 lines per millimeter, but it does become a consideration in any form of microoptics, from fiber excitation to optical disk readout. Indeed, Rayleigh was a farsighted man to derive this result at the turn of the twentieth century. As we soon will see, this result also applies to problems of partially coherent propagation, as well as to the case of monochromatic propagation that we have just considered.

Before proceeding to partial coherence, let us first consider an example of a Rayleigh limited coherent imaging system, as depicted in Figure 7.20. An infinity-corrected objective works with the object at its front focal plane and its image at infinity, so no specific length is needed. A secondary lens forms a real image viewed with the eye. The magnification of the objective and the secondary lens pair equals the ratio of their focal lengths. The arrangement facilitates focal adjustment.

For an objective that is free of aberration (i.e., "diffraction-limited"), its numerical aperture and the wavelength determine the limitation of the image. An estimate of the resolution is given by the Rayleigh criterion, according to which two coherent points of equal intensity can be regarded as distinct when their separation equals the radius of the first zero of the diffraction patterns. This separation is roughly $d = \lambda/2NA$, as was discussed above. Objectives that work in air have a numerical aperture up to about 0.95, corresponding to a half angle of 71°.

Any geometrical magnification is possible, but a microscope can enlarge an image until there is no further relevant detail. For example, if the first diffraction orders of a periodic object fall beyond the collection cone of the objective, then regardless of the magnification, the image is uniform, consisting only of the zeroth order. In other words, object amplitude Fourier components with spatial frequencies greater than $V_{\max}$ do not contribute to image modulation where $V_{\max} = 2NA/\lambda$.

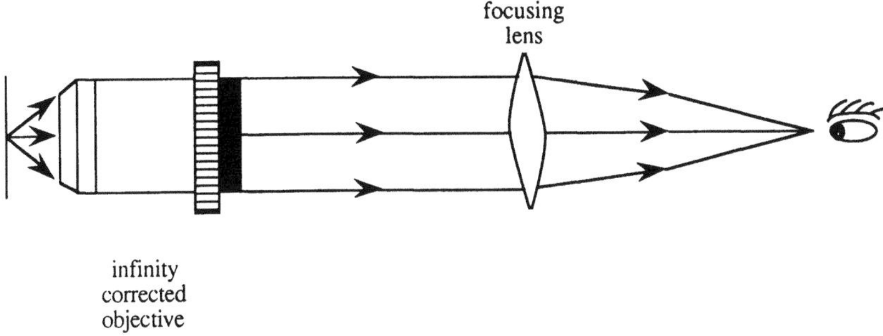

FIGURE 7.20.   Depiction of a Rayleigh limited coherent imaging system.

7.4 VAN CITTERT-ZERNICKE THEOREM

We will now apply diffraction theory to partially coherent propagation as is discussed in a number of sources such as Beran and Parent (1964), Goodman (1968) and Goodman (1985). To remind us of earlier discussions as well as to introduce some new notation, an archetypical diffractometer is illustrated in Figure 7.21. As one will recall, the intensity in the plane at $z = L$ is given by (6-68)

$$I(x_2, y_2, L) = \Gamma_{11}(x_1, y_1, x_2, y_2) + \Gamma_{22}(\alpha_1, \beta_1, x_2, y_2)$$
$$+ 2\,\mathrm{Re}\,[\Gamma_{12}(x_1, y_1, \alpha_1, \beta_1, x_2, y_2)] \tag{7-78}$$

where all of the interference effects were contained in the Γ_{12} term. As the whole diffractometer argument was based on paraxial approximations, it was generally assumed that the dependences of Γ_{11} and Γ_{22} on x_2, y_2 could be ignored. These dependences could not quite be ignored in Γ_{12}, as there was a grating term in the Γ_{12}. In the one-dimensional case that we considered in Chapter 6, Γ_{12} was given by

$$\Gamma_{12}(d) = \sqrt{\Gamma_{11}\left(\frac{d}{2}\right)}\sqrt{\Gamma_{22}\left(-\frac{d}{2}\right)}\,\hat{\gamma}_{12}(d) \tag{7-79}$$

where the complex degree of coherence was given by

$$\hat{\gamma}_{12}(d) = \gamma_{12}(d)e^{i\alpha(d)}e^{i(kxd)/L} \tag{7-80}$$

In the present case, there will also be a term like the complex exponential, which we could generically express as

$$\hat{\gamma}_{12}(x_1, y_1, x_2, y_2) = \gamma_{12}(x_1, y_1, \alpha_1, \beta_1)e^{i\alpha(x_1, y_1, \alpha_1, \beta_1)}e^{i\varphi(x_2, y_2)} \tag{7-81}$$

It is therefore clear that one could define quantities that belong to the x_1, y_1, α_1, β_1 plane and are completely independent of where the observation plane is. These quantities are the mutual intensity J_{12} and the complex degree of coher-

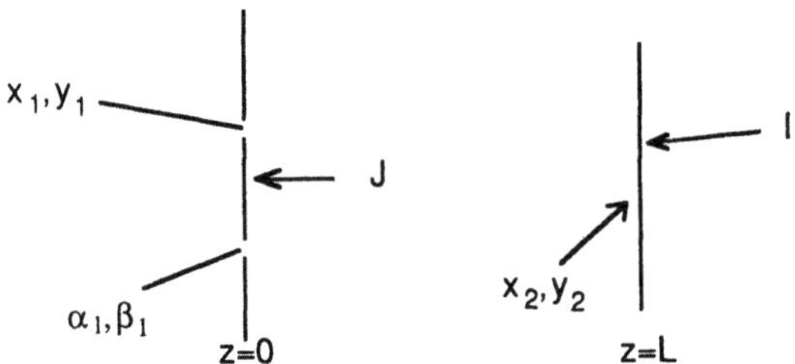

FIGURE 7.21.   Depiction of an archetypical diffractometer.

ence $\hat{\mu}_{12}$ and, in the present case, are given by

$$J_{12}(x_1, y_1, \alpha_1, \beta_1) = \Gamma_{12}(x_1, y_1, \alpha_1, \beta_1, x_2 = y_2 = 0) \qquad \text{(a)}$$

$$\hat{\mu}_{12}(x_1, y_1, \alpha_1, \beta_1) = \hat{\gamma}_{12}(x_1, y_1, \alpha_1, \beta_1, x_2 = y_2 = 0) \qquad \text{(b)}$$

$$(7\text{-}82)$$

An alternative definition for J_{12} is

$$J_{12}(x_1, y_1, \alpha_1, \beta_1) = \frac{1}{2\eta} \langle U^*(x_1, y_1)U(\alpha_1, \beta_1)\rangle \qquad (7\text{-}83)$$

where the U's are the fields, and now, explicitly, J_{12} is dependent only on the fields defined in the two apertures of the plane at $z = 0$. The function J_{12} contains the full information about the coherence length in the plane $z = 0$. The problem we want to solve here is the problem of, given a knowledge of J_{12} in some plane $z = z_1$, finding J_{12} in another plane $z = z_2$.

To determine the propagation characteristics of the mutual intensity function J_{12}, we can use the results of scalar diffraction theory, at least in the paraxial limit. Certainly by using (7-83) as an example case, we can write

$$J_{12}(x_2, y_2, \alpha_2, \beta_2) = \frac{1}{2\eta} \langle U^*(x_2, y_2)U(\alpha_2, \beta_2)\rangle \qquad (7\text{-}84)$$

where the U's can be expressed by

$$U^*(x_2, y_2) = \frac{-1}{i\lambda(z_2 - z_1)} \int U^*(x_1, y_1)e^{ikr_{x_1 y_2}} \, dx_1 \, dy_1 \qquad \text{(a)}$$

$$(7\text{-}85)$$

$$U(\alpha_2, \beta_2) = \frac{-1}{i\lambda(z_2 - z_1)} \int U^*(x_1, y_1)e^{ikr_{\alpha_1 \alpha_2}} \, d\alpha_1 \, d\beta_1 \qquad \text{(b)}$$

where the $r_{x_1 x_2}$ and $r_{\alpha_1 \alpha_2}$ notations are used to represent the quantities

$$r_{x_1 x_2} = \sqrt{(x_2 - x_1)^2 + (y_2 - y_1)^2 + (z_2 - z_1)^2} \qquad \text{(a)}$$

$$r_{\alpha_1 \alpha_2} = \sqrt{(\dot{\alpha}_2 - \alpha_1)^2 + (\beta_2 - \beta_1)^2 + (z_2 - z_1)^2} \qquad \text{(b)}$$

$$(7\text{-}86)$$

Using the definitions (7-84) and (7-85), one can write that

$$J(x_2, y_2, \alpha_2, \beta_2)$$

$$= \frac{1}{\lambda^2(z_2 - z_1)^2} \int J(x_1, y_1, \alpha_1, \beta_1)e^{ik(r_{\alpha_1 \alpha_2} - r_{x_1 x_2})} \, dx_1 \, dy_1 \, d\alpha_1 \, d\beta_1 \qquad (7\text{-}87)$$

Equation (7-87) is the general result of which we shall soon see the Van Cittert-Zernicke theorem is a special case.

Before going on to the Van Cittert-Zernicke theorem, let us consider a couple of special examples of the intensity pattern radiated by sources of various coherences. Clearly, equation (7-87) predicts that the paraxial intensity in a plane z_2 due to a partially coherent source in a plane z_1 is given by

$$I(x_2, y_2) = \frac{1}{\lambda^2(z_2 - z_1)^2} \int J(x_1, y_1, \alpha_1, \beta_1) e^{ik(r_{\alpha_1\alpha_2} - r_{x_1x_2})} \, dx_1 \, dy_1 \, d\alpha_1 \, d\beta_1$$

$$(7\text{-}88)$$

where the identifications of the r's should be self-evident from the relations of equation (7-88). A first example of (7-88) could be the intensity resulting from an incoherent source placed in the plane z_1. A totally incoherent source would have a mutual intensity given by

$$J_{12}(x_1, y_1, \alpha_1, \beta_1) = L_s^2 \, I(x_1, y_1) \, \delta(x_1 - \alpha_1) \, \delta(y_1 - \beta_1) \qquad (7\text{-}89)$$

where L_s is the source's linear extent and $\delta(a - b)$ is Dirac's delta function. Equation (7-89) is a mathematical idealization of an incoherent source, as the delta functions indicate a correlation length of zero, which clearly is not physically possible. If each infinitesimal element of a source were radiating with a phase that was totally uncorrelated with the phase of the next infinitesimally close infinitesimal element, the sum over the source would have to equal zero on the average. A totally incoherent source, therefore, could not radiate. A better approximation of the mutual intensity of an incoherent source might well be

$$J_{12}(x_1, y_1, \alpha_1, \beta_1) = I(x_1, y_1) \frac{1}{4L_{cx}L_{cy}} \, e^{-(|x_1 - \alpha_1|/L_{cx})} \, e^{-(|y_1 - \beta_1|/L_{cy})} \qquad (7\text{-}90)$$

where (7-89) is the limit of (7-90) when L_{cx} and L_{cy} go to zero. Allowing ourselves the luxury of using (7-89), we find that the intensity in a plane z_2 due to an incoherent source in a plane z_1 is given by

$$I(x_2, y_2) = \frac{L_s^2}{\lambda^2(z_2 - z_1)^2} \int I(x_1, y_1) \, dx_1 \, dy_1 \qquad (7\text{-}91)$$

Equation (7-79) is not too surprising a result. We recall that everything we are doing is in the paraxial approximation. A Lambertian source radiates in a pattern defined by $\cos \theta$, which is equal to 1 in the paraxial limit. Further, incoherent sources must obey the inverse square law as was discussed in Chapter

4. Indeed, equation (7-91) has just such a square law dependence in its denominator.

A second enlightening example of the radiation prediction of equation (7-88) could be the example of a totally coherent source. A coherent source's mutual intensity could be expressed in the form

$$J_{12}(x_1, y_1, \alpha_1, \beta_1) = \frac{1}{2\eta_0} U^*(x_1, y_1)U(\alpha_1, \beta_1) \tag{7-92}$$

where no averaging brackets are necessary because of the monochromaticity of the fields. Plugging into equation (7-88), one finds that

$$I(x_2, y_2) = \frac{1}{2\eta_0} U^*(x_2, y_2)U(x_2, y_2) \tag{7-93}$$

where the U's are defined by

$$U(x_2, y_2) = \frac{1}{i\lambda(z_2 - z_1)} \int U(x_1, y_1)e^{ikr_{21}} \, dx_1 \, dy_1 \tag{7-94}$$

Equation (7-93) tells us that the intensity of a monochromatic field is just the magnitude squared of the field, apart from a constant. The result is not especially surprising.

We now have all the machinery in place to consider the Van-Cittert Zernicke theorem in some detail. The theorem predicts the coherence length at a plane z_2 due to an incoherent source in a plane z_1, at least in the limit of Fraunhofer diffraction. Using (7-89) in (7-87), one finds immediately that

$$J_{12}(x_2, y_2, \alpha_2, \beta_2) = \frac{L_s^2}{\lambda^2(z_2 - z_1)^2} \int I(x_1, y_1)e^{ik(r_{\alpha_1\alpha_2} - r_{x_1x_2})} \, dx_1 \, dy_1 \tag{7-95}$$

Assuming that the source is a disk of radius a, expressible as

$$I(x_1, y_1) = I_0 \, \text{circ}\left(\frac{r}{a}\right) \tag{7-96}$$

and that the disk is sufficiently far away from the plane z_2 that Fraunhofer diffraction applies and therefore the r's can be expanded as

$$r_{x_1\alpha_1} = (z_2 - z_1)\left[1 + \frac{\alpha_2^2 + \beta_2^2}{2(z_2 - z_1)^2} - \frac{(x_1\alpha_2 + y_1\beta_2)}{(z_2 - z_1)^2}\right] \tag{7-97}$$

then using (7-96) and (7-97) in (7-95), one finds that

$$J_{12}(x_2, y_2, \alpha_2, \beta_2) = \frac{L_s^2 e^{i\psi}}{\lambda^2(z_2 - z_1)^2} \, \mathcal{F}\left[\mathrm{circ}\left(\frac{r}{a}\right)\right]_{\substack{f_x = (\alpha_2 - x_2)/(\lambda(z_2 - z_1)) \\ f_y = (\beta_2 - y_2)/(\lambda(z_2 - z_1))}} \tag{7-98}$$

where the ψ is given by

$$\psi = k\left[\frac{\alpha_2^2 + \beta_2^2}{2(z_2 - z_1)} - \frac{(x_2^2 + y_2^2)}{2(z_2 - z_1)}\right] \tag{7-99}$$

From the previous section, we know that the transform of the circ function is given by a Bessel function, and indeed we can write out that

$$J_{12}(x_2, y_2, \alpha_2, \beta_2) = \frac{L_s^2 e^{i\psi}}{\lambda^2(z_2 - z_1)^2} \, \pi a^2 \frac{J_1(2\pi a f_\rho)}{\pi a f_\rho} \tag{7-100}$$

where the f_ρ is given by

$$f_\rho = \sqrt{\frac{(\alpha_2 - x_2)^2 + (\beta_2 - y_2)^2}{\lambda^2(z_2 - z_1)^2}} \tag{7-101}$$

Equation (7-100) is the main result of the Van Cittert-Zernicke theorem, which is basically embodied by equation (7-95). The result, (7-100), is most interesting as it indicates that an incoherent source in the plane z_1 produces a partially coherent disturbance in the plane z_2, with a coherence length given by

$$L_c = \frac{\lambda(z_2 - z_1)}{2a} = \frac{\lambda}{2NA} \tag{7-102}$$

where NA refers to the effective numerical aperture of the viewing system. An interesting point here is that this minimum coherence length is identical to the diffraction limit. This means that, among other things, a diffraction-limited spot must be completely coherent, and, therefore, a limit is placed on the efficiency with which an incoherent field can be focused down. Also, this result states that the coherence will increase with propagation length. As an example, let us consider the sun as an incoherent source and consider the coherence length of the solar rays that impinge on the earth. The radius of the sun is roughly .7 million kilometers, and it is at a distance of 150 million kilometers. The effective numerical aperture that it subtends is, therefore, roughly .005. We can see

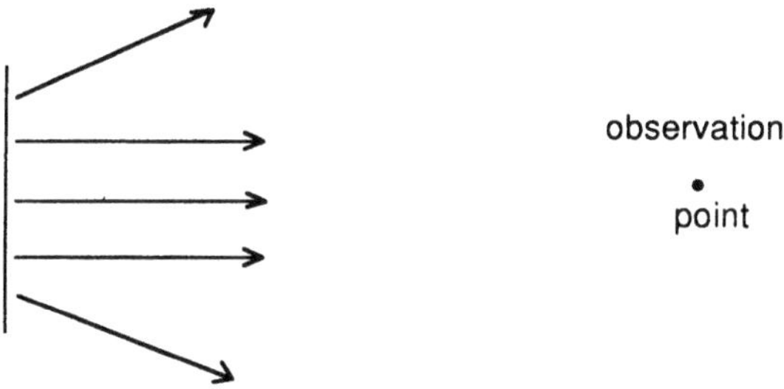

FIGURE 7.22.   Illustration of the rays that will be seen at an observation point far from the source of the radiation.

from (7-102) that the coherence length of the light has increased from nearly zero on the face of the sun to about 100 λ at the surface of the earth. The effect is essentially a selection effect, as is illustrated in Figure 7.22. The farther away the point is from the source, in units of the source radius, the fewer rays from the source that will impinge on the observation point. Further, the rays that do reach the observation point will be essentially all parallel and therefore will present a (more) coherent phase front. Another way to consider this problem is via the diffractometer. Consider the situation depicted in Figure 7.23. Here a diffractometer is illuminated by several independent (i.e., incoherent) point sources. The fields in the pinholes due to a point source at coordinate x_0 will be given by

$$V\left(\frac{d}{2}\right) = V_0 \frac{e^{iks_{12}}}{s_{12}} e^{i\varphi} \qquad \text{(a)}$$

$$V\left(-\frac{d}{2}\right) = V_0 \frac{e^{iks_{12}}}{s_{12}} \qquad \text{(b)}$$

$$(7\text{-}103)$$

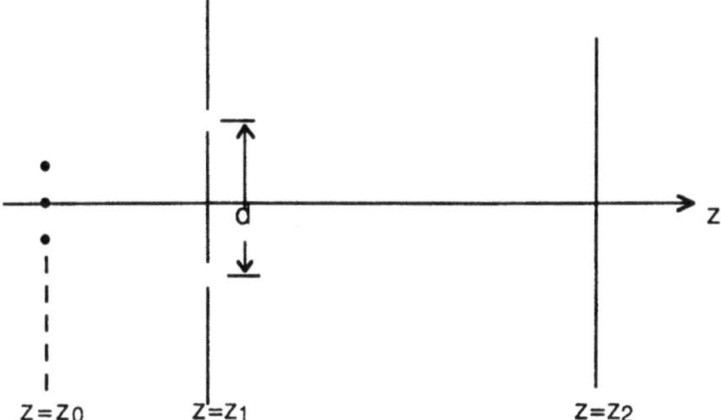

FIGURE 7.23.   Illustration of a diffractometer illuminated by several point sources.

where the φ is given paraxially by $kx_0 d/(z_1 - z_0)$. The intensity pattern on the screen due to one point source will thus be given by

$$I(x_2) = \frac{1}{\eta_0} \frac{V_0^2}{(z_2 - z_1)^2} \left[1 + \cos\left(\frac{kx_2 d}{z_2 - z_1} + \frac{kx_0 d}{z_1 - z_0} \right) \right] \qquad (7\text{-}104)$$

Because the source was a single point source and therefore subtends zero angle (has zero effective numerical aperture), the fringe pattern has a visibility of unity. Let us say, however, that there are two mutually incoherent point sources at coordinates x_{01} and x_{02}. Here the pattern observed in z_2 would be given by

$$I(x_2) = \frac{V_0^2}{\eta_0} \frac{1}{(z_2 - z_1)^2} \left[2 + \cos\left(\frac{kx_2 d}{z_2 - z_1} + \frac{kx_{01} d}{z_1 - z_0} \right) \right.$$
$$\left. + \cos\left(\frac{kx_2 d}{z_2 - z_1} + \frac{kx_{02} d}{z_1 - z_0} \right) \right] \qquad (7\text{-}105)$$

and clearly here, unless the source locations are chosen very specially, the fringe visibility will be less than unity because the maxima and minima of the cosine functions will not be at the same place. For a finite incoherent source extending from $-L/2$ to $+L/2$, the intensity pattern would be in general

$$I(x_2) = \frac{V_0^2}{\eta_0} \frac{1}{(z_2 - z_1)^2} \left[1 + \frac{1}{L} \int_{-1/2}^{1/2} dx_0 \cos\left(\frac{kx_2 d}{z_2 - z_1} + \frac{kx_0 d}{z_1 - z_0} \right) \right]$$
$$(7\text{-}106)$$

and the fringe contrast is lowered from unity. Indeed, one can see in (7-106) that the larger the solid angle subtended by the source, the poorer the fringe visibility is.

Indeed, Michelson used the effect of the interference between incoherent points to make his stellar interferometer. Consider the situation depicted in Figure 7.24. A star has an angular intensity dependence of $I(\alpha)$ where α is the angle to the interferometer input. The interferometer itself has a configuration

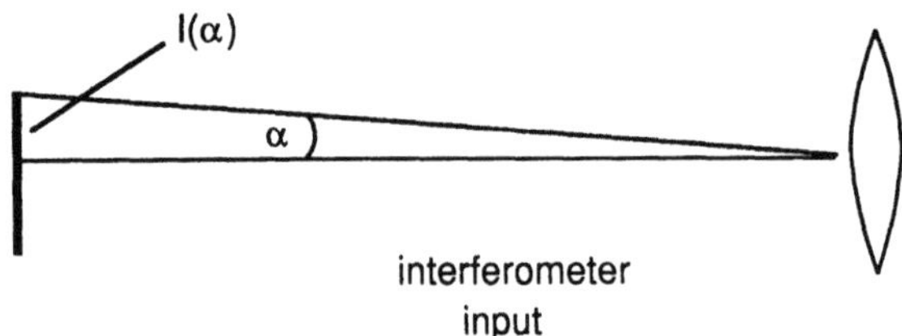

FIGURE 7.24.   Schematic depiction of the use of a Michelson stellar interferometer.

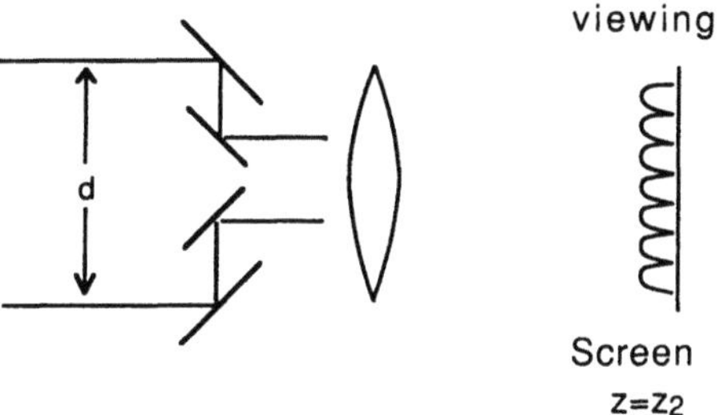

FIGURE 7.25.   The configuration of a Michelson stellar interferometer.

as illustrated in Figure 7.25. The outside mirrors were mounted on tracks so that they could be moved, thereby scanning the distance d. The intensity pattern that is observed in the plane $z = z_2$ would be given by

$$I(x_2) = \frac{V_0^2}{2\eta_0} \frac{1}{(z_2 - z_1)^2} \left[1 + \int d\alpha I(\alpha) \cos\left(\frac{kx_2 d}{z_2 - z_1} + k\alpha d \right) \right] \quad (7\text{-}107)$$

As d is increased from zero, the fringe visibility will decrease from one to zero, as is depicted in Figure 7.26. The zero value should occur at the point where the $k\alpha d$ term is equal to π, or

$$d = \frac{\lambda}{2\alpha} \quad (7\text{-}108)$$

Michelson was able to scan his mirrors over roughly 6 meters, thereby obtaining an angular sensitivity of roughly 5×10^{-8} radian. For a star with the radius of the sun, 7×10^{10} cm, this would restrict one to a viewing range of 1.4×10^{18} cm or 1.4 light years. Unfortunately, our nearest star, Alpha Centauri, is at a distance of more than 4 light years. However, Alpha Ori, at a distance of 10 light years is 300 times as large as the sun, thereby subtending an angle of

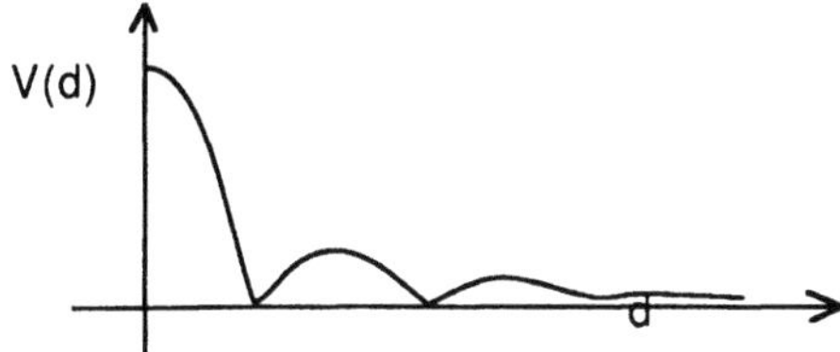

FIGURE 7.26.   The fringe visibility of a Michelson stellar interferometer as a function of the mirror separation d.

2×10^{-6} radian, and Michelson was able to resolve this stellar disk with his interferometer.

7.5 DIFFRACTION GRATINGS AND SPECTROMETERS

Our final topic is that of diffraction gratings. To begin the discussion, let us consider the situation depicted in Figure 7.27, the problem of the diffraction of a single slit. Let us say that the transmission of the slit is expressible as a pupil function

$$P(x, y) = \text{rect}\left(\frac{2x}{L}\right) \tag{7-109}$$

where the rect is defined in equation (6-19) as being unity for the absolute value of the argument less than $1/2$ and zero otherwise, and where the incident illumination is given by

$$U(x, y, z < 0) = U_0 e^{ikz \cos \theta} e^{ikx \sin \theta} \tag{7-110}$$

We soon will see that, in our oversimplified model, the angle of the incident wave has only a trivial effect, a fact that is not borne out in practice and that we will come back to later. With the substitutions of (7-109) and (7-110) into (7-50), one can write that

$$U(x, y, z) = U_0 \frac{e^{ikr} e^{(ik/2z)(x^2 + y^2)}}{i\lambda z} \int_{-\infty}^{\infty} dy' \int_{-L/2}^{L/2} dx' \, e^{i(2\pi/\lambda)x' \sin \theta} e^{-i2\pi(f_x x' + f_y y')} \tag{7-111}$$

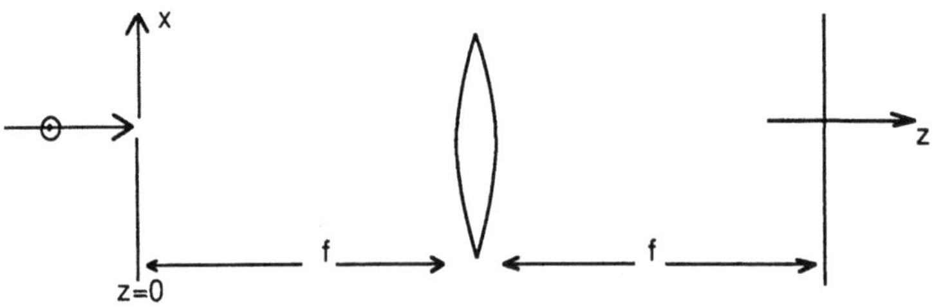

FIGURE 7.27.   Depiction of a y-polarized wave incident on a slit with an optical system designed to Fourier-transform the diffraction pattern.

Because of the y symmetry of the problem, the integral gives a delta function. This results in

$$U(x, y, z) = U_0 \frac{e^{ikz} e^{(ik/2z)(x^2 + y^2)}}{i\lambda z} \delta(y) \int_{-L/2}^{L/2} e^{i2\pi x'[(\sin\theta/\lambda) - f_x]} \, dx' \quad (7\text{-}112)$$

The remaining integral is trivial, and the result can be written as

$$U(x, y, z) = U_0 \frac{e^{ikz} e^{(ik/2z)(x^2 + y^2)}}{i\lambda z} \delta(y) \frac{\sin \pi \left(f_x - \dfrac{\sin\theta}{\lambda} \right) L}{\pi \left(f_x - \dfrac{\sin\theta}{\lambda} \right)} \quad (7\text{-}113)$$

with the corresponding intensity

$$I(x, y, z) = \frac{U_0^2 L^2}{\lambda^2 z^2} \delta^2(y) \left[\frac{\sin \pi \left(f_x - \dfrac{\sin\theta}{\lambda} \right) L}{\pi \left(f_x - \dfrac{\sin\theta}{\lambda} \right) L} \right]^2 \quad (7\text{-}114)$$

where the f_x is given by $x/\lambda f$. The result is plotted in Figure 7.28. It is immediately noted here that the effect of the input angle is only to shift the whole pattern up and down the x-axis. In a "real" grating, a change of the input angle has a much more profound effect than this, but we will return to this point later. For now, we will ignore angles until we can take them into account correctly. A second point is the y dependence of the pattern. Because of the y symmetry of the problem, one would expect there to be no y variation. Instead, we get a

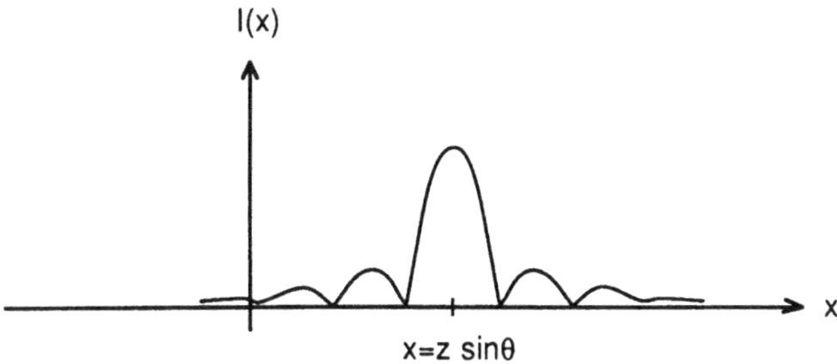

FIGURE 7.28.   A plot of the x dependence of the function in (7-114).

delta function, which is due to the Fraunhofer approximation. An ideal lens will take a y-independent incident wave and focus it down to a line. That is just what happened here. The interesting dependence is the x dependence. Rewriting (7-114) without the θ or y dependence, one sees that

$$I(x, y, z) = \frac{U_0^2 L^2}{\lambda^2 z^2} \left[\frac{\sin \pi \frac{x}{\lambda z} L}{\pi \frac{x}{\lambda z} L} \right]^2 \tag{7-115}$$

It is rapidly seen that (7-115) will have a set of equally spaced zeros at points defined by

$$x_N = N \frac{\lambda z}{L} \quad N \text{ integer} \tag{7-116}$$

The maximum at zero will be the largest, as the denominator grows monotonically with x. As is easily seen from (7-116), the width of the central maximum is inverse in its dependence on L, the slit opening.

We now wish to consider the effect of an array of slits on an incident plane wave. Here we can write that the pupil function is given by

$$\begin{aligned} P(x) &= \sum_{n=0}^{N} P_n(x) = \sum_{n=0}^{N} \text{rect}\left(\frac{x - x_N}{L}\right) \\ &= \sum_{n=0}^{N} \text{rect}\left(\frac{x - nd}{L}\right) \end{aligned} \tag{7-117}$$

where d is the distance between adjacent slits. Putting (7-117) into the diffraction integral, one finds that

$$U(x, z) = U_0 \frac{e^{ikz} e^{(ik/2z)x^2}}{i\lambda z} \int dx'\, e^{-i2\pi f_x x'} \sum_{n=0}^{N} \text{rect}\left(\frac{x - nd}{L}\right) \tag{7-118}$$

Making a change of variables and rearranging, one finds

$$U(x, z) = \sum_{n=0}^{N} e^{-i2\pi f_x nd} \left[\frac{U_0 e^{ikz} e^{ikx^2/2z}}{i\lambda z} \int dx'\, e^{-i2\pi f_x x} \text{rect}\left(\frac{2x'}{L}\right) \right] \tag{7-119}$$

which is a rather interesting result. The term in the last bracket is just exactly the pattern due to one aperture. In fact, we would have gotten an exactly similar result no matter what form we assumed for the aperture. We would have gotten an array term times a single aperture pattern. The lead factor is generally called the array factor, and it can be simplified from its sum form by using the identity

$$\sum_{n=0}^{N} x^n = \frac{1 - x^{N+1}}{1 - x} \tag{7-120}$$

to get the form

$$U(x, z) = e^{-i\pi f_x Nd} \frac{\sin \pi f_x (N + 1)d}{\sin \pi f_x d} L \frac{U_0 e^{ikz} e^{(ik/2z)x^2}}{i\lambda z} \frac{\sin \pi f_x L}{\pi f_x L} \tag{7-121}$$

with the corresponding intensity

$$I(x, z) = U_0 \frac{U_0^2 L^2}{\lambda^2 z^2} \left(\frac{\sin \pi f_x (N + 1)d}{\sin \pi f_x d} \right)^2 \left(\frac{\sin \pi f_x L}{\pi f_x L} \right)^2 \tag{7-122}$$

Note that, as $L < d$, the last term in (7-122) is going to be broader than the first term. Further, if

$$\frac{xd}{\lambda z} \gg 1 \tag{7-123}$$

then the first bracketed term will have multiple maxima where the denominator and numerator have zeros. The situation is as it is depicted in Figure 7.29. It should be noted here that for $d < \lambda$ there is only one peak, as x/z is like the sine of an angle and cannot be greater than one. (That we have x/z at all is an artifact of the paraxial approximation. The exact result would only have had $\sin \theta$.) Practically, we will see that d is a couple of wavelengths and a few times L, the slit width. This means, of course, that L is on the order of a wavelength, and our Kirchhoff boundary conditions cannot be very good. The effect of the

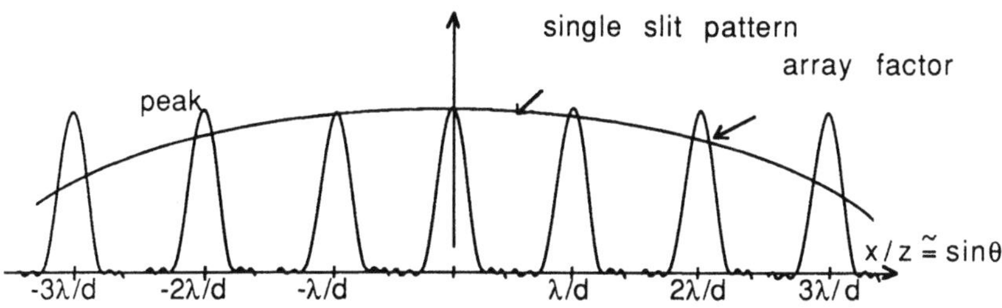

FIGURE 7.29.   Schematic depiction of the Fraunhofer pattern of an array of slits.

correct boundary condition, however, can be figured out from the following argument. Because the field must smoothly go to zero on the surface of the screen rather than having a rectangular field in the slit, the field will be smooth and roughly half as wide as its Kirchhoff counterpart. Being smaller, it will actually diffract more, and the broad peak in Figure 7.29 will become even broader. Therefore, even more diffraction orders will have appreciable values than with the ideal fields. It should be noted here that the peak at $n\lambda/d$ is in general called the nth diffracted order.

We now wish to consider what happens to a polychromatic wave incident on such a diffraction grating as the one just discussed. Clearly, we can write the grating transmission as a transmission function, which is

$$T(x, \lambda) = \frac{\sin^2\left(\pi \dfrac{x(N + 1)d}{\lambda z}\right)}{\sin^2\left(\pi \dfrac{xd}{\lambda z}\right)} [I(x)]_{1\,\text{slit}} \qquad (7\text{-}124)$$

and the intensity transmitted by the grating could be expressible as

$$I_t(x) = \int T(x, \lambda)I(\lambda)\,d\lambda \qquad (7\text{-}125)$$

where $I(\lambda)$ is the intensity distribution of the incoming radiation. For the moment, let us make the approximation that N is very large, and therefore the half width of the diffraction peaks, which is given by

$$X_{HW} = \frac{\lambda z}{(N + 1)d} \qquad (7\text{-}126)$$

is very narrow indeed. With this, we can roughly approximate T with the form

$$T(x, \lambda) = \sum a_n \delta\left(\frac{x}{z} - n\frac{\lambda}{d}\right) \qquad (7\text{-}127)$$

and thereby obtain I_t in the form

$$I_t(x) = \sum a_n' I\left(\frac{d}{n}\sin\theta\right) \qquad (7\text{-}128)$$

The situation described by (7-128) is illustrated in Figure 7.30. The zeroth-order pattern has essentially zero width and therefore contains no information.

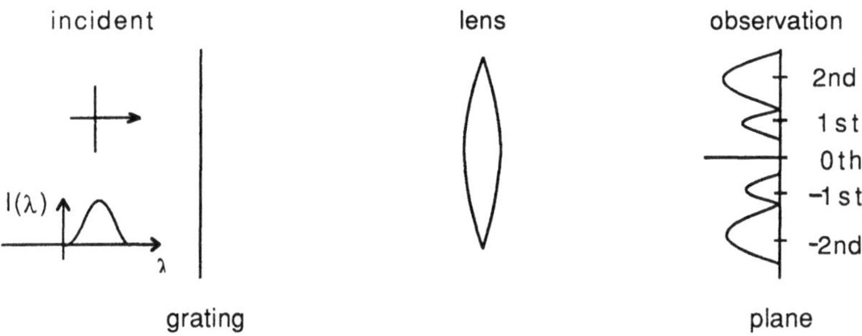

FIGURE 7.30.   Illustration of how a grating spectrometer works.

The first orders, however, carry a reproduction of the incident spectrum; that is, the spectrum is repeated as a function of θ where the relation $\lambda = d \sin \theta$ applies. In higher orders, the spectrum is also reproduced by a magnification that is given by the number of the order. For example, if the spectrum contained just two lines, these lines would be unresolved in the zeroth order, located at a relative distance of 1 in the first order, at a relative distance of 2 in the second order, and so on. Therefore, one sees that as far as resolution goes, it can be advantageous to operate in as high a grating order as is possible.

Let us briefly consider the resolving power of the grating spectrometer. Say that the incident spectrum is of the form

$$I(\lambda) = \delta(\lambda - \lambda_0) + \delta(\lambda - \lambda_0 - \Delta\lambda) \tag{7-129}$$

such that the output of the spectrometer in some given order would look like the intensity pattern sketched in Figure 7.31. The depiction there is of the situation in which the two lines are just resolvable; that is, their separation is given by their half-widths. The half-width $\Delta\theta$ of an order is given by

$$\Delta\theta = \frac{m\Delta\lambda}{d} \tag{7-130}$$

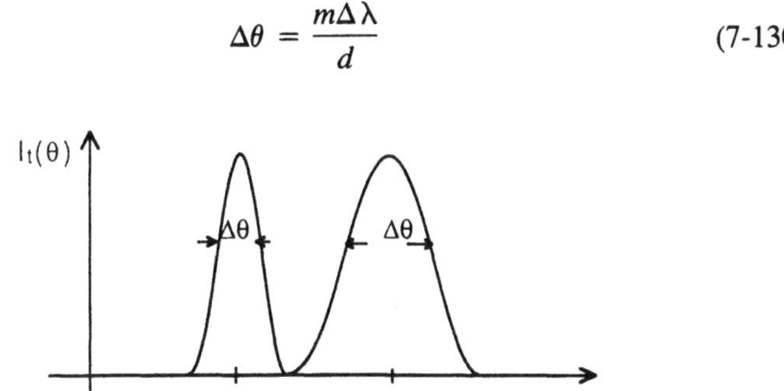

FIGURE 7.31.   The output of a grating spectrometer for an incident spectrum of the form of (7-129).

where m is the grating order. Now, the spacing between centers of the lines $\delta\theta$ will be given by

$$\delta\theta = \frac{\lambda}{Nd} \tag{7-131}$$

Combining equations (7-130) and (7-131), one finds that

$$\frac{\lambda}{\Delta\lambda} = mN \tag{7-132}$$

If one were to use the tenth order of a 1-cm-long grating with 1 μm spacing between the grooves, one could obtain a grating resolving power of 100,000 points, which means that for a center wavelength of 5000 Å one could resolve down to 0.05 Å or about 5 GHz. It must be borne in mind, however, that the total $\Delta\lambda$ must satisfy

$$\Delta\lambda = \frac{\lambda z}{md} \tag{7-133}$$

lest there be aliasing as there was in the case of Fabry-Perot. This can set a limit on how high an order one may pick.

The above-derived resolving power is obtainable in practice, but it is in general quite expensive to achieve. In spectroscopy experiments where one wants to resolve some spectra, the incoming beam usually is a spatially incoherent wave. To use a grating spectrometer, the beam need be spatially coherent. Other methods are necessary if one cannot collimate the beam. Let us consider a spectrophotometer, for example. Spectrophotometry is an experimental method used to measure quantitatively the absorption spectrum of materials. When there is enough energy present, a beam could be shown through a material, recollimated and have its spectrum taken in a grating spectrometer. Other techniques are by necessity more painstaking.

In a spectrophotometer, a beam falls on the absorbing specimen, and its transmission factor τ is measured by means of a photometric device. If ρ is the reflection coefficient, 2α is the absorption coefficient, and d is the thickness of the specimen, α can be obtained from the relation:

$$\tau = (1 - \rho)^2 \, e^{-2\alpha d} \tag{7-134}$$

A spectrophotometer consists of the following main components: (1) sources of radiation, (2) the monochromator, (3) photometric devices, and (4) radiation detectors, described below.

1. *Source of radiation.* Radiation sources with a continuous spectrum are used for spectrophotometers. Different sources are used for measurement in different spectral regions. Typical sources are:

UV region: rare-gas discharge lamps (He, Ar, Kr, Xe); high-pressure mercury lamp.

UV–visible region: high-pressure xenon discharge lamp; deuterium discharge lamp; Lyman continuum source.

IR region: Nernst filament; the G lobar; very high-pressure mercury lamp.

UV–visible–IR region: tungsten-filament lamp; quartz–iodine lamps.

2. *The monochromator.* A prism or a diffraction grating may be used to obtain a more monochromatic incident radiation flux. If a diffraction grating is used, different orders of the spectra may overlap. Suitable filters may be used to filter out unwanted wavelengths, but more generally a prism monochromator is used in series with the grating monochromator.

3. *Photometric devices.* There are three main classes of photometric instruments, characterized by different methods of measuring and comparing the absorption of specimens:

   Class I: The instrument has two beams, in one of which is the absorbing specimen and in the other a photometric device—polarizing prisms, adjustable aperture, and so forth—for matching the intensity of the two beams. *T* is obtained from the status of the photometric device, which has been calibrated.

   Class Ia: This is a modification of Class I, in which there is only one beam and one detector, with the absorbing specimen and the photometric device being used in succession in the beam.

   Class II: A single beam and a single detector are used, the transmission of a specimen being the ratio of the detector currents with and without the absorbing specimen. It is assumed that the response of the detector is proportional to the intensity falling on it.

4. *Radiation detectors.* Four main classes of detectors are used:

   Class I. Photoemissive cells: photomultipliers
   Class II. Photovoltaic cells: photodiodes, PIN photodiodes
   Class III. Photoconductive cells: PbS, PbSe, InSb, and so on.
   Class IV. Thermal detectors: thermopiles, bolometers, the Golay detector.

A main point here is that the wave from the primary source is spatially incoherent and will not diffract from the monochromatic grating. This wave must be made spatially coherent, at a large loss of signal intensity, for it to be diffracted by the grating. The grating discussed above, however, splits up the wave into a large number of diffracted orders, thereby further subdividing an already limited signal power and causing signal-to-noise problems. For this reason, one needs a grating much different from the slit type discussed; so we next consider some different types of gratings. We will begin the discussion with examples

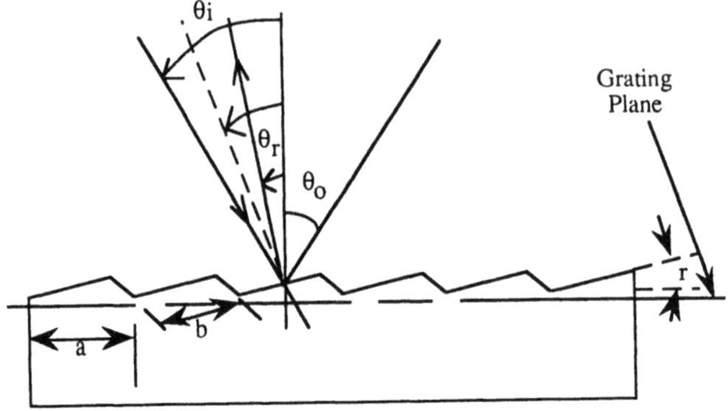

FIGURE 7.32.   Depiction of an Echeller grating.

of typical gratings before turning our attention to photographically produced gratings.

An Echeller (blazed) grating is illustrated in Figure 7.32. If the facets in the figure had not been tilted, specular reflection would have carried most of the energy off into the zeroth order at θ_0; instead it now goes out at θ_r. At normal incident $\theta_i = 0$, the zeroth order ($m = 0$) is at $\theta_0 = 0$. Most of the diffracted radiation is now concentrated at $\theta_r = 2r$, and this will correspond to the mth order when

$$a \sin 2r = m\lambda \tag{7-135}$$

In this case the mth-order interference peak will reside at the central maximum of the single slit diffraction pattern. A blazed grating has shaped grooves that can concentrate energy in a particular order.

For example, let us compute the angle r needed to strongly channel normally incident radiant energy at a wavelength of 633 nm into the third order by a grating with 500 lines/mm. Since $a \sin 2r = m\lambda$,

$$\begin{aligned}
r &= \frac{1}{2} \sin^{-1} \left(\frac{m\lambda}{a} \right) \\
&= \frac{1}{2} \sin^{-1} \left(\frac{3 \times 633}{\dfrac{10^6}{500}} \right) = 35.86°
\end{aligned} \tag{7-136}$$

A Littrow grating, depicted in Figures 7.33 and 7.34, is a blazed reflection grating in a specially designed configuration that allows for high compactness and high resolving power. As with any blazed grating, the general grating equation holds: $a(\sin \theta_m = \sin \theta_1) = m\lambda$ where a is the width of a line on the

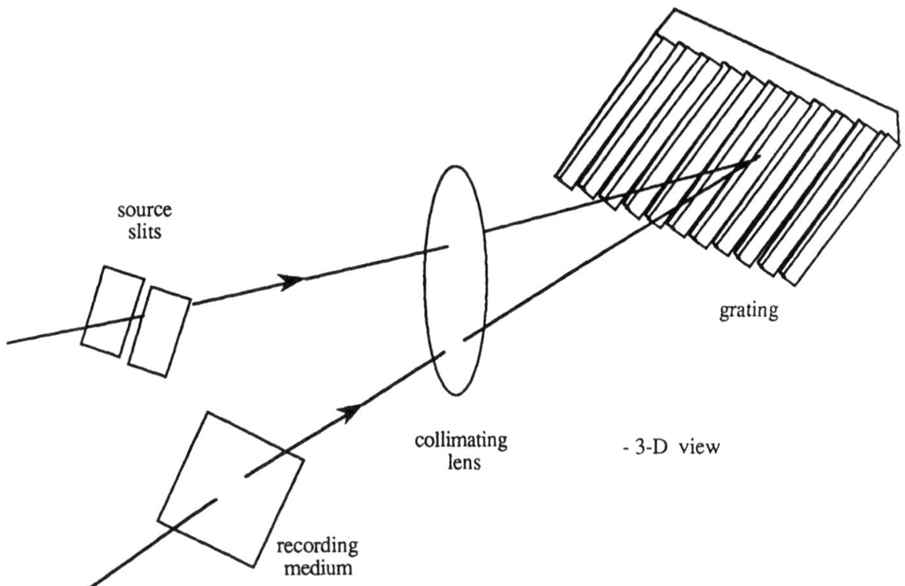

FIGURE 7.33.   Depiction of a Littrow grating with a three-dimensional perspective.

grating, λ is the incident wavelength, θ_1 is the angle of incidence on the grating, and qm is the angle of the mth-order reflection. In the special case of a Littrow grating, $\theta_1 \approx \theta_m$, which makes the grating equation read as $2a \sin \theta_1 = m\lambda$. This is known as the autocollimation condition. The resolving power of a grating is defined as

$$R = \lambda/\Delta\lambda_{\min} = \frac{Na(\sin \theta_m - \sin \theta_i)}{\lambda}$$

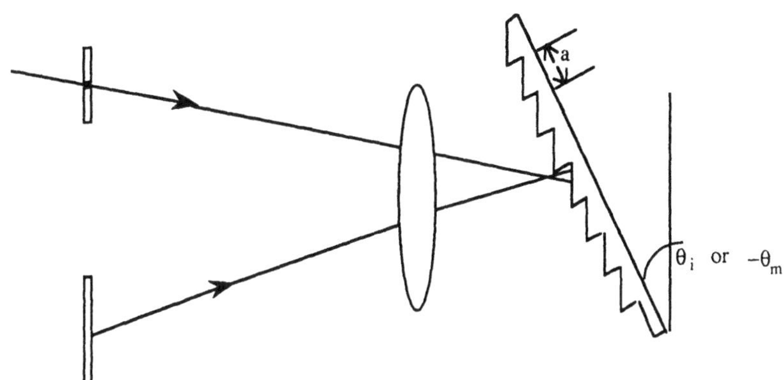

FIGURE 7.34.   Depiction of a Littrow grating in two-dimensional perspective.

where N is the number of lines per width, and Na is the width of the grating. In a Littrow grating, R is maximized because $\theta_m \approx \theta_i$; thus

$$R = \frac{2Na(\sin \theta_i)}{\lambda} \tag{7-137}$$

Quantitatively that means that for a grating 300 mm wide, at an angle of 60° in a Littrow mount, with $\lambda = 550$ nm the resolving power is $9.45 \cdot 10^5$.

Now let us turn our attention to some photographically produced gratings. We will first consider a so-called thin amplitude grating. To form such a grating, one could employ such a setup as that depicted in Figure 7.35. Here, a beam of coherent monochromatic light is split into two approximately equal paths and brought back together again with an angle 2θ between them. The resulting field distribution on the film plane could be expressed in the form

$$U(x) = 2U_0 \cos (kx \sin \theta) \tag{7-138}$$

If one assumes that the development process can be made completely linear, then one could assume that the transmittance of the resulting developed film plate would be proportional to the intensity incident on the plate during exposure. With this, one could write the transmittance as

$$t(x) = \left(\frac{1}{2} + \frac{m}{2} \cos 2\pi f_0 x\right) \text{rect}\left(\frac{2x}{L}\right) \tag{7-139}$$

where the m denotes a modulation index, and L is the linear extent of the film plate. If we were to stick this grating into a spectrometer type apparatus, as was illustrated in Figure 7.30, we would find that

$$U(x, y, z) = \frac{e^{ikz}}{i\lambda z} e^{(ik/2z)(x^2 + y^2)} \mathcal{F}[t(x)] \tag{7-140}$$

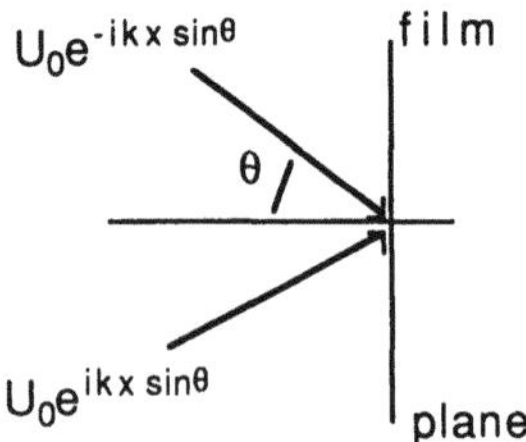

FIGURE 7.35.   Schematic depiction of a setup for producing amplitude gratings.

where the Fourier transform can be expressed as

$$\mathcal{F}\left[t(x)\right] = \frac{L}{2}\operatorname{sinc}\left(\frac{xL}{\lambda z}\right) + \frac{mL}{4}\operatorname{sinc}\left(\frac{x}{\lambda z} + f_0\right)L$$

$$+ \frac{mL}{4}\operatorname{sinc}\left(\frac{x}{\lambda z} - f_0\right)L \tag{7-141}$$

where the sinc function is defined as follows:

$$\operatorname{sinc}(x) = \frac{\operatorname{sinc}\pi x}{\pi x} \tag{7-142}$$

Now if $f_0 L > \pi$, that is, the grating contains at least a period, then one can write approximately that

$$I(x) \cong \frac{L^2}{4(\lambda z)^2}\left[\operatorname{sinc}^2\left(\frac{xL}{\lambda z}\right) + \frac{m^2}{4}\operatorname{sinc}^2\left(\frac{x}{\lambda z} - f_0\right)L\right.$$

$$\left. + \frac{m^2}{4}\operatorname{sinc}^2\left(\frac{x}{\lambda z} + f_0\right)L\right] \tag{7-143}$$

which is plotted in Figure 7.36. Although the number of grating lobes has been lowered to three, the energy efficiency of the grating still is not very good. The maximum fractional energy transmission is given by

$$\begin{array}{c}\text{fraction}\\\text{of incident}\\\text{transmitted}\end{array} \cong \frac{1}{4} + \frac{m^2}{8} \tag{7-144}$$

which means that as much as 75% of the incident energy can be absorbed by the grating. Further, as we are primarily interested in using one of the first orders only, the actual usable energy is only $m^2/16$. Thus, at best, 6% of the incident energy is usable. We would like to do better than this.

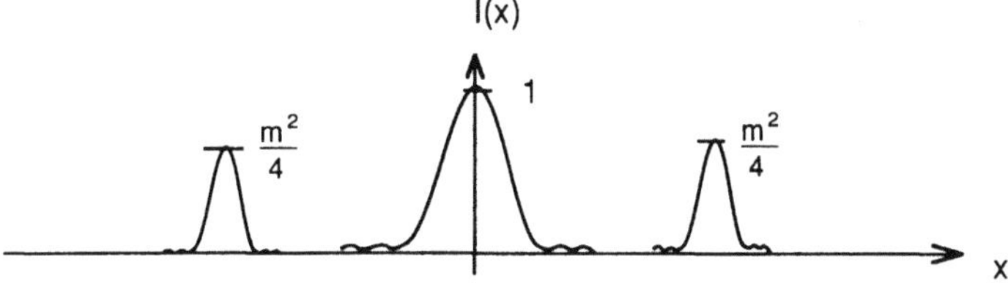

FIGURE 7.36.   Fraunhofer diffraction pattern of a thin amplitude grating.

The easiest way to improve on the above grating might well be to remove the absorption of the film grains. There is a process for doing this, known as bleaching. In this process, one actually removes the metallic silver in the commonly used silver halide emulsion and replaces the silver with an agent that causes these regions to take on a higher index than the average index of the emulsion. With this, one can generate a transfer function that looks like

$$t(x) = \exp\left\{i\,\frac{m}{2}\,\sin 2\pi f_0 x\right\}\,\text{rect}\left(\frac{2x}{L}\right) \tag{7-145}$$

Recalling that

$$e^{i(m/2)\sin 2\pi f_0 x} = \sum_{q=-\infty}^{\infty} J_q\left(\frac{m}{2}\right) e^{i2\pi q f_0 x} \tag{7-146}$$

one can show for a grating with several periods that

$$I(x) = \frac{L^2}{\lambda z} \sum_{q=-\infty}^{\infty} J_q^2\left(\frac{m}{2}\right)\,\text{sinc}\,(f_x + q f_0)L \tag{7-147}$$

Now, implicit in assuming that the grating is thin is the assumption that m must be small. For small m, the resulting Fraunhofer pattern might appear as depicted in Figure 7.37. As the zeroth-order Bessel function is unity at the origin and all others are zero at the origin, for small m the vast majority of the power will stay in the zeroth order. Now, the sum of the square of the Bessel functions is unity, which implies that no energy is absorbed by the plate here. One therefore would think that by increasing $m/2$ to the first zero of J_0 or to the first maximum of J_1, one could generate a much higher diffraction efficiency. However, the thin grating approximation is buried everywhere in our derivation, and further improvement is necessary before we can continue.

Perhaps the most important thing that was lost in making the thin grating approximation is the Bragg effect. Consider a thick grating, as depicted in Fig-

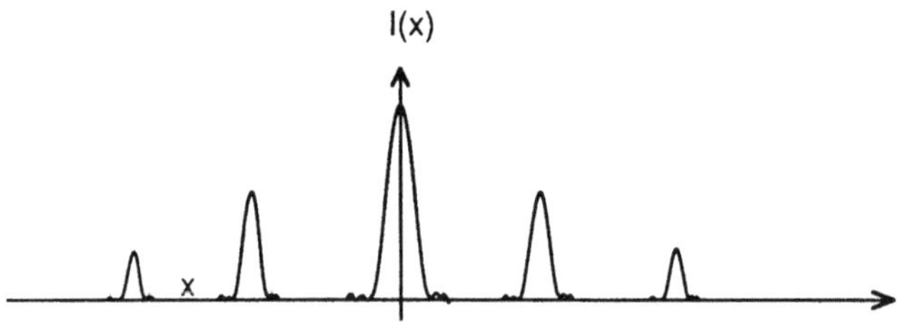

FIGURE 7.37.   Depiction of the Fraunhofer diffraction pattern represented by equation (7-147).

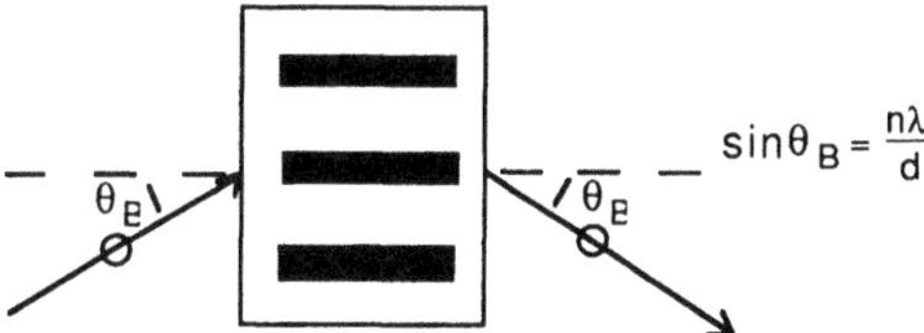

FIGURE 7.38.   Depiction of a thick grating, for which incidence at the Bragg angle will yield much greater diffraction efficiency than that at any other angle.

ure 7.38. In all of our work up to now, we have seen no essential effect of changing the incident angle. This was buried in the transmission function approximation. But there must be a profound effect of rotation of the input angle for a thick grating. For a wave incident at the Bragg angle, all the reflections from the grating planes will add up in phase for the wave exiting at the Bragg angle. This diffracted order can have essentially a 100% diffraction efficiency, and one can choose which order will have maximal efficiency through the parameters chosen for grating construction. It is in this way that people can design high signal-to-noise, efficient grating spectrometers. A rough overview of how such a thick grating works is given in the following paragraphs. [Thick gratings are discussed in various places, including Kogelnik (1969), Collier, Burckhardt, and Lin (1981, Chapter 9) and Filmore and Tynan (1971).] Primary attention in the following argument will be given to describing the Bragg angle diffraction process, a process well known from propagation in the periodic structures which make up matter (Brillouin 1956; Kittel 1971), in which 100% diffraction efficiency can be achieved.

To consider the diffraction grating problem as a boundary value problem, one needs to consider the form of the solution of the wave equation in the striated region, then match this solution at the boundaries (assumed to be $z = 0$ and $z = L$) to the (known) solutions in the regions $z < 0$ and $z > L$. The coordinate system relative to relevant parameters is illustrated in Figure 7.39. Symmetry dictates that the polarization in all regions will be y-directed.

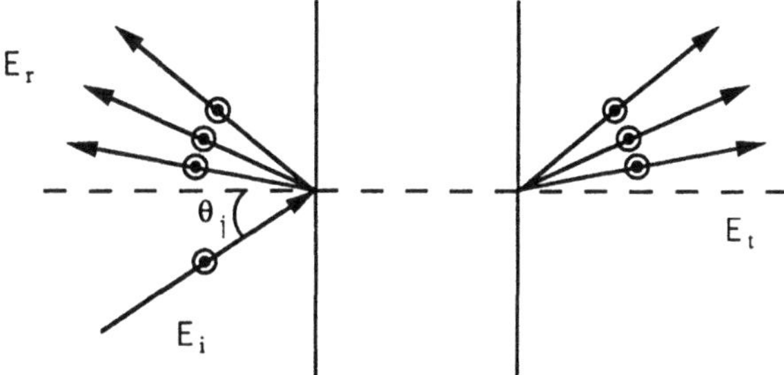

FIGURE 7.39.   Depiction of the thick grating problem as a boundary value problem.

As the solution within the x-varying region will in general be of a quite complicated form, one must assume the transmitted and reflected waves to take the forms of general plane wave expansions

$$\mathbf{E}_t = \int_{-1}^{1} A_t(\alpha)\, e^{ik\alpha x}\, e^{ikz\sqrt{1-\alpha^2}}\, d\alpha\; \hat{\mathbf{e}}_y \qquad \text{(a)}$$

$$\tag{7-148}$$

$$\mathbf{E}_r = \int_{-1}^{1} A_r(\alpha)\, e^{ik\alpha x}\, e^{-ikz\sqrt{1-\alpha^2}}\, d\alpha\; \hat{\mathbf{e}}_y \qquad \text{(b)}$$

Assuming, in region II, a refractive index variation of the form

$$n^2(x) = \epsilon_r + \epsilon_r \Delta\epsilon \, \cos Kx \tag{7-149}$$

where $K = 2\pi f_0$, one can express the wave equation in this region in the form

$$\frac{\partial^2 E_y}{\partial x^2} + \frac{\partial^2 E_y}{\partial z} + k_0^2 \epsilon_r E_y + k_0^2 \epsilon_r \Delta\epsilon \, \cos Kx E_y = 0 \tag{7-150}$$

subject to continuity conditions of E_y, $\partial E_y/\partial z$ at the boundaries $z = 0$ and $z = L$. To solve equation (7-150), one can try separation of variables—that is, assume a solution $E_y = X(x)Z(z)$—to find

$$\frac{1}{X}\frac{\partial^2 X}{\partial x^2} + \frac{1}{Z}\frac{\partial^2 Z}{\partial z^2} + k_0^2 \epsilon_r + k_0^2 \Delta\epsilon \, \cos Kx = 0 \tag{7-151}$$

Separating (7-151) into two equations, one finds

$$\frac{\partial^2 Z}{\partial z^2} + k_z^2 Z = 0 \qquad \text{(a)}$$

$$\tag{7-152}$$

$$\frac{\partial^2 X}{\partial x^2} + k_x^2 X + k_0^2 \epsilon_r \Delta\epsilon \, \cos Kx X = 0 \qquad \text{(b)}$$

The k_z can be determined from elementary considerations. Consider that $\Delta\epsilon \rightarrow 0$. In this case, one must have that $k_x^2 + k_z^2 = k_0^2 \epsilon_r$, which has the obvious solution $k_x = k_0 \sqrt{\epsilon_0}\, \sin\theta$ and $k_z = k_0\sqrt{\epsilon_r}\, \cos\theta$, where θ is the angle in medium II and not in the free-space media of I and III. But if $\Delta\epsilon$ becomes greater than zero, this can have no effect on k_z, and therefore the solution to (7-152) (a) must be

$$Z(z) = Z_f e^{ik_z z} + Z_b e^{-ik_z z} \tag{7-153}$$

where $k_z = k_0 \sqrt{\epsilon_r} \cos \theta$, and where the Z_f and Z_b must be determined from the boundary conditions at $z = 0$ and $z = L$. The solution to (7-152) (b) also is known, at least in terms of Floquet's theorem (Morse and Feshbach 1953, Whittaker and Watson 1973). The form of the solution is

$$X(x) = e^{ik_x x} \sum_{n = -\alpha}^{\infty} a_n e^{-inKx} \qquad (7\text{-}154)$$

where the k_x is unknown, and the a_n's are unknown constants. To find relations for the k_x and a_n's, one needs to plug back the results of (7-153) and (7-154) into (7-150). To do this, one can without loss of generality take $Z_b = 0$ to obtain

$$[-(k_x - nK)^2 - k_z^2 + k_0^2 \epsilon_r] e^{ik_x x} \sum a_n e^{-nKx}$$

$$+ \frac{k^2 \epsilon_r \Delta \epsilon}{2} e^{ik_x x} \sum a_n e^{-i(n+1)Kx} + \frac{k^w \epsilon_r \Delta \epsilon}{2} e^{ik_x x} \qquad (7\text{-}155)$$

$$\cdot \sum a_n e^{-i(n-1)Kx} = 0$$

where the common Z_p term has been eliminated. One can obtain a matrix equation by multiplying through by e^{imKx} and integrating the resulting expression over one period of e^{imKx}. Performing this operation and eliminating the common $e^{ik_x x}$ term, one finds that

$$\mathbf{Ma} = 0 \qquad (7\text{-}156)$$

where $\mathbf{a}$ is an infinite vector given by

$$\mathbf{a} = [\; \cdots \; a_{-n} \; \cdots \; a_{-1} \;\; a_0 \;\; a_1 \; \cdots \; a_n \; \cdots \;]^T \qquad (7\text{-}157)$$

where T denotes the transpose, and where M is given by

$$M = \text{Tridiag} \left[\frac{k_0^2 \epsilon_r \Delta \epsilon}{2}, \;\; k_0^2 \epsilon_r - (k_x^2 - nK)^2 - k_z^2, \;\; \frac{k_0^2 \epsilon_r \Delta \epsilon}{2} \right] \qquad (7\text{-}158)$$

where Tridiag denotes that M is a tridiagonal matrix whose nth-row diagonal element is given by the central argument in the brackets, and whose nth row therefore is given by the row in the brackets, centered on the diagonal.

Equation (7-156) is a homogeneous matrix equation and therefore has solutions only when

$$\det M = 0 \qquad (7\text{-}159)$$

Equation (7-159) can be considered to be the determinential equation for k_x. As M is an infinite matrix, (7-159) describes an infinite-order polynomial equation. This equation should have an infinite number of roots k_x. Perusal of (7-155), however, indicates that for each value of k_x found, $k_x - nK$ will also be solutions. Therefore, as it turns out, there is really only one solution for k_x, as can be illustrated on the Brillouin diagram (Brillouin 1956) of Figure 7.40. The point is that, as the perturbation becomes small, k_x as a function of $k_0 \sqrt{\epsilon_r}$, essentially will be a straight line. As $k_0 \sqrt{\epsilon_r}$ only shows up squared in (7-155), this line can have either positive or negative slope. The positive slope lines correspond to positive phase velocity waves, and the negative slope lines to negative phase velocity waves. In fact, were one to plug these solutions for k_x back into M, one could solve the resulting system of equations for the **a** vector as a one-parameter string (i.e., one always has to decide on the normalization of an eigenvector) to find that

$$M(k_x)\mathbf{a} = 0 \tag{7-160}$$

has solution $\mathbf{a}_f$ and $\mathbf{a}_b$, where $\mathbf{a}_f$ is a forward-traveling eigenvector and $\mathbf{a}_b$ is a backward-traveling eigenvector. Denoting components of $\mathbf{a}_f$ by a_{fn} and $\mathbf{a}_b$ by a_{bn}, one could form the complete solution to the wave equation inside the medium by

$$\mathbf{E}_{\mathrm{II}} = [Z_f e^{ik_0\sqrt{\epsilon_r}z\cos\theta} e^{ik_x x} \sum a_{fn} e^{-inKx} + Z_b e^{-ik_0\sqrt{\epsilon_r}z\cos\theta} e^{ik_x x} \sum a_{bn} e^{-inKx}]\hat{\mathbf{e}}_y \tag{7-161}$$

A complete solution of the boundary value problem thus could be effected by equating (7-161) and its derivative at $z = 0$ with an incident wave plus (7-148) (a) and its derivative, and equating (7-161) and its derivative with (7-148) (b) and its derivative at $z = L$.

A major problem with this strategy is that it requires at least two solutions of infinite sets of equations. Obtaining even a single solution of an infinite set

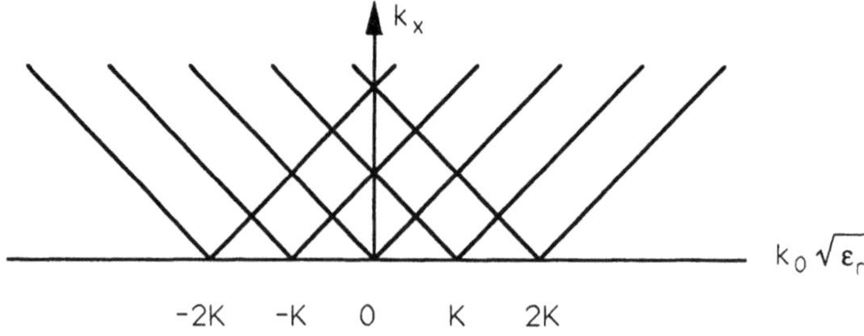

FIGURE 7.40.   A Brillouin diagram of the unperturbed solutions of equation (7-159).

of equations is a formidable task. A possible simplification of this task would be to truncate the system at a given value of n and solve the resulting finite sets of equations. Such a solution has been effected by Burckhardt (1966). Such a treatment is approximate, and it is not clear exactly what the approximation is. In the following paragraph, we will give an alternative approximate treatment that leads to a very simple treatment of the Bragg incidence problem.

We now have solved the problem of the optical flat twice, once in Chapter 2 when we considered optical filters and once in Chapter 6 when the Fabry-Perot interferometer was discussed. One way to solve the problem was to consider it as a boundary value problem, as in Chapter 2, whereas the other was to consider it as a multiple-bounce problem. Here we wish to consider the thick hologram problem as a multiple-bounce problem. In the glass plate problem, we found that the higher-order bounces could have quite an effect on the magnitude of the reflected wave, although by comparison this was only a 16% maximum effect on the magnitude of the transmitted wave. Here we are much more interested in the direction of the transmitted wave than in its absolute magnitude, so we will consider the first transit and ignore the rest. In practice we could go further, but in most cases the corrections will not be too great. You can make the multiple-bounce calculations when you have finished the chapter.

In the Fabry-Perot multiple-bounce argument, the procedure was to calculate the reflection at the first interface, propagate the wave at the second facet, calculate reflection and transmission, propagate back to the first facet, and so on. In our problem, the important step is to calculate the propagation effects, as we know what happens at the interfaces (assuming that the perturbation is weak and we can use the Fresnel coefficients for an interface with a medium with $\epsilon_r = n_r^2 \epsilon_0$). If we believe Floquet's theorem, we will accept that for any plane inside the medium the y-directed field can be expanded in the form

$$E_y(\text{II}) = e^{ik_1 \sqrt{\epsilon_r} x \sin\theta}\, e^{ik_0 \sqrt{\epsilon_r} z \cos\theta} \sum a_n(z) e^{-inKx} \tag{7-162}$$

where we have now stuffed the eigenvalue problem into the z variation of the coefficients $a_n(z)$. Now the expansion of (7-162), which is analogous to the expansion of (3-37), will not be of any value unless some approximation is valid. As with (3-37), we will assume that the slowly varying approximation, this time slowly varying with coordinate z, is a valid simplification and that we can therefore write that

$$\frac{\partial a_n}{\partial z} \ll k_z a_n \tag{7-163}$$

Assuming an E-field of the form of (7-162), one can use Maxwell's curl equation (2-1) (a) to find that

$$\mathbf{H} \cong \frac{1}{i\omega\mu_0} [-ik_z E_y \hat{\mathbf{e}}_x + i(k_x - nK)E_y \hat{\mathbf{e}}_z] \tag{7-164}$$

where the higher-order (derivative of a_n) terms have been dropped. Plugging (7-148) and (7-162) into Maxwell's curl equation (2-1) (b) and using $k_x^2 + k_z^2 = k_0^2 \epsilon_r$, one finds that

$$\begin{aligned} -ik_z \sum \frac{\partial a_n}{\partial z} e^{-inKz} &+ (2k_x nK - n^2 K^2) \sum a_n e^{-inKx} \\ &= \frac{k_0^2 \epsilon_r \Delta\epsilon}{2} \sum a_n [e^{-i(n+1)kz} + e^{-i(n-1)Kx}] \end{aligned} \tag{7-165}$$

where it should be noted that the zeroth-order terms exactly canceled, that is, $k_x^2 + k_z^2 = k_0^2 \epsilon$, and, therefore, unlike the situation in (7-164), there were no higher-order terms to ignore. Employing the previously used trick of multiplying by e^{imKx} and integrating over a period, one finds that

$$\frac{\partial \mathbf{a}}{\partial z} + \mathbf{Ma} = 0 \tag{7-166}$$

where M is given by

$$M = \text{Tridiag} \left[-i\frac{k_0^2 \epsilon_r}{2k_z} \Delta\epsilon, \quad i\frac{2k_x nK - n^2 K^2}{k_z}, \quad -i\frac{k_0^2 \epsilon_r}{k_z} \Delta\epsilon \right] \tag{7-167}$$

Equation (7-166) defines the characteristics of the propagation through the transversely striated medium. Given initial conditions at either interface, one can use (7-166) to compute the amplitudes and phases of the "partial" waves arriving at the other interface. Now, qualitatively, we could describe the form of the solutions of (7-166) much as we described the WKB solution to the matrix equation (5-18). The solution (5-18) showed that nothing much happened to the solution except near singular points of the matrix M, but at these points it was necessary to find a canonical solution to a simplified problem to connect the unperturbed solutions on either side of the singular point. We know that the singularities of M lie near the Bragg angles as illustrated in Figure 7.38. Presently we will proceed to expand the system of (7-166) about a Bragg point and try to find a canonical solution that represents the amount of coupling out of the fundamental spatial mode and into the higher modes. That is, in our approximation, the incident wave will couple only to the zeroth-order harmonic, leaving us the system of equations

$$\frac{\partial a_n}{\partial z} + i\left(\frac{2k_x nK - n^2 K^2}{k_z}\right) a_n = -i\frac{k_0^2 \epsilon_r}{k_z} \Delta\epsilon (a_{n+1} + a_{n-1}) \tag{7-168}$$

subject to the initial condition that

$$a_0(0) = 1 \qquad \text{(a)}$$

$$a_i(0)|_{i \neq 0} = 0 \qquad \text{(b)}$$

$$(7\text{-}169)$$

As in the boundary value problem previously discussed, we cannot solve (7-168) for an infinite number of terms, but we must at some point truncate. For Bragg angle incidence, where

$$k_x = \frac{K}{2} \qquad (7\text{-}170)$$

this is not a hard truncation to decide upon. The right-hand side of (7-168) is in essence the pumping term—that is, the source term for the nth partial wave. In our case, the a_0 will act as a source for a_1 and a_{-1}, which will build up and then act as sources for a_2 and a_{-2} as well as coupling back to a_0, and so on. The coupling process will be most efficient if the second term on the right-hand side of (7-168) is small, as can be seen from noticing that the solution to the equation

$$\frac{df}{\partial z} + i\alpha f = g \qquad (7\text{-}171)$$

is

$$f(z) = f_0 e^{i\alpha z} \int g(z') e^{-i\alpha z'} \, dz' \qquad (7\text{-}172)$$

which for large α and roughly constant g will be roughly zero. However, for zero α and roughly constant g, f will grow monotonically in z. The $n = 0$ phase term (second term on the left-hand side) of (7-168) is automatically zero. For Bragg incidence, the $n = 1$ phase term is also zero, indicating that the zeroth- and first-order modes will be most strongly coupled. Truncating (7-168) to these two partial waves, one finds that

$$\frac{\partial a_1}{\partial z} = -i\,\frac{k_0^2 \epsilon_r}{kz}\,\Delta\epsilon a_0 \qquad \text{(a)}$$

$$\frac{\partial a_0}{\partial z} = -i\,\frac{k_0^2 \epsilon_r}{kz}\,\Delta\epsilon a_1 \qquad \text{(b)}$$

$$(7\text{-}173)$$

subject to initial conditions

$$a_0(0) = 1 \qquad \text{(a)}$$
$$a_1(0) = 0 \qquad \text{(b)}$$
$$(7\text{-}174)$$

with the attendant solutions

$$a_1(z) = \sin\left(\frac{k_0^2\epsilon_r}{kz}\,\Delta\epsilon z\right) \qquad \text{(a)}$$
$$a_0(z) = \cos\left(\frac{k_0^2\epsilon_r}{kz}\,\Delta\epsilon z\right) \qquad \text{(b)}$$
$$(7\text{-}175)$$

Equations (7-173) through (7-175) comprise the major results of the so-called coupled mode theory. As can be seen explicitly from the solutions, full-power transfer can occur between the zeroth- and first-order modes. Unfortunately, after a coupling length L_c, defined by

$$L_c = \frac{kz\pi}{2k_0^2\epsilon_r\Delta\epsilon} \qquad (7\text{-}176)$$

the power will begin to flow back again from the first order to the zeroth order. Therefore, complete coupling will require some rather stringent fabrication tolerances. Also, although (7-175) seems to indicate that the power will couple back and forth between the modes ad infinitum, the ignored terms will become even more important at lengths beyond l_c. Even at l_c, the power carried by those modes, although small, will prevent true 100% coupling efficiency.

References

Beran, M. J. and G. B. Parent, Jr., *Theory of Partial Coherence*, Prentice Hall, Englewood Cliffs, NJ (1964).

Born, M. and E. Wolf, *Principles of Optics*, Fifth edition, Pergamon Press, New York (1975).

Brillouin, L., *Wave Propagation in Periodic Structures*, Dover, New York (1956).

Burckhardt, C. B., Diffraction of a plane wave at a sinusoidally stratified dielectric grating, *J. Opt. Soc. Am.* **56**, 601 (1966).

Carrier, G. F., M. Krook, and C. E. Pearson, *Functions of a Complex Variable*, McGraw-Hill, New York (1966).

Collier, R. J., C. B. Burckhardt, and L. H. Lin, *Optical Holography*, Academic Press, New York (1981).

Fillmore, G. L. and R. F. Tynan, The Sensitometric characteristics of hardened dichromated gelatin films, *J. Opt. Soc. Am.* **61**, 199–203 (1971).

Goodman, J. W., *Introduction to Fourier Optics*, McGraw-Hill, New York (1968).

Goodman, J. W., *Statistical Optics*, Wiley, New York (1985).

Hecht, E., *Optics*, Addison-Wesley, Reading, MA (1987).

Jackson, J. D., *Classical Electrodynamics*, Second edition, John Wiley & Sons, New York (1975).

Kittel, C., *Introduction to Solid State Physics*, Fourth edition, John Wiley & Sons, New York (1971).

Klein, M. V. and T. E. Furtak, *Optics*, Second edition, John Wiley & Sons, New York (1986).

Kogelnik, H. Coupled wave theory for thick hologram gratings, *Bell Sys. Tech. J. 28*, 2909 (1969).

Levine, H. and J. Schwinger, On the theory of diffraction by an aperture in an infinite plane screen, Part I, *Phys. Rev. 74*, 958 (1948).

Levine, H. and J. Schwinger, On the theory of diffraction by an aperture in an infinite plane screen II, *Phys. Rev. 75*, 1423–1432 (1949).

Levine, H. and J. Schwinger, On the theory of electromagnetic wave diffraction by an aperture in an infinite plane conducting screen, *Comm. Pure Appl. Math. 3*, 355–391 (1950).

Meixner, J. and W. Andrejewski, Strenge Theorie der Beugung ebener electromagnetischer Wellen an der volkommen leitenden Kreisscheibe und an der kreisförmigen Oeffnung im vollkommen leitenden ebenen Schiron, *Ann. d. Physik 7*, 157–168 (1950).

Morse, P. M. and H. Feshbach, *Methods of Theoretical Physics*, McGraw-Hill, New York (1953).

Rayleigh, J. W. S. *The Theory of Sound II*, Second edition, Dover, New York (1945).

Smythe, W. R., *Phys. Rev. 72*, 1066 (1947).

Smythe, W. R., *Static and Dynamic Electricity*, Third edition, McGraw-Hill, New York (1969).

Sommerfeld, A. *Math. Ann. 47*, 317 (1896).

Sommerfeld, A. *Optics*, Volume III of *Lectures on Theoretical Physics*, translated from German, Academic Press, New York (1964).

Wolf, E. and E. W. Marchand, Comparison of the Kirchhoff and the Rayleigh—Sommerfeld theories of diffraction at an aperture, *J. Opt. Soc. Am. 54*, 587–590 (1964).

Whittaker, E. T. and G. N. Watson, *A Course of Modern Analysis*, Fourth edition, Cambridge University Press, Cambridge (1973).

Problems

1. In the text, Fresnel diffraction was considered for a plane wave incident on aperture. Here, we wish to consider the source to be at a finite distance l_s. In (a) through (d), find the form of the Fresnel diffracted field on the axis behind the aperture for:

    (a) A point source on the axis.

    (b) A point source off axis at $x = d_s$.

    (c) Symmetric point sources at $x = \pm d_s$.

    (d) A line source defined by $|X| < d_s$.

2. Consider the Fresnel zone of an aperture with a plane wave incident on it.

    (a) What is the $\lim_{z \to 0} U(x, y, z)$ if one considers a rectangular aperture that is open for $l_{y-} < x < l_{x+}$ and $l_{y-} < y < l_{y+}$? Comment on your answer.

(b) Find the $U(x, y, z)$ in the Fresnel zone (not the limit of small z), behind a straight edge, that is an aperture that transmits positive x coordinate. Sketch the magnitude of your result.

(c) Find $U(r = 0, z)$, the on-axis field behind a circular aperture of radius a.

(d) Find $U(r = 0, z)$, the on-axis field behind an annulus of inner radius a_1 and outer radius a_2.

3. Recall that the one-dimensional Fresnel diffraction pattern of the field in a plane $z = 0$ can be expressed in the form

$$U(x, L) = \frac{e^{ikL}}{i\lambda L} [U^{(i)}(x) * e^{(ik/2L)x^2}]$$

where the convolution operator $*$ is defined by

$$F(x) * g(x) = \int_{-\infty}^{\infty} f(x - x')g(x') \, dx' = \int_{-\infty}^{\infty} g(x = x')f(x') \, dx'$$

In (a) through (d), find the Fresnel pattern of the given $z = 0$ field distribution, and try to interpret your results in terms of the associated Fraunhofer diffraction pattern.

(a) $U^{(i)} = \delta(x - x_i)$.

(b) $U^{(i)} = e^{(-ik/2L)(x - x_i)^2}$

(c) $U^{(i)} = e^{ik(x - x_i) \sin \theta}$

(d) $U^{(i)} = (1/\sqrt{2\pi\sigma}) \, e^{-(x - x_i)^2/2\sigma}$

4. Let $g(r, \theta)$ be separable in polar coordinates to the form $g_r(r)g_\theta(\theta)$. Then one can show that (Goodman 1968, p. 27):

$$\mathcal{F}\{g(r, \theta)\} = \sum_{k = -\infty}^{\infty} c_k(-j)^k e^{jk\varphi} H_k\{g(r)\}$$

where

$$c_k = \frac{1}{2\pi} \int_0^{2\pi} g_\theta(\theta)e^{jk\theta} \, d\theta$$

Use the result above to find and sketch the intensity pattern of a distance f behind a lens that is a distance f in front of a circular aperture with a transmittance function given by:

(a) $g_\theta(\theta) = \cos l\theta$, l integer.

(b) $g_\theta(\theta) = \text{rect} (\cos l\theta)$ where, as per problem 9, page 135, rect is defined as

$$\text{rect} (x) = \begin{cases} 1 & |x| < \frac{1}{2} \\ 0 & |x| > \frac{1}{2} \end{cases}$$

5. Define the angular spectrum A_z of U by

$$A_z\left(\frac{\alpha}{\lambda}, \frac{\beta}{\lambda}\right) = \int_{-\infty}^{\infty}\int U(x, y, z)e^{i2\pi((\alpha/\lambda)x + (\beta/\lambda)y))}\, dx\, dy$$

where

$$(\nabla^2 + k^2)U = 0$$

(a) Find the differential equation satisfied by A.
(b) Given $A_0\,(\alpha/\lambda, \beta/\lambda)$, what is A_z in terms of A_0?
(c) How does one interpret those values of A with $\alpha^2 + \beta^2 \geq 1$?
(d) What is the angular spectrum of a constant? A delta function?

6. Consider the transformation of a Gaussian beam by a thin lens as shown in Figure 7.41. Knowing that $d_0'^2[(s_2 - f)^2 + s_R^2] = f^2 d_0^2$ where $d_0 = 4\lambda/\pi\theta$ and $s_R = d_0/\theta$

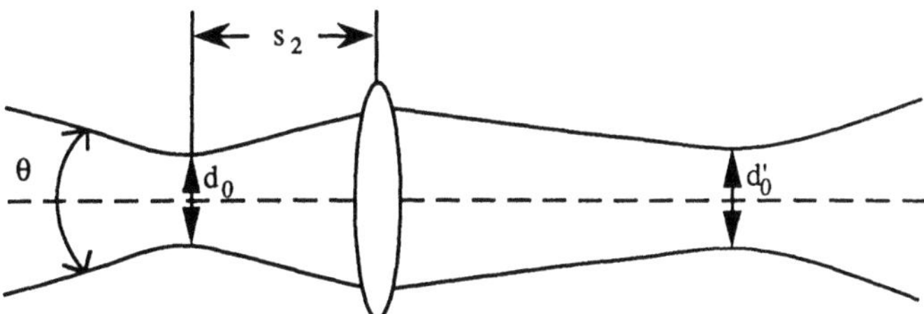

FIGURE 7.41.   Figure for problem 6.

(s_R is the Rayleigh range), one wants to image a laser beam ($\lambda = 628$ nm) with a divergence θ of 0.8 mrad is focused with a lens of $f = 0.1$ m. What are the minimum beam diameters d_0' for the following cases?
(a) $s_2 \ll s_R$ ($s_2 = s_R/10$).
(b) $s_2 = s_R$.
(c) $s_2 \gg s_R$ ($s_2 = 10 \cdot s_R$).

7. Recall that the mutual intensity propagates according to the rule

$$J(x_2, \alpha_2) = \frac{1}{\lambda(z_2 - z_1)}\int J(x_1, \alpha_1)e^{ik(r_{\alpha_1\alpha_2} - r_{x_1 x_2})}\, dx_1\, d\alpha_1$$

where

$$r_{x_1 x_2} = \sqrt{(x_2 - x_1)^2 + (z_2 - z_1)^2}$$

$$r_{\alpha_1\alpha_2} = \sqrt{(\alpha_2 - \alpha_1)^2 + (z_2 - z_1)^2}$$

Consider the optical setup depicted in Figure 7.42, where paraxial propagation can be easily assumed.

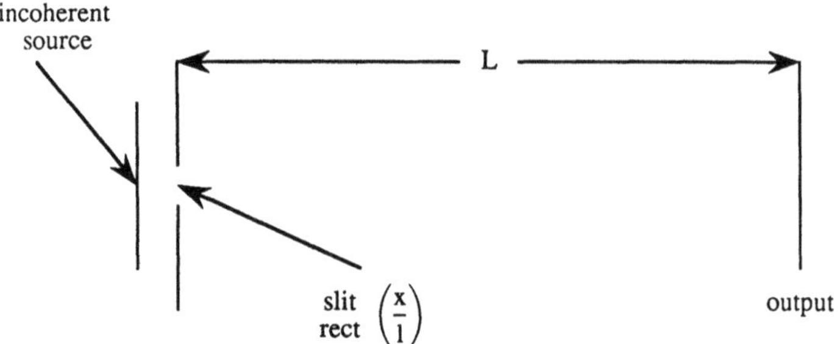

FIGURE 7.42.   Figure for problem 7.

(a) Find the mutual intensity in the output plane.
(b) Find the coherence length in the output plane.
(c) Find and sketch the intensity in the output plane.

8. Consider the optical arrangement in Figure 7.43, masked by an aperture with transmittance

$$T_0(x_0, y_0) = \text{circ}\left(\frac{r_0}{a_{s0}}\right) - \text{circ}\left(\frac{r_0}{a_{si}}\right)$$

where

$$r_0 = \sqrt{x_0^2 + y_0^2} \quad \text{and} \quad a_{si} < a_{s0}$$

(a) Find the mutual intensity in plane z_1 as a function d_1, the distance between two pinholes in z_1. In the text, d_1 is written as

$$d_1 = \sqrt{(x_1 - \alpha_1)^2 + (y_1 - \beta_1)^2}$$

Sketch the mutual intensity function.
Say that the aperture stop in $z = z_1$ has transmittance given by $t_1(x_1, y_1) = \text{rect}(r_1/a)$ where $r_1 = \sqrt{x_1^2 + y_1^2}$.

(b) Find and sketch the intensity in plane $z = z_1$.

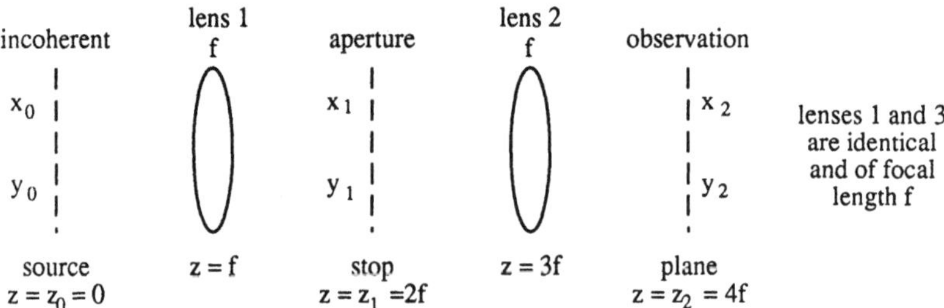

FIGURE 7.43.   Figure for problem 8.

    (c) Say that a_1 is "large." Find the expression for the intensity and mutual intensity in place $z = z_2$.

    (d) Say that a_1 is "small." Find and sketch the intensity and mutual intensity in the plane $z = z_2$.

9. Consider a rectangularly shaped arc lamp source, as depicted in Figure 7.44, placed at plane z_0 in a set-up like the one depicted in figure 7.43, but without lenses. The source radiates as a blackbody (Lambertian) source.

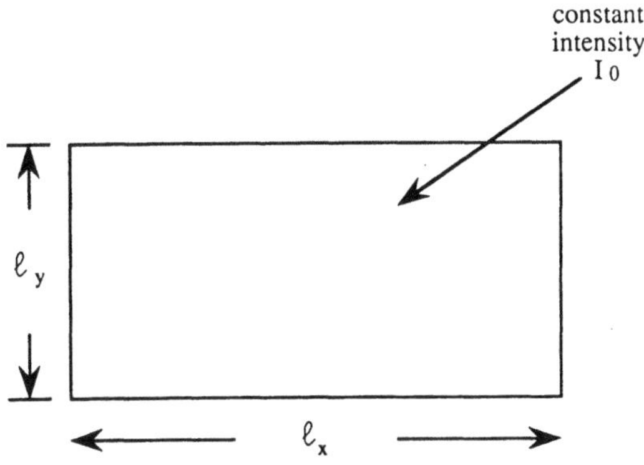

FIGURE 7.44.   Figure for problem 9.

    (a) Find the complex degree of coherence of light propagated a distance z_1 from this source.

    (b) Say that a rectangular aperture of dimensions d_x, d_y is placed at z_1. Find the complex degree coherence at a plane z_2 behind the aperture. The result can be expressed as a convolution integral. In (c) and (d) one will evaluate the integral in two special cases.

    (c) Evaluate the integral found in (b), given that z_1 is very short in comparison with other length scales.

    (d) Repeat (c) but in the case where z_1 is very long.

10. Find and sketch the mutual intensity function and optical intensity that exist in a plane a distance f in front of a lens that in turn is a distance f in front of an incoherent source, where the incoherent source has the following spatial distributions:

    (a) $I(x_s) = I_0 \sum_{i=-N}^{N} \delta(x_s - x_i)$, $x_i = i/2N \, l_s$.

    (b) $I(x_s) = I_0 \cos 2\pi f x_s$.

    (c) $I(x_s) = I_0 \operatorname{rect}(\sin 2\pi f x_s)$.

    (d) $I(x_s) = I_0 e^{-x_s^2/2\sigma_s^2}$.

11. Consider a partially coherent source in a plane A, which is located at a great distance R from an observation plane B, as illustrated in Figure 7.45. In (a) through (d) we wish to find the intensity $I(Q)$ in plane B, given the source shape and normalized mutual intensity function. In (a) and (b) assume that the intensity in A is expressible as $I_0 \operatorname{rect}(2x/L_x) \operatorname{rect}(2y/L_y)$, where I_0 is suitably normalized. In (c) and (d) assume that the intensity in A is expressible as $I_0 \operatorname{circ}(\sqrt{x^2 + y^2}/a)$.

FIGURE 7.45.   Figure for problem 11.

(a) Find $I(Q)$ for $\mu(P_1, P_2)$ if A is unity.
(b) Find $I(Q)$ for $\mu(P_1, P_2) \rightarrow \delta(d_{12})$ where d_{12} is the distance from P_1 to P_2.
(c) Repeat (a) and (b) for the circ function.
(d) Find $I(Q)$ for $\mu_{12} = 1/N \, e^{-d_{12}^2/\sigma}$ where N is a suitable normalization (σ dependent). What happens when $\sigma \rightarrow \infty$? $\sigma \rightarrow 0$?

12. Consider the optical setup depicted in Figure 7.46, where the aperture stop is assumed to fit right up against the face of the incoherent source. Consider both s and L to be sufficiently long and the viewing area sufficiently small that the paraxial, far-field approximation holds.

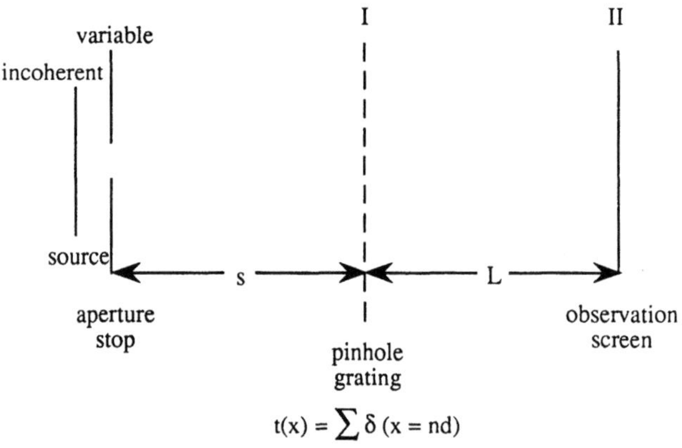

FIGURE 7.46.   Figure for problem 12.

(a) Find the mutual intensity directly in front of I.
(b) Give an expression for the mutual intensity at plane II.
(c) What should the intensity at II be for an incoherent illumination of I? Show that your result in (b) contains this result as a limit.
(d) What should the intensity at II be for coherent illumination of I? Show that your result in (b) contains the answer as a limit.

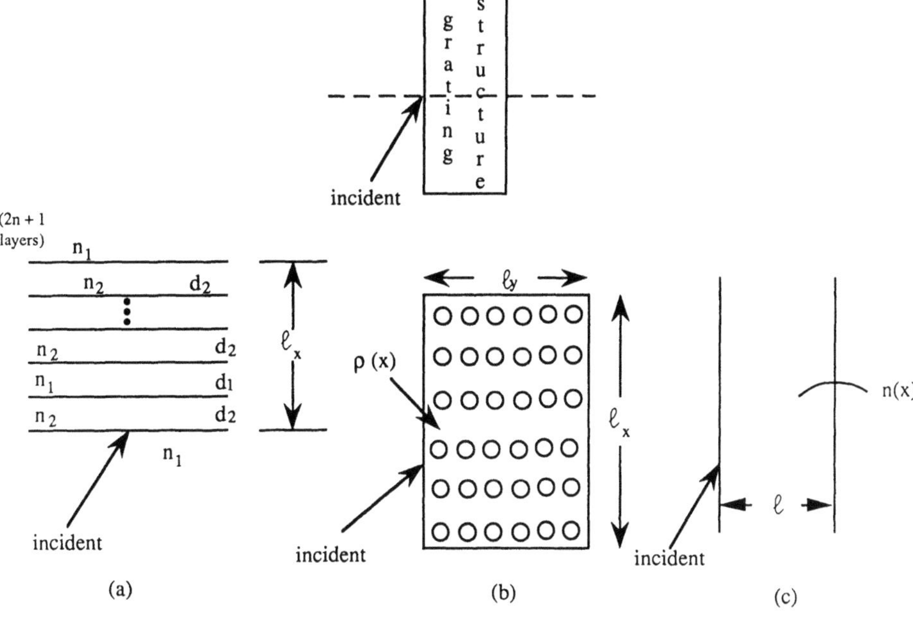

FIGURE 7.47.   Figure for problem 13.

13. Three models, (a), (b), and (c), of a grating, as depicted in Figure 7.47, correspond to parts (a), (b), and (c) of the problem, respectively. In (a) through (c) you are to discuss how sharp the Bragg peak is in angle and what kind of diffraction efficiencies are achievable by the given model.

   (a) Here the grating is modeled as a dielectric multilayer. Discuss what d_1 and d_2 must have to be in order for the structure to exhibit a strong Bragg effect.

   (b) Here the grating is modeled by a set of dipole radiators of density $\rho(x) = (1 + \cos(2\pi/d)\,x)$. You can assume that the dipoles are driven in phase synchronism with the incident wave, as if the incident wave acted as a driving current. That is, each dipole radiator is a point source $e^{i(kr+\varphi)}/r$, where φ is determined by the incident. One can find the far field by integrating over this distribution. Find the far-field pattern, and comment in some detail how it varies with l_x and l_z.

   (c) Here one wishes to attack the grating problem as a boundary value problem. We will assume, however, that the input and output facets are so well index-matched that one need not worry about reflections. Assume that

$$n^2(x) = \epsilon_1 + \Delta\epsilon \cos(2\pi f_0 x)$$

where $\Delta\epsilon \ll \epsilon_1$, and assuming the solution inside the medium looks like

$$x(x, z) = \sum_{q=-\infty}^{\infty} a_q(z)e^{i2\pi q f_0 x} e^{i\beta z}$$

Discuss how Bragg's effect exhibits itself, and, as well as discussing the sharpness of the output Bragg peak, discuss the effect of changing the input angle away from that of Bragg.

14. Consider a "loaded" diffraction grating, that is, one in which each slit is filled with a chunk of dielectric such that the grating transmission function is of the form

$$t(x) = \sum_{n=1}^{N} \text{rec} \, [(x - nd)/l] e^{in\alpha}$$

where

$$\text{rect} \, (x) = \left\{ \begin{matrix} 1 & |x| < \tfrac{1}{2} \\ 0 & \text{otherwise} \end{matrix} \right\}$$

Assume $N \geq 20$.
(a) Find the Fraunhofer diffraction pattern of this grating.
(b) Sketch the diffraction pattern of this grating for $\alpha = 0, \pi/4, \pi$.
(c) Sketch the diffraction pattern for $\alpha = \alpha_r = i\alpha_i$, $\alpha_r = 0, \pi/4$, $\alpha_1 = 0.01, 0.1$, and 10.

15. Consider a grating (Ronchi ruling) where the transmission function is given by

$$t(x) = \text{rect} \left(\frac{x}{x_0/2} \right) * \sum_{n=-\infty}^{\infty} \delta(x - nx_0)$$

where $*$ denotes convolution as is defined in equation 7-42, where rect (x) is defined in equation (6-19), and can be sketched as in Figure 7-48.

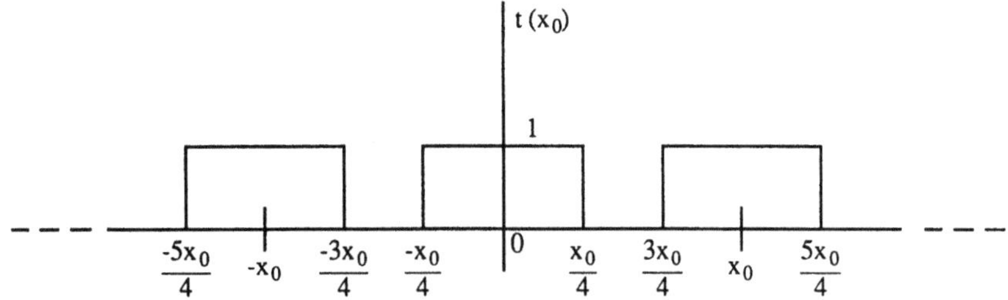

FIGURE 7.48.   Figure for problem 15.

(a) If the ruling is illuminated by a normally incident plane wave, calculate the angular spectrum of the transmitted wave.
(b) Use the angular spectrum to write the transmitted field as a Fourier series.
(c) Show that after propagating a distance $Z = (X_0^2/(2m + 1)^2\lambda) \, 2k$ the field returns to its original value (i.e., self-focuses), and that it appears with reversed contrast $(U_z(x) = 1 - U_z(x))$ after propagating $z = ((2k + 1)X_0^2)/((2m +)^2\lambda)$ in the paraxial limit.

X = 0

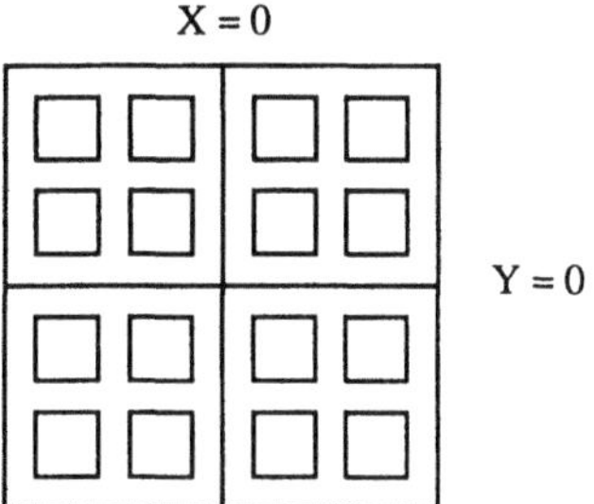

Y = 0

FIGURE 7.49.   Figure for problem 16.

16. Consider a rectangular array of rectangular apertures (only the first four apertures of the $N \times N$ array have been sketched in Figure 7.49).
The transmittance function of the array is expressible in the form

$$t(x, y) = \sum_{l=-N/2}^{N/2} \sum_{k=-N/2}^{N/2} \mathrm{rect}\left(\frac{x - (2k + 1)2\lambda}{\lambda}\right) \mathrm{rect}\left(\frac{y - (2l + 1)2\lambda}{\lambda}\right)$$

where $\mathrm{rect}(x)$ is defined as per equation (6-19). Say that a wave $e^{ikz\cos\theta}\,e^{ikx\sin\theta}$ is incident from $-z$ on this aperture array in the plane $z = 0$. Assume that (miraculously) the Kirchhoff boundary conditions apply.
   (a) Find an expression for the far-field radiation pattern of the array. What is the array factor? What is the element factor?
   (b) Plot this pattern along the x and y axes.
   (c) Assume that a second wave $e^{ikz\cos\theta - ikx\sin\theta}$ is also incident on the structure. What is the resulting radiation pattern?
   (d) Sketch the transmission factor (fraction of incident power transmitted) as a function of the half angle θ at which the two incident beams interfere.
17. Find and sketch the diffraction pattern produced by gratings with the following transmittances:

   (a) $t(x) = \mathrm{rect}(\cos 2\pi f_0 x)\, \mathrm{rect}\left(\dfrac{2x}{l}\right)$ where $\mathrm{rect} = (x) = \begin{pmatrix} 1 & |x| < \frac{1}{2} \\ 0 & |x| > \frac{1}{2} \end{pmatrix}$.

   (b) $t(x) = \exp\left[i((\sin 2\pi f_1 x)/2))\right] \exp(i\pi \sin 2\pi f_2 x)\, \mathrm{rect}(2x/l)$.
   (c) $t(x) = \frac{1}{2}[1 + \cos 2\pi f_1 x] \exp[i\pi \sin 2\pi f_2 x]\, \mathrm{rect}(2x/l)$.
   (d) $t(x) = \mathrm{rect}(\cos 2\pi f_1 x) \exp[i\pi\, \mathrm{rect}(\sin 2\pi f_2 x)]\, \mathrm{rect}(2x/l)$.
      where $\mathrm{rect}(x)$ is defined as per equation (6-19).
18. Bragg's law states that the maximum diffraction occurs for a grating structure with spacing d, when the incident angle θ_B satisfies the relation $\sin\theta_B = h\lambda/2d$ where n is an arbitrary integer. Assume a wave $\exp[i\pi\, \mathrm{rect}(\sin 2\pi f_x x)]\, \mathrm{rect}(2x/l)$ incident on a grating in the plane $z = 0$. Discuss whether Bragg's law is satisfied for
   (a) A transmission grating with $t(x) = \frac{1}{2}(1 + \cos 2\pi f_g x)$.
   (b) A phase grating with $t(x) = \exp[i(m/2)\sin(2\pi f_g x)]$.

19. Consider a wave with intensity spectrum $I(\lambda)$ incident on a grating.
    (a) If the grating has amplitude transmission function $t(x) = \frac{1}{2} + \frac{1}{2}\cos 2\pi f_g x$, what is the maximum achievable resolution?
    (b) For a phase grating, what is the maximum achievable resolution? Are practical limitations here more severe than they were in (a)?
    (c) Write an expression for the intensity pattern recorded behind the grating in (a) for an $I(\lambda)$ that is flat between λ_1 and λ_2 and collimated into a plane wave.
    (d) Write an expression for the intensity pattern recorded behind the grating in (a) for a quasi-monochromatic incoherent source of spatial extent $l(<z_s)$ located a distance z_s in front of the grating.

20. Say we wish to analyze the spectrum of a process that we know emits radiation between 3000 Å and 7000 Å. Design a grating spectrometer to do this. The design should include details of the grating, that is, the grating spacing, number of lines, and orientation. The grating should be less than about 1 cm. The form of the viewing screen and length from the grating also should be specified. The number of resolvable points of accuracy in the pattern should be specified.

21. The coupled waves solution to the transverse grating structure predicted that, for Bragg angle incidence, the wave amplitudes propagate according to

$$a_1(z) = \sin\left(\frac{k_0^2 \epsilon_r}{k_z} \Delta\epsilon z\right)$$

$$a_2(z) = \cos\left(\frac{k_0^2 \epsilon_r}{k_2} \Delta\epsilon z\right)$$

When $z = l$, these equations predict the value at the output, given that a_0 was at the input. However, to truly solve the boundary value problem, we would need to sum a series of partial reflections. Let us say that the grating structure lies within an emulsion plate with an index roughly equal to that of glass, as depicted in Figure 7.50. We now wish to solve the boundary value problem by summing the partial waves in the following limits (do not bother to find the intensity):
    (a) l arbitrary and $\Delta\epsilon \to 0$.
    (b) $l = (\pi/2)(k_z/k_0^2\epsilon_r\Delta\epsilon)$.
    What goes wrong for other values of l? *Hint:*

$$\sum_{n=0}^{\infty} x^n = \frac{1}{1-x}$$

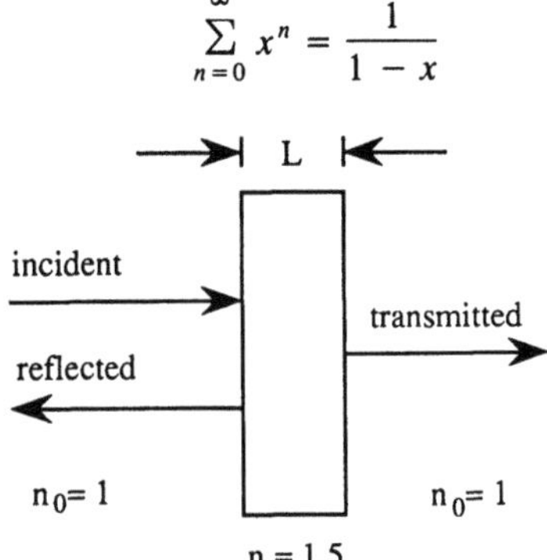

FIGURE 7.50.   Figure for problem 21.

Index